Thermal Management System for Lithium-Ion Batteries

Thermal Management System for Lithium-Ion Batteries

Editors

Jinsheng Xiao
Hengyun Zhang
Sousso Kelouwani

MDPI

Basel • Beijing • Wuhan • Barcelona • Belgrade • Novi Sad • Cluj • Manchester

Editors

Jinsheng Xiao
Wuhan University of Technology
Wuhan
China

Hengyun Zhang
Shanghai University of Engineering Science
Shanghai
China

Sousso Kelouwani
Université du Québec à Trois-Rivières
Trois-Rivieres, QC
Canada

Editorial Office
MDPI AG
Grosspeteranlage 5
4052 Basel, Switzerland

This is a reprint of articles from the Special Issue published online in the open access journal *Batteries* (ISSN 2313-0105) (available at: https://www.mdpi.com/journal/batteries/special_issues/035JWFFAC4).

For citation purposes, cite each article independently as indicated on the article page online and as indicated below:

Lastname, A.A.; Lastname, B.B. Article Title. *Journal Name* **Year**, *Volume Number*, Page Range.

ISBN 978-3-7258-2421-2 (Hbk)
ISBN 978-3-7258-2422-9 (PDF)
doi.org/10.3390/books978-3-7258-2422-9

Contents

About the Editors

Jinsheng Xiao

Jinsheng Xiao is a full professor at the School of Automotive Engineering, Wuhan University of Technology, China, and a a visiting professor at the Hydrogen Research Institute, University of Quebec at Trois-Rivieres, Canada. He obtained his M.S. in 1986 in Marine Engineering, Wuhan University of Technology, and his Ph.D. in 1999 in Engineering Thermophysics, Tsinghua University. His recent research interests include the simulation and optimization of hydrogen production, storage, and fuel cell systems, and battery thermal management systems.

Hengyun Zhang

Hengyun Zhang is a distinguished professor of Shanghai University of Engineering Science, a Senior Member of IEEE, and the Co-Chairman of the ICEMAT2023 International Conference Organizing Committee. He obtained a B.E. from the University of Science and Technology of China in 1994 and a PhD from Nanyang Technological University in 2001. He has led and strongly participated in four National Natural Science Foundation projects and three provincial projects, among others. His main research interests include the characterization of thermal and electrical parameters and the thermal management of new energy vehicle batteries. He has published over 150 papers, and one key national book. He has been granted more than a dozen invention patents, and won the Innovation and Entrepreneurship Award from the China Invention Association. Prof. Zhang was also recognized as a China highly cited scholar (Elsevier) in 2023.

Sousso Kelouwani

Sousso Kelouwani is a Full Professor of Mechatronics with the Department of Mechanical Engineering at the Université du Québec à Trois-Rivières (UQTR), a Senior Member of IEEE, a Canada Research Chair in the Energy Optimization of Intelligent Transport Systems, and a Research Chair in the Intelligent Navigation of Autonomous Industrial Vehicles. He was the Co-President and President of the Technical Committee of the IEEE International Conferences on Vehicular Power and Propulsion in Chicago in 2018 and in Hanoi in 2019. His research interests include optimizing energy systems for vehicle applications, advanced driver assistance techniques, and intelligent vehicle navigation. He holds four patents in the USA and Canada, in addition to 100 scientific articles.

Preface

Lithium-ion batteries (LIBs) have been widely used as power sources for the automotive and energy storage industries due to their high energy density, high power output, low self-discharge rate and minimal memory effect. However, the performances of LIBs are greatly affected by their operating temperature. Various studies on the thermal management and thermal performance enhancement of LIBs have thus been carried out to address their ever-increasing application scenarios. This Special Issue, "Thermal Management System for Lithium-Ion Batteries", aims to present and disseminate the most recent advances in the thermal management of LIBs under various application conditions. Ten high-quality papers on the different aspects of the thermal management of lithium ion batteries are included in this Special Issue. Among them, eight papers are devoted to the thermal cooling of battery modules, with five on the single cooling of battery modules in either liquid channel cooling or direct contact cooling forms; three papers consider hybrid cooling in combination with phase change materials (PCM) and liquid cooling or air cooling. In contrast to thermal cooling techniques, another paper is devoted to the development of the pulse heating of a battery under subzero temperature conditions. In addition to the above battery level studies, one paper focuses on the development of a new material coating made of polyacrylonitrile, which is used as a shutdown film to prevent thermal runaway in LIBs.

The editors would like to congratulate all the authors for publishing their excellent original work in this Special Issue. The editors would also like to express special gratitude to the reviewers, who selflessly devoted their time and expertise to provide comments and suggestions for the improvement of the quality of papers. The second editor would like to acknowledge the support from the Natural Science Foundations of China (52476079, 12272217), as well as the Qinghai Kunlun Talent Plan. We hope that the papers in this Special Issue will benefit readers and add new knowledge to the academic community in this important area.

Jinsheng Xiao, Hengyun Zhang, and Sousso Kelouwani

Editors

batteries

MDPI

Article

Comparison of Cooling Performance in a Cylindrical Battery with Single-Phase Direct Contact Cooling under Various Operating Conditions

Minjun Kim [1], Jeonggyun Ham [2], Donghyeon Shin [3] and Honghyun Cho [2,*]

1 Department of Mechanical Engineering, Graduate School of Chosun University, 309 Pilmundaero, Dong-gu, Gwangju 61452, Korea
2 Department of Mechanical Engineering, Chosun University, 309 Pilmundaero, Dong-gu, Gwangju 61452, Korea
3 Korea Automotive Technology Institute, 303 Pungse-ro, Pungse-Myeon, Cheonan 31214, Korea
* Correspondence: hhcho@chosun.ac.kr; Tel.: +82-62-230-7769

Abstract: This study compares the performance according to a working fluid, the number of battery cooling block ports, and header width required for cooling according to the application of the direct contact single-phase battery cooling method in a 1S16P battery module and examines the battery cooling performance according to the flow rate under the standard and summer conditions based on an optimized model. The analysis result verified that R134a showed low-pressure drop and high cooling performance as the working fluid of the direct contact single-phase cooling system in the 1S16P battery module, and R134a showed the best cooling and stability when applied with three ports and a 5 mm header. In addition, under 25 °C outdoor conditions, the maximum temperature of the battery and the temperature difference between the batteries at 3 and 5 lpm excluding 1 lpm are 30.5 °C, 4.91 °C, and 28.7 °C, 3.28 °C, indicating that the flow rate of refrigerant was appropriate for battery safety. In contrast, in the summer condition of 35 °C, the maximum temperature of the battery and temperature difference between the batteries were 38.8 °C and 3.27 °C at the R134a flow rate of 5 lpm or more, which was verified as a stable flow condition for battery safety.

Keywords: battery thermal management; direct cooling; cylindrical battery; cooling performance; single phase cooling

Citation: Kim, M.; Ham, J.; Shin, D.; Cho, H. Comparison of Cooling Performance in a Cylindrical Battery with Single-Phase Direct Contact Cooling under Various Operating Conditions. *Batteries* **2022**, *8*, 195. https://doi.org/10.3390/batteries8100195

Academic Editors: Jinsheng Xiao, Hengyun Zhang and Sousso Kelouwani

Received: 6 September 2022
Accepted: 13 October 2022
Published: 20 October 2022

1. Introduction

As greenhouse gas emissions and the depletion of natural resources have become major issues in recent years, there is increasing interest in developing and applying eco-friendly energy to solve energy problems. Now, in the transportation industry, conventional internal combustion engine vehicles, which account for 49% of the existing fossil fuel oil resources and is the largest energy consuming and fastest growing sector in the world, are being replaced by eco-friendly energy vehicles [1,2]. In addition, many countries are investing much effort in achieving the goal of zero carbon emissions by 2050. In particular, the electric cars and battery industry have grown significantly to contribute to achieving zero carbon emissions through international agreements in the automobile industry [3]. Electric cars use lithium-ion batteries mainly for power sources, and lithium-ion batteries are a key element in energy storage and automobiles, having advantageous properties, such as high energy density, lightweight, no memory effect, and low-self-discharge rate [4]. However, in lithium-ion batteries, the operating temperature changes significantly due to electrochemical reactions during charging and discharging time, which directly affects the performance and lifespan of the battery. Hence, a cooling system to control the heat generation of the battery is essential [5].

In general, the safe operation and management temperature of a battery should be 20–40 °C and the temperature difference (TD) between batteries should be maintained at

5 °C or less [6]. Failure to adequately cool or control the heat generation of lithium-ion batteries results in reduced capacity and performance degradation of battery batteries due to the increase in temperature non-uniformity between battery cells and pack temperatures, which may lead to thermal runaway [7,8]. Wang et al. [9] experimentally tested the performance of the lithium-ion batteries according to the charge/discharge cycle when the ambient air temperature was 25 °C and 45 °C, respectively, and reported that the battery charge capacity decreased significantly over time, and specifically at high temperatures. In addition, it was verified that after 200 cycles, the capacity of the battery decreased to 21.3% and 22.6%, respectively, under the condition that the ambient air temperature was 25 °C and 45 °C. Li et al. [10] analyzed the change in internal structure due to mechanical deformation of 21700 lithium-ion batteries using X-ray computer tomography technology. As a result, as the mechanical deformation and the charge/discharge cycle increase, the internal resistance of the lithium-ion battery increases, and the capacity and state of health of the lithium-ion battery decrease significantly. Sun et al. [11] conducted experiments and analyzed the mechanical deformation characteristics against overcharging in 72 pouch battery modules. They reported that it is very important to prevent mechanical deformation of lithium-ion batteries because the pouch cell is deformed even at a low temperature (below −60 °C), and the volume change offset is 277% under internal pressure of 32,434 Pa. Supported by various study results, a battery thermal management system (BTMS) that maintains a constant temperature inside the battery, prevents mechanical deformation, and dissipates heat is essential for the normal operation of the battery.

The current study of battery cooling systems consists mainly of air cooling [12,13], liquid cooling [14,15], phase change material (PCM) cooling [16,17], and heat pipe cooling [18,19]. Air cooling poses no risk of leakage, such as liquid cooling. In addition, since cooling is performed using air, there are advantages, including the low maintenance cost, simple structure without a cooling loop, and low weight and energy consumption. However, due to the low heat capacity and thermal conductivity of air, the long-term high C-rate charge and discharge, high ambient temperature, and the like may cause problems such as battery malfunction or overheating [20]. Compared to air cooling, liquid cooling is more complex, heavy, and expensive to build but exhibits excellent cooling performance with a compact structure, high thermal capacity, and thermal conductivity [12]. In addition, the PCM-based cooling system has a significant advantage in terms of temperature distribution uniformity of the battery to be utilized as a cooling system. However, developing a single-PCM-BTMS is limited due to low thermal conductivity, structural instability due to leakage, and phase change. However, the PCM-BTMS system can be applied to batteries as a composite PCM [5]. Finally, due to its flexibility, low maintenance, and high thermal conductivity, the heat pipe cooling system rapidly expands within industries, including BTMS [21]. Such heat pipe-based cooling has not been applied in the battery thermal management field due to inefficient heat pipe and immaturity in the production standards and equipment, and remains in the research stage [22].

Other than the thermal management system presented above, a study on a refrigerant cooling system using a refrigerant is being conducted, and in the case of an electric vehicle, it can be applied as a system to expand a refrigeration system of a conventional air conditioning system, which has the advantage of reducing the cost. However, the refrigerant used in the heating, ventilation, and air conditioning (HVAC) system has to be maintained at a higher pressure than the liquid-cooling system. It is not easy to apply due to the design or control of the system. Shen et al. [23] conducted a study on improving the cooling performance of battery packs through indirect cooling using R134a and a cooling plate to study the refrigerant cooling system. As a result of the change in the flow rate and C-rate of the refrigerant, the maximum temperature (MXT) of the battery can be controlled within the safe operating temperature range, and a method for improving the thermal performance of the large battery pack to prevent from thermal runway expansion of the cell was proposed. Hong et al. [24] conducted a study on the thermal characteristics of batteries by comparing the performance of liquid cooling and refrigerant cooling systems under actual vehicle conditions. It was suggested that the two-phase refrigerant cooling method is more efficient than liquid cooling as refrigerant cooling provides 16.1% higher battery charge capacity and 15.0% lower internal resistance than liquid cooling, even in extreme environments.

Liquid cooling or refrigerant cooling system may be classified into direct-contact and indirect-contact cooling methods. Manufacturers widely use indirect cooling in the EV field in various BTMSs, including Audi, Tesla, GM, BMW, and Ford [25]. Xie et al. [26] analyzed the indirect liquid cooling system using an aluminum plate on the top and bottom parts of the battery module. They reported that it is an effective cooling technology that can optimize the thickness and flow rate of the cooling plate and maintain the MXT of the battery below 31.8 °C at a 3 C-rate. Chen et al. [27] proposed a 7-channel-based parallel liquid cooling system at the bottom of the battery and analyzed thermal performance and energy costs through an analytical study. They reported through analysis that the MXT, TD, and energy cost of the battery at 2.5 C-rate are 33.1 °C, 0.9 °C, and 17.29 J, respectively, which are effective cooling technology. Lu et al. [28] proposed an indirect liquid cooling method through a one-stage Tesla valve on the side of a battery and analyzed the valve angle, channel distance, and coolant effect at a 3 C-rate. Based on the analysis result, they suggested that it is an efficient cooling technology that can maintain the MXT of the battery below 30.5 °C. As such, indirect liquid cooling is performed using a working fluid obtained by mixing liquid with ethylene glycol to avoid electrically short-circuited and chemical erosion. However, these devices increase the non-uniformity of temperature distribution inside the battery pack and the cost and complexity of the battery due to a decrease in thermal conductivity. In addition, the battery performance and life cycle can be degraded significantly due to the non-uniformity of temperature distribution between the battery and the other battery [29].

Contrarily, direct-contact cooling does not require a channel design process of the working fluid and provides compactness and high efficiency compared to indirect contact cooling with high heat transfer performance. Nevertheless, direct contact BTMS has not been widely used [30]. In direct contact cooling, liquid usage is limited due to electrical and chemical reactions caused by direct contact with the battery and problems such as liquid leakage. However, various studies are currently conducted to apply direct contact cooling to battery systems. Tan et al. [25] designed a direct contact cooling system adopting HFE-6120 dielectric coolant under 3 C-rate conditions and analyzed the effect through system optimization. The study presented the results of improving the battery's MXT and temperature deviation by 18.1% and 25.0%, respectively, and provided direct contact cooling design guidelines. In addition, Wu et al. [29] designed a direct contact cooling system using silicone oil and reported that the MXT increment rate of a battery was significantly reduced by 20% to 30% compared to indirect contact cooling, which showed more efficient performance. Liu et al. [30] conducted theoretical and experimental verification with a direct contact cooling battery system using transformer oil (TO) at a 2 C-rate discharge rate, analyzed the effect of change in flow rate on performance, and Al Zareer et al. [31] established a direct contact refrigerant-based thermal management system using R134a refrigerant-based analyzed under 5 C-rate conditions. The result verified that when the working fluid covers 40% or more of the battery, the battery system is maintained below 30 °C to design a stable system.

In an actual battery, the working fluid for direct contact cooling is very limited, and fluid ensuring electrical insulation, non-toxicity, temperature resistance, and chemical stability must be used [30]. However, notwithstanding such disadvantages, direct contact cooling minimizes thermal resistance between the battery and the cooling fluid compared to indirect contact, thus showing an efficient cooling performance, and particularly, it is an effective method for dissipating heat from the battery even at a high battery heating value. Moreover, it is a cooling system that can be a based model for utilizing heat pipes and PCM cooling methods.

However, despite these shortcomings, direct contact cooling exhibits efficient cooling performance by minimizing the thermal resistance between the battery and the cooling fluid compared to indirect contact and is an effective way to dissipate heat from the battery, especially even at a high battery heating value. To increase the mileage of the electric vehicle, a large number of batteries must be installed in a limited space than before, thus excellent battery cooling performance is essentially required. Due to the dense arrangement of the battery, it is necessary to micronize the cooling plate and the cooling structure in order to utilize the indirect contact cooling method. Micronization of the cooling structure increases the pressure drop and requires high pump power consumption. On the other hand, the direct contact cooling method can efficiently manage battery heat due to a decrease in thermal resistance between the battery and the working fluid.

In this study, to evaluate the possibility of introducing the direct contact cooling method to the electric vehicle thermal management system, candidates for the working fluid (silicone oil, transformer oil, HFE-6120, R134a), considering the cost and their cooling performances, were investigated. In addition, the influence of the number and arrangement of inlet and outlet ports was evaluated to improve the battery cooling performance by optimization of the limited space of space. Finally, since the external air temperature affects the battery cooling system, the battery cooling performance according to the external air temperature was evaluated. This study is expected to contribute to the practical and efficient design for the BTMS of electric vehicles through a comprehensive analysis of the cooling performance and cost of working fluid of the direct contact cooling method.

2. Numerical Methodology

2.1. Battery and Cooling System Modeling

This study analyzed the cooling performance of the battery using ANSYS Fluent 2021 R2 [32]. Figure 1 shows a battery module model and fluid volume for analysis according to the change in the number of ports of the working fluid in the battery module. In this study, the cooling performance was studied by changing the number of inlet and outlet ports of the battery module to 1, 2, and 3. The battery model used for the analysis was a cylindrical 21700 battery, which has a capacity of 4.95 Ah and was designed with a total number of 96. The battery module in Figure 1 was designed to have a battery size of 21 mm (D) $\times$ 70 mm (H), a battery holder size of 181.6 mm (L) $\times$ 276.4 mm (W) $\times$ 14.5 mm (H), and a battery cooling block size of 183.6 mm (L) $\times$ 273.4 mm (W) $\times$ 72 mm (H). The gap between the batteries is maintained at 1.2 mm. For the analysis under the same flow conditions, the flow rate flowing into each port was designed as $\dot{Q}$ in Figure 1a, and the flow rate entering from Figure 1b,c to each port was designed as $\dot{Q}/2$ and $\dot{Q}/3$ of the flow rate flowing into in Figure 1a, respectively, to perform the analysis. Figure 1d shows the volume of fluid flowing through the battery cooling block in the model with three ports. In addition, Table 1 shows the thermal properties of each component of the battery module applied in this study.

The cooling performance and battery temperature distribution depending on the change of working fluid were analyzed to select a suitable working fluid for cooling in a battery module with a single port before evaluating the cooling performance according to the change of inlet and outlet ports of the battery block. The types of working fluids applied to the analysis are silicone oil, TO, HFE-6120, and R134a, and each working fluid was selected based on the working fluids used in the previous study [25,29–31]. R1234yf and R1234ze as new generation refrigerants are used as alternative refrigerants for R134a. The GWP for R1234yf, R1234ze, and R134a is 4, 7, and 1430, respectively, thus R134a has a higher GWP than R1234yf and R1234ze. However, R1234yf and R1234ze have flammability. Moreover, they are very expensive compared to other refrigerants such as R134a. In addition, the price of the working fluid is important to utilize the direct contact cooling method because the direct contact cooling method has a wider space for the fluid to flow compared to the existing indirect contact cooling method. Therefore, R1234yf and R1324ze as working fluids in the direct contact cooling method are excluded in this study [34]. Accordingly, R134a was selected as the working fluid candidate group for direct contact cooling.

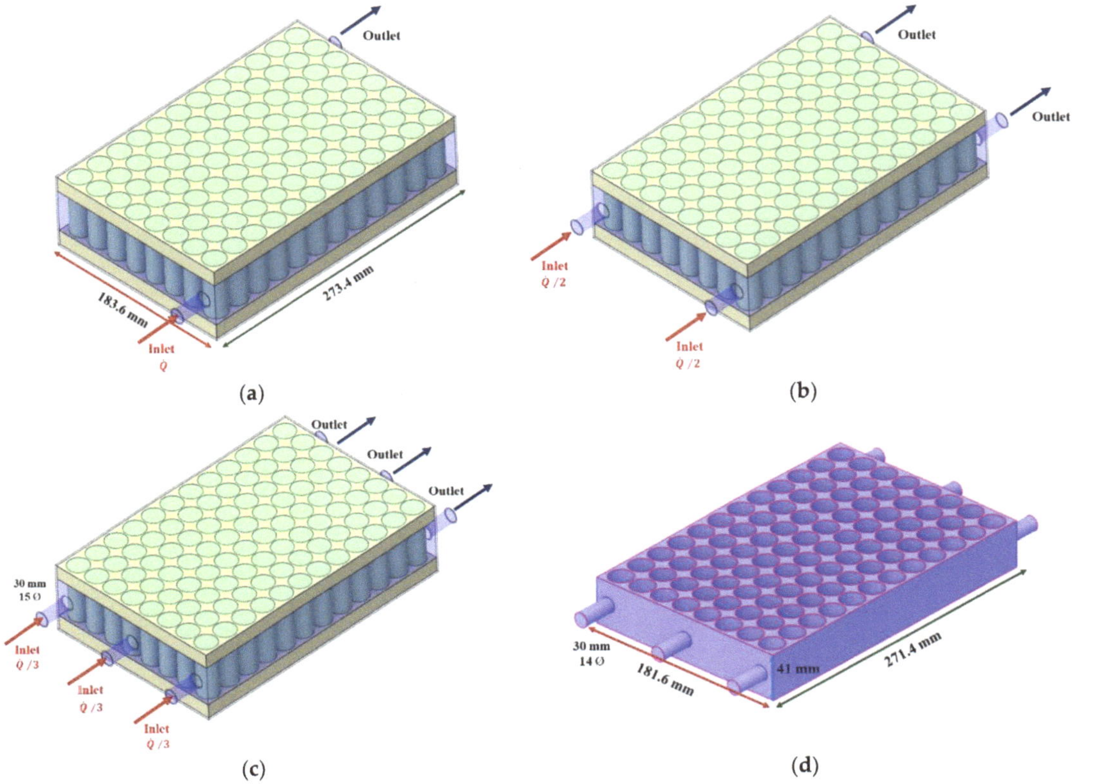

Figure 1. Modeling of the battery module according to the port number: (**a**) Port = 1; (**b**) Port = 2; (**c**) Port = 3; (**d**) Fluid volume at port 3.

Table 1. Thermo-physical properties of the battery module [33].

Parameters	Battery	Battery Holder	Battery Cooling Block
Density (kg m^{-3})	2525.5	1200	2719
Specific heat (J kg^{-1} °C^{-1})	$1040.5 + 2.29 \times T$	1250	871
Thermal conductivity (W m^{-1} °C^{-1})	23.6	0.2	202.4

Table 2 shows the typical thermal properties and cost of working fluids applied in this study. The silicone oil [29] is obtained at an operating temperature of 17 °C, TO [30] is obtained at an operating temperature of 20 °C, and HFE-6120 [25] is obtained at an operating temperature of 21 °C. In general, the thermal properties of working fluids change according to temperature. Thus, thermal properties should be applied according to temperature. In addition, the cost of working fluid was confirmed as R134a showed the lowest cost at $41/kg [35], while silicon oil showed the highest cost at $245/kg [36]. The selected working fluids for the batter thermal system were analyzed based on the model presented in Figure 1a to select the recommendable working fluid for battery cooling. A model optimized for battery thermal management was derived with the analysis of the cooling performance of various ports and header width. The width of the header applied in this study was applied equally to both the inlet and outlet, and the header width was changed to 0 mm, 5 mm, and 10 mm to be analyzed. Figure 2 shows a model with a header width of 0 mm and 5 mm to consider the battery cooling performance according to the header width change.

Table 2. Thermo-physical properties and cost of the working fluid [25,29,30,35–38].

Parameters	Silicone Oil	TO	HFE-6120	R134a
Density (kg m^{-3})	965	820	1600	$(-1.20 \times 10^{-2}) \times T^2 - 3.28 \times T + 1297.94$
Specific heat (J kg^{-1} °C^{-1})	2490	2530	1170	$(3.62 \times 10^{-2}) \times T^2 + 2.52 \times T + 1336.14$
Thermal conductivity (W m^{-1} °C^{-1})	0.16	0.3875	0.23	$(-1.42 \times 10^{-6}) \times T^2 - (4.29 \times 10^{-4}) \times T + 0.10$
Viscosity (Pa s)	50	0.0082	0.0019	$(1.83 \times 10^{-8}) \times T^2 + (-3.40 \times 10^{-6}) \times T + (2.69 \times 10^{-4})$
Cost ($ kg^{-1})	245	81.68	217	41

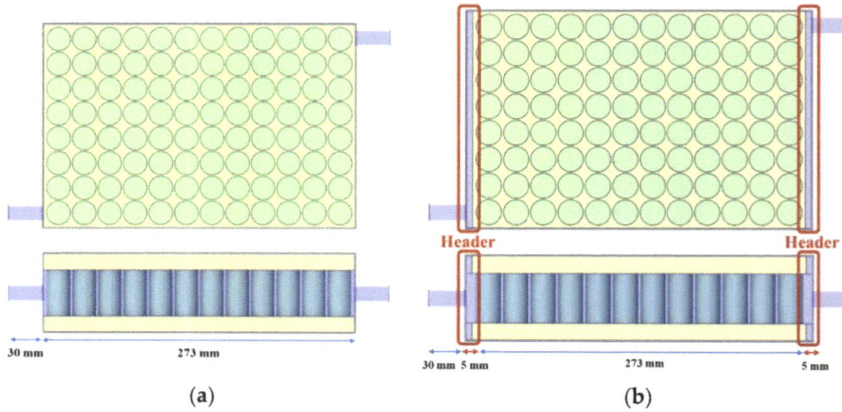

Figure 2. Modeling of the battery module according to the changing header width: (**a**) Header width = 0 mm; (**b**) Header width = 5 mm.

2.2. Governing Equation and Boundary Conditions

In the direct cooling system, the heat dissipation of the battery is achieved by direct heat exchange between the battery and working fluid and has a very complex structure. Therefore, in order to simplify the analysis, the following assumption was made.

(1) There is no phase change in the interpretation of this analysis.
(2) It is assumed that the working fluid is filled with the cooling model, and there is no air.
(3) Thermal resistance and radiation are not considered.
(4) The working fluid is incompressible.

The equation of energy generated in a battery cell is as follows [39].

$$\rho_{ba} c_{ba} \frac{\partial T_{ba}}{\partial t} = -\lambda_{ba} \nabla^2 T_{ba} + \dot{q} \quad (1)$$

where T, ρ, and c are temperature, density, and specific heat capacity of the battery, respectively. λ and $\dot{q}$ are thermal conductivity and volumetric heat generation rate of battery (subscript *ba*), respectively.

The equation of heat generated in a battery (Q_{gen}) is as follows [40,41].

$$Q_{gen} = Q_{ir} + Q_{re} = I^2 R - IT \frac{\partial U_{OCV}}{\partial T} \quad (2)$$

where Q_{gen}, Q_{ir}, and Q_{re} represent total heat generation, irreversible heat and reversible heat of battery cell, respectively. I and R represent discharge/charge current and equivalent internal resistance of the battery cell, respectively. T represents the battery temperature. $\partial U_{OCV}/\partial T$ represents the temperature coefficient of open-circuit voltage.

The mass, momentum, and energy balance equations of the governing equations of working fluid in the battery module are as follows [31].

$$\frac{\partial \rho_{wf}}{\partial t} + \nabla\cdot(\rho_{wf}\vec{u}) = 0 \tag{3}$$

$$\frac{\partial}{\partial t}(\rho_{wf}\vec{u}) + \nabla\cdot(\rho_{wf}\vec{u}\vec{u}) + \nabla P = \rho g \tag{4}$$

$$\frac{\partial}{\partial t}(\rho_{wf}c_{wf}T_{wf}) + \nabla(\rho_{wf}c_{wf}\vec{u}T_{wf}) = -\nabla(\lambda_{wf}\nabla T_{wf}) \tag{5}$$

where $\vec{u}$, P, and g are velocity, static pressure, and gravitational acceleration (subscript wf), respectively.

In addition, the cooling capacity of the cooling system for controlling the battery heating value is calculated as follows.

$$Q = \dot{m}c_{wf}(T_{out} - T_{in}) \tag{6}$$

where $\dot{m}$ is mass flow rate, Q is cooling capacity, while T_{out} and T_{in} are outlet temperature and inlet temperature of working fluid, respectively.

In the battery cooling block, the theoretical pump power is assumed to be 100% efficiency, and the pump power required at this time can be expressed by the following simple equation.

$$W_{pump} = \Delta P\dot{Q} \tag{7}$$

where W_{pump} is pump power, ΔP is the pressure drop between the inlet and the outlet pressure of the cooling plate, and $\dot{Q}$ is the flow rate.

In this study, there are four types of fluids considered as working fluids in the battery cooling block, among which silicone oil, TO, and HFE-6120 are shown as Reynolds number Re < 2300, in which a Laminar flow model is applied. An R134a, Re > 2300, is applied to a realizable k-ε turbulence model. In addition, this study applied 1 lpm (L/min), 3 lpm, and 5 lpm to the flow rates of working fluid and set the outdoor air temperature, the temperatures of the initial battery system, and the inlet of working fluid to 25 °C. Here, the convective heat transfer coefficient for the ambient air was 10 W/m·K [42]. The battery charge/discharge rate applied to the analysis and verification experiments was applied at a 2 C-rate. In addition, when a direct contact battery cooling block was applied to the summer, the temperature distribution characteristics and cooling performance of the battery were additionally reviewed under outdoor air and inlet temperature of working fluid of 35 °C [43,44] based on the developed analysis model.

2.3. Grid Independence Test

The analysis for evaluating grid independence was conducted by applying refrigerant R134a fluid at initial battery temperature, outdoor air temperature, and coolant inlet temperature at 25 °C when the number of the port was 1 and the flow rate was 1 lpm. Figure 3 shows the change in pressure drop in the battery module according to the number of meshes. When the initial number of meshes was 3,885,580, the pressure drop decreased rapidly from 33.68 Pa, and even if the number of meshes increased further from 4,063,590 to 33.14 Pa, the pressure drop did not increase and converge as a result. Therefore, this study analyzed the battery temperature and cooling performance of the battery cooling block using 4,063,590 meshes.

2.4. Verification Experiment Setup and Method

An experiment was conducted on cooling during battery charging and discharging in order to verify the reliability of the analysis model used, and a schematic diagram of the experimental apparatus was shown in Figure 4. As shown in Figure 4a, the experimental apparatus for verification consists of Power supply (PNE Solution Co., Suwon, Korea, Uncertainty: ±0.1%), Data acquisition (GRAPHTEC Co., Yokohama, Japan, Uncertainty: ±0.5%), computer, temperature chamber, 1 serial × 16 parallel (1S16P) battery module, T-type thermal couple (Omega Co., Techlink, Singapore, Uncertainty: ±0.5). As a condition for the battery charge/discharge experiment, the experiment was carried out under 2 C-rate conditions after maintaining the ambient temperature at 25 °C. 1S16P battery module has a capacity of 80 Ah and consists of a battery holder, a busbar, and a 21,700 battery. Figure 4a shows the location of the T-type thermocouple attached to the battery module. A total of nine thermocouples were used. Three thermocouples were installed at the top, middle, and bottom of the battery, and the average temperature of the battery was used. The presented experiment results are the average value of 5 repeated experiments, and the error of the obtained temperature is 6.43%.

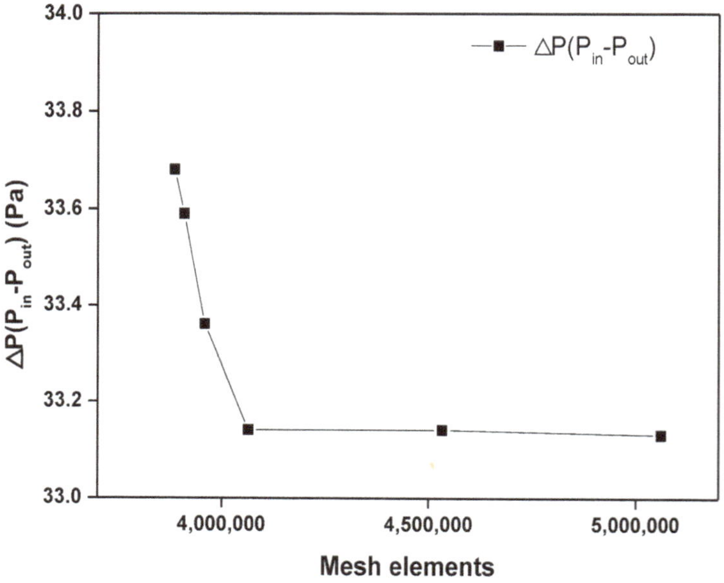

Figure 3. Grid independence according to mesh elements.

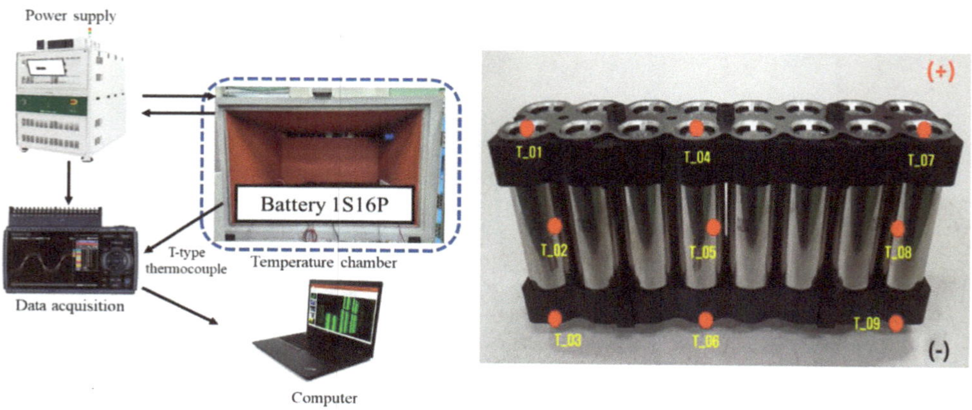

(**a**) (**b**)

Figure 4. Verification experimental setup: (**a**) Schematic diagram of experimental setup; (**b**) 1S16P battery module.

3. Results and Discussion

3.1. Verification of Battery Analytical Model

A change in temperature of the battery according to the time under the 2 C-rate discharging condition is shown in Figure 5. Figure 5a shows a change in the average temperature of each of the three thermocouples located at the upper end (T_01, 04, 07), middle end (T_02, 05, 08), and bottom end (T_03, 06, 09) under natural convection condition. When discharging was performed for 1800 s under 2 C-rate conditions, the battery temperature continued to rise over time, and the average temperature at the top, middle, and bottom ends of the battery was 52.3 °C, 60.2 °C, and 54.2 °C, respectively. Heat is generated by a chemical reaction internal cell during the charging and discharging of the battery cell. The temperature in the middle of the battery during discharging was the highest, and the top and bottom parts of the battery showed almost the same temperature. Since the heat is generated by a chemical reaction in the battery cell, the temperature farther from the center of the battery cell is lower than that at the center. It is confirmed that a similar temperature distribution was also observed in the verification experiment on the heat generation of the battery cell. To compare the experimental and analysis results of the 1S16P battery module when the natural convection cooling method was applied under the same conditions, the MXT of the battery was compared and presented in Figure 5b. The MXT of the battery was 58.3 °C and 60.5 °C, respectively, in the analysis and experiment when the time of discharge reached 1800 s. In addition, the change in the MXT of the battery over time showed a similar tendency, and the average error rate of the MXT over time was 2.09%, verifying the reliability of the battery analysis model used in this study.

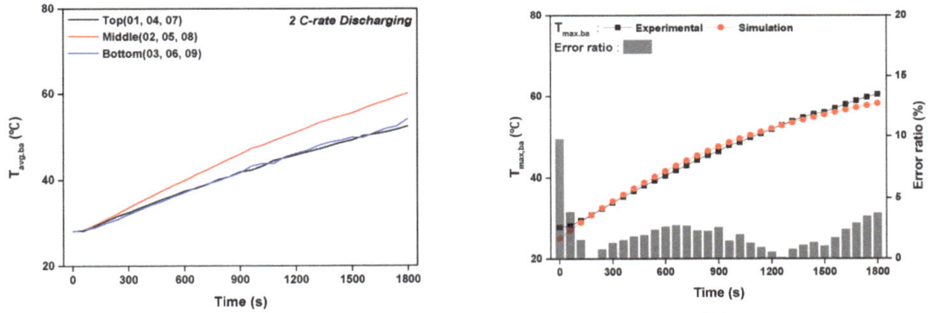

Figure 5. Battery temperature variation under 2 C-rate discharge condition: (**a**) Average temperature according to time; (**b**) Comparison of MXT according to time between experimental and simulation results.

3.2. Cooling Performance according to the Working Fluid

The cooling performance and temperature distribution according to the change of working fluid were analyzed by applying a battery module with a single port, as illustrated in Figure 1a, to select a suitable working fluid necessary for cooling the battery module. The initial, ambient, and inlet temperature was 25 °C, and the flow rate of the working fluid was 1 lpm, and Figure 6 illustrates the battery temperature, pressure drop of the battery cooling block, and cooling performance depending on the change of the working fluid. Figure 6a,b show the MXT and a TD between the batteries according to the change of the working fluid. When discharging for 30 min under natural convection conditions without cooling using working fluid, the MXT of the battery was 58.3 °C, and in the battery cooling block, when the working fluids were silicone oil, TO, HFE-120, and R134a, the MXT of the battery was 35.3 °C, 35.1 °C, 34.9 °C, and 35.9 °C, respectively. In addition, as presented in Figure 6b, the TD between batteries were 8.56 °C, 9.20 °C, 9.6,3 °C, and 10.30 °C, respectively, and the TD between batteries was the largest in the order of R134a, HFE-6120, TO and silicone oil. Based on the analysis results, HFE-6120 was the most suitable fluid for managing the

MXT of the battery, and silicone oil was the most suitable for managing the TD between batteries.

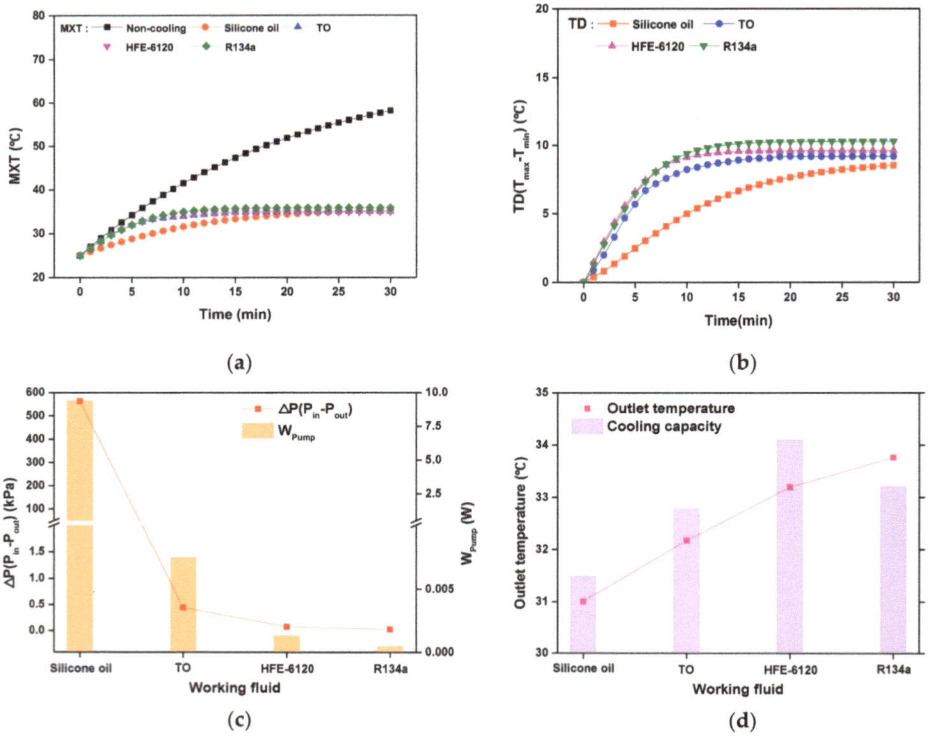

Figure 6. Variation of cooling performance according to working fluids: (**a**) MXT of battery according to time; (**b**) TD of battery according to time; (**c**) Pressure drop and pump power for different working fluids; (**d**) Outlet temperature and cooling capacity for different working fluids.

Figure 6c shows a comparison of pressure drop and pump power for each working fluid in the battery module. Silicone oil, TO, HFE-6120, and R134a were applied, and the pressure drop of the battery cooling block was 564 kPa, 0.45 kPa, 0.08 kPa, and 0.03 kPa, respectively, and the power required for the pump was 9422 mW, 0.08 mW, 1.33 mW, and 0.55 mW. The pressure drop for each working fluid is in the order of silicon oil, TO, HFE-6120, and R134a. Since the silicone oil and TO, has a higher viscosity than R134a and HFE-6120, the use of such a large viscosity leads to a decrease in the performance and capacity of an electric vehicle or battery system. Figure 6d shows the outlet temperature and cooling capacity of the battery cooling block for each working fluid. When silicon oil, TO, HFE-6120, and R134a are applied, the outlet temperatures of the battery cooling block are 31.0 °C, 32.2 °C, 33.2 °C, and 33.8 °C, respectively, followed by order of R134a HFE-6120, TO, and silicone oil. In addition, the cooling capacity of the battery cooling block was 207.5 W, 213.9 W, 220.6 W, 216.1 W, respectively, in the order of HFE-6120, R134a, TO, and silicone oil. This difference is caused by the thermal properties of each working fluid, and it is determined that HFE-6120 and R134a, excluding silicone oil and TO, which consume a large pump power, are appropriate cooling fluids. As a result, the cooling capacity of HFE-6120 and R134a was 4.47 W, indicating that HFE-6120 showed slightly greater cooling performance, but HFE-6120 generated 0.8 mW more power. However, HFE-6120 is a newly developed material, and the price of HFE-7000 is $217/kg [37], 5.29 times higher than R134a, which is $41.0/kg [35]. Therefore, this study comprehensively considered cooling performance, the power required, and the economic aspect and selected R134a as a working fluid to cool the direct contact single phase of the battery module. The cooling performance according to variables is considered analytically.

3.3. Cooling Performance according to the Number of Port and Header Width Using R134a

The cooling performance according to the number of inlet and outlet ports of the battery cooling block and the change in the width of the inlet and outlet headers was studied using R134a selected through the analysis results as the working fluid. The number of ports of the battery cooling block was changed to 1, 2, and 3, and the width of the inlet/outlet header was changed to 0 mm, 5 mm, and 10 mm (as shown in Figure 2) to perform the analysis. The results are shown in Figure 7. Figure 7a,b show the MXT of the battery and a TD between the batteries according to the change of width. In addition, Figure 7c,d show the pressure drop and cooling capacity of R134a in the battery cooling block. As the header width increased from 0 mm to 10 mm in a single port, the MXT of the battery and the TD between batteries were higher in the order of the models with header widths of 10 mm, 5 mm, and 0 mm, whereas the pressure drop of the battery cooling block and the cooling capacity showed the largest value in the order of header width 0 mm, 5 mm, and 10 mm. The pressure drop of the battery cooling block in a single port decreases to 10.2 mm/s, 9.19 mm/s, and 8.43 mm/s, respectively, when the header widths are 0 mm, 5 mm, and 10 mm, the maximum temperature difference between the batteries increases, while the pressure drop decreases to 33.1 Pa, 20.1 Pa, and 18.9 Pa, respectively. Thus, when the battery cooling block has a single port, the increase in the width of the inlet/outlet header has a negative effect on the battery thermal management. The best cooling performance was shown when the header width was 0 mm. In addition, when there are two ports, the MXT of the battery, the TD between batteries, and the cooling capacity of the battery cooling block were significantly higher in the order of 10 mm, 0 mm, and 5 mm. At the same time, the pressure drop was more significant in the order of 0 mm, 5 mm, and 10 mm. For a single port, the header width of 0 mm showed the best cooling performance, whereas for two ports, the header width of 5 mm showed the best cooling performance. With two ports and 0 mm header width, the cooling fluid does not flow in the center of the outlet and shows a dead zone where the cooling fluid stagnates. Battery cooling by increasing the number of headers is related to the fluid inflow characteristics according to the location of the header. In the battery cooling structure, the distance between the inlet and the battery is very close. Therefore, a smooth distribution of working fluid does not occur when the working fluid enters the cooling area. Therefore, the working fluid flows to the side with relatively little flow resistance, and the battery located in the center cannot be cooled sufficiently. Due to the difference in the number of ports, there is a difference in entering velocity from the inlet port to the battery cooling area. Therefore, the effect of reducing the flow resistance by the header width is different. As the number of ports increases, entering velocity is slow. The flow resistance through the batteries decreases and the flow of fluid into the center becomes smooth. Due to uniform flow velocity, the cooling performance increases. On the other hand, when the number of ports is two, the inlet port is located on the side. Therefore, the fluid flows smoothly due to the collision of the fluid flow in the center. When the header width of the cooling structure with 3 ports exceeds 5 mm, the flow resistance to the side is lower than the flow resistance in the center. Thus, the battery cooling performance is reduced due to the non-uniformity of the fluid flow distribution.

When the number of inlet/outlet ports is two, the flow phenomenon at the outlet side can be identified in Figure 8. Figure 8 shows the temperature and velocity distribution according to the change in header width in the battery cooling block with two ports. In header widths 0 mm, 5 mm, and 10 mm, the MXT of the battery all showed in the red zone of the outlet, and the temperatures were 39.0 °C, 38.7 °C, and 41.0 °C, respectively. At header widths 0 mm, 5 mm, and 10 mm, the velocity of R134a in the battery cooling block was 88.2 mm/s, 76.4 mm/s, and 75.7 mm/s, respectively. When header width is 0 mm, a space in the inlet and outlet is narrow, so the flow rate of cooling fluid flows along the wall of the blue zone is higher than that of an internal battery, so the fluid velocity on the wall is faster than in a case where a header width is 5 mm or 10 mm. This phenomenon has been verified in Figure 8D–F. These figures enlarged the velocity distribution of the red zone, showing the MXT of the battery. The maximum velocity in the red zone at header widths of 0 mm, 5 mm, and 10 mm were 15.3 mm/s, 18.8 mm/s, and 18.9 mm/s, respectively, which were much lower than the average velocity. As the header width increases from 0 mm to 5 mm, the maximum velocity of R134a in the battery cooling block decreases to 11.8 mm/s. Still, the maximum velocity in the red zone increases to 3.50 mm/s, so the cooling effect between the batteries increases. As a result, the MXT of the battery and the TD between the batteries decreased by 0.32 °C and 0.47 °C, respectively. However, when the header width increases from 5 mm to 10 mm, the maximum velocity of R134a in the battery cooling block decreases by 0.66 mm/s, and the maximum velocity of the red zone increases by 0.14 mm/s. This is because the inlet/outlet space increases due to an increase in header width, and a stagnation section at the rear surface of the battery increases due to a low velocity of R134a, so the MXT of the battery increases by 2.36 °C and 2.44 °C, respectively. In addition, the cooling capacity of the battery cooling block was decreased by 3.84 W, which adversely affected the battery thermal management. These results verified that a model with a header width of 5 mm showed the best cooling performance in a model with two ports of the battery cooling block. A model with two ports showed a better maintenance cost because of a smaller pressure drop compared to a model with a single port, but a model with two ports and a battery with a maximum temperature difference showed a smaller cooling capacity in the same header width than a model with a single port and three ports.

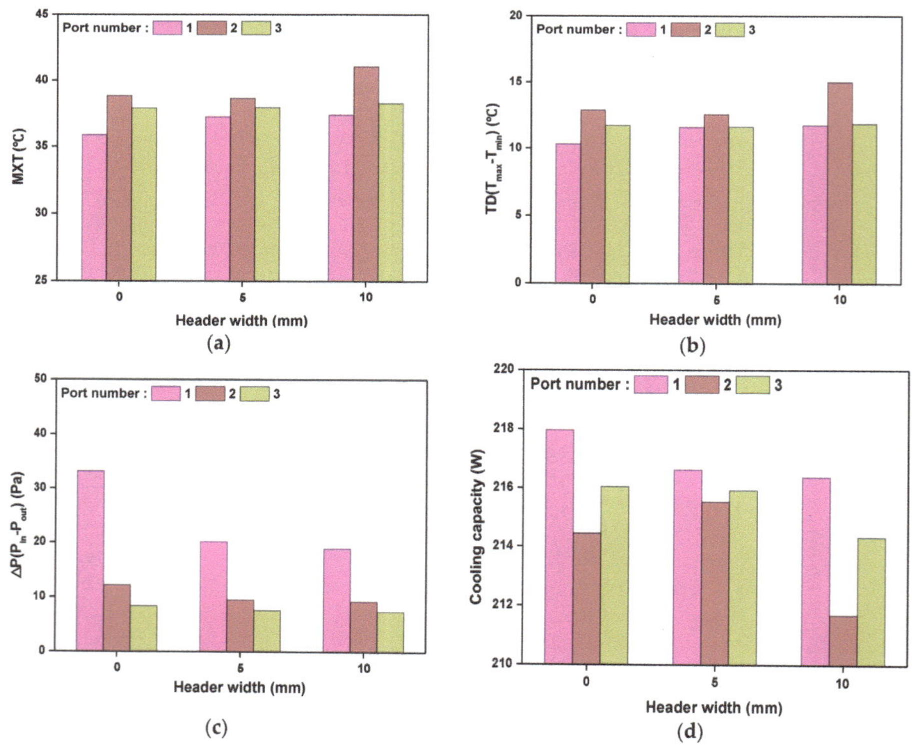

Figure 7. Comparison of cooling performance according to header width: (**a**) MXT of battery for different header width; (**b**) TD of battery for different header width; (**c**) Pressure drop for different header width; (**d**) Cooling capacity for different header width.

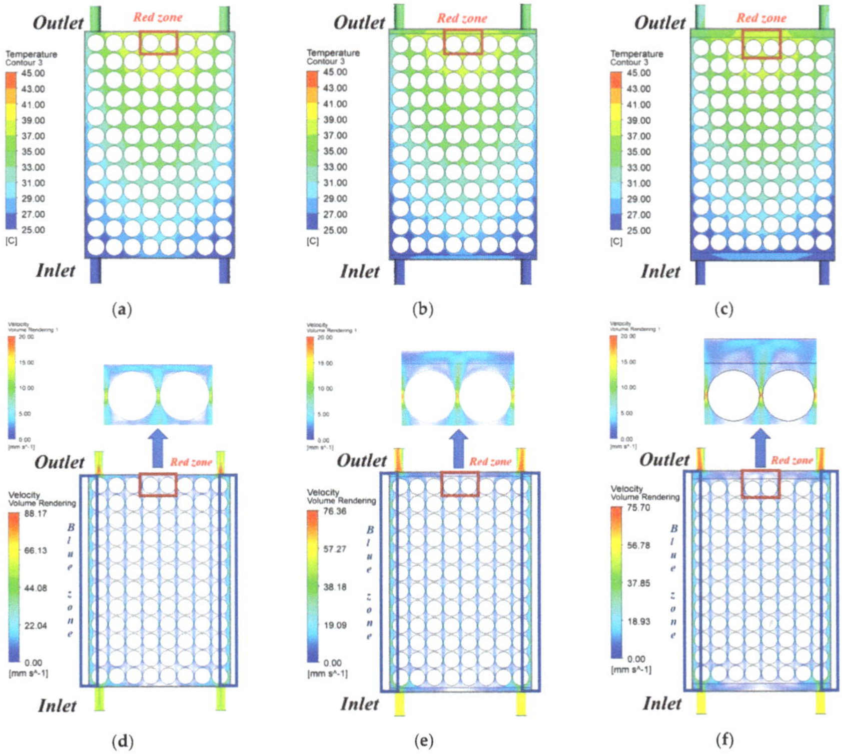

Figure 8. Temperature and velocity distribution according to header width at the battery cooling block with port 2: (**a**) Temperature contour at 0 mm; (**b**) Temperature contour at 5 mm; (**c**) Temperature contour at 10 mm; (**d**) Velocity contour at 0 mm; (**e**) Velocity contour at 5 mm; (**f**) Velocity contour at 10 mm.

When the number of ports in the battery cooling block is three and the header width increases from 0 to 10 mm, the MXT of the battery is higher in the order of the models with header widths of 10 mm, 5 mm, and 0 mm, and the TD between the batteries is 10 mm, 0 mm and 5 mm model showed the highest temperature, and the pressure drop and cooling capacity of the battery cooling block showed the largest values in the order of 0 mm, 5 mm, and 10 mm models. As the header width increased from 0 mm to 5 mm, the MXT of the battery increased by 0.03 °C, while the cooling capacity decreased by 0.12 W, which had a negative effect on the battery performance. However, the TD and pressure drop between the batteries was 0.11 °C and 0.79 Pa decreased, showing excellent cooling performance. This is because, as illustrated in Figure 8, the increase in header width from 0 mm to 5 mm resolves the flow of fluid that has been stagnant at the center of the outlet and reduces the TD and pressure drop between batteries, thereby having a positive effect on BTMS. On the other hand, as the header width increases from 5 mm to 10 mm, the MXT of the battery and the TD between the batteries increase by 0.33 °C and 0.24 °C, and the cooling capacity of the battery cooling block decreases by 1.60 W, which degrades the cooling performance. As a result, it was found to have an adverse effect on battery thermal management. A pressure drop of the battery cooling block decreased by 0.2 Pa from 7.57 Pa, but as the MXT of the battery and the TD between the batteries increases, and the cooling capacity decreases, under the condition of having three ports, a model with a header width of 10 mm is relatively inefficient, and a header width of 5 mm having three ports is the most effective model for the battery thermal management. As a result of analysis according to various condition changes, the model with the best battery cooling performance was found to have a header width of 0 mm for a single port and a header width of 5 mm for three ports, and when comparing the performance under two conditions, the MXT of the battery and the pressure drops of the battery cooling block was 35.9 °C, 33.1 Pa, and 37.9 °C, 7.57 Pa, respectively. In other words, under the condition of having a single port, the MXT of the battery was lower than in the condition of having three ports, but the pressure drop of the battery cooling block increased by approximately 4.38 times In both cases, a battery cooling block with a header width of 5 mm was selected as the optimal model considering a pressure drop that significantly affects maintenance cost, because the MXT of the battery is maintained under a safe operating temperature, which was below 40 °C. Based on the selected model, the battery temperature and cooling performance were compared depending on the change of the working fluid flow rate under the standard conditions of 25 °C and summer conditions of 35 °C, and the flow rate conditions were analyzed to maintain the safe operation management temperature of the battery in summer.

3.4. Comparison of Cooling Performance according to the Flow Rate of R134a under Standard and Summer Conditions

This study selected R134a as the working fluid through analysis according to changes in the working fluid, number of inlet and outlet ports, and header width. It was verified that a model with three ports of battery cooling block and 5 mm header width had the best cooling performance. The analysis was performed assuming that the initial temperature of the battery cooling block and inlet temperature of the working fluid, and the outside air temperature are 25 °C and 35 °C, respectively, to compare the battery cooling performance under standard and summer conditions. In addition, by changing the flow rate of R134a, which is the working fluid, the flow rate of the working fluid to maintain the MXT of the battery, which is the safe operation management temperature of the battery, 40 °C, and the TD between the batteries of 5 °C or less were verified. Figure 9 shows the comparison of cooling performance according to the flow rate change of R134a in the model with three inlet and outlet ports. Figure 9a shows the MXT of the battery according to the discharging time under standard and summer conditions. When the flow rate of R134a was 1, 3, and 5 lpm under standard and summer conditions, the MXT of the battery at 30 min of discharging time was 38.0 °C, 30.5 °C, 28.7 °C, and 48.0 °C, 40.5 °C, and 38.8 °C, respectively. In the case of standard conditions, the MXT of the battery was below the safe operating temperature of 40 °C under all flow conditions, indicating stable performance for the battery thermal management. However, in summer conditions, the MXT of the battery was 40 °C or higher at 1 lpm and 3 lpm except for 5 lpm, confirming that the battery cooling system was not suitable for thermal management. Figure 9b shows the MXT of the battery and a TD between batteries according to outdoor air conditions. When the discharging time reaches 30 min, and the flow rate of R134a is 1 lpm, 3 lpm, and 5 lpm under standard and summer conditions, the TD between the batteries was 11.6 °C, 4.91 °C, 3.28 °C, and 11.6 °C, 11.6 °C, 4.85 °C, 3.27 °C, respectively. Depending on the change in the outdoor air conditions, the variation in the TD between batteries at 3 lpm flow rate of the same working fluid was at the maximum of 0.06 °C. In addition, at 1 lpm, the TD between batteries in both standard and summer conditions was 6.63 °C and 6.61 °C, which is significantly higher than 5 °C, verifying that it is not suitable for battery thermal management. However, under the condition that the flow rate of R134a is 3 lpm or more, the TD between batteries is maintained at 5 °C or lower, verifying that the flow rate condition is suitable for the battery thermal management.

Figure 9c shows the pressure drop of the battery cooling block according to the change in the outdoor air condition. When the discharge time reached 30 min, as the flow rate of R134a increased to 1 lpm, 3 lpm, and 5 lpm under standard and summer conditions, the pressure drop of the battery cooling block was 7.57 Pa, 45.5 Pa, 124 Pa, and 6.82 Pa, 41.7 Pa, and 103 Pa, respectively, and the required power of the pump was 0.13 mW, 2.27 mW, 10.4 mW, and 0.11 mW, 2.09 mW, and 8.58 mW, respectively. When the flow rate of R134a was 1 lpm, 3 lpm, and 5 lpm, the difference in pressure drop of the battery cooling block was 0.75 Pa, 3.75 Pa, and 21.4 Pa as the ambient temperature changed from 25 °C to 35 °C. The pressure drop varies greatly depending on the change in ambient temperature because the density of R134a, which is directly proportional to the pressure drop, decreases as the temperature increases [45]. Therefore, an increase in the flow rate of working fluid leads to a pressure drop in the battery cooling block, which is proportional to an increase in the power required by the pump, so the flow rate of appropriate cooling fluid for cooling has to be controlled The outside temperature increases the inlet temperature of the battery cooling system. Accordingly, the viscosity of the working fluid decreases due to the increased temperature of the working fluid, and the pressure drop also decreases. However, it does not significantly affect the cooling performance of the battery cooling system. In Figure 9d, the cooling capacity of the battery cooling block according to the outdoor conditions is shown according to the change in the flow rate of R134a. When the discharging time reached 30 min, the cooling capacity of the battery cooling block was 215.9 W, 221.5 W, 222.8 W, and 215.9 W, 221.5 W, 222.8 W, as the flow rate increased to 1 lpm, 3 lpm, and 5 lpm under standard and summer conditions. This is because the outdoor temperature affects the temperature of the surface of the battery cooling structure, but does not significantly affect the temperature of the working fluid that affects cooling performance. Therefore, the cooling performance is not significantly reduced.

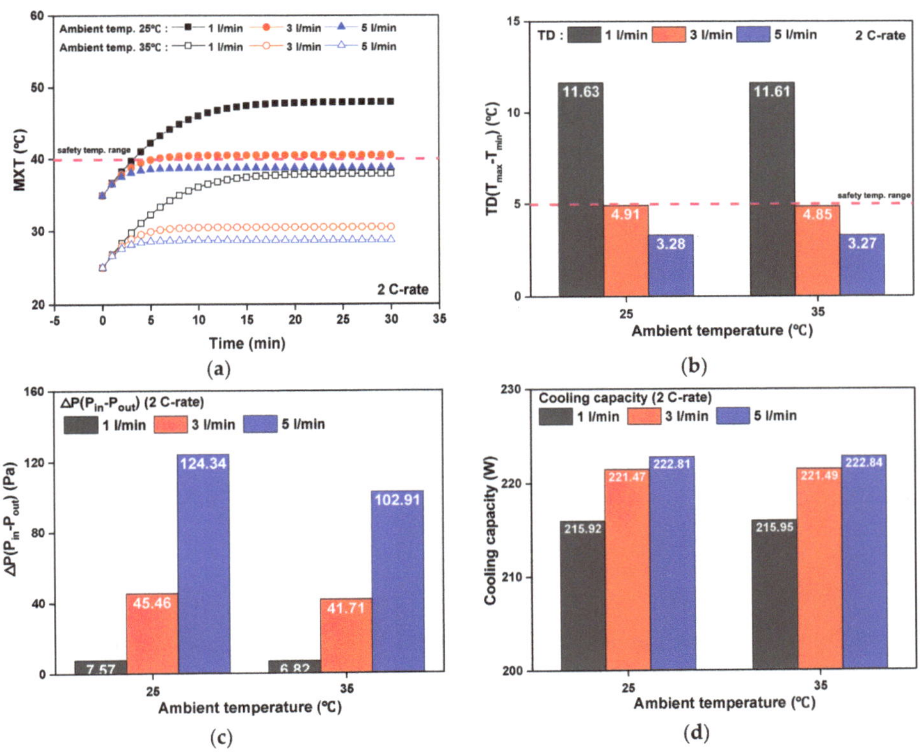

Figure 9. Comparison of cooling performance according to flow rate under 25 °C and 35 °C outdoor air conditions: (**a**) MXT of the battery according to time; (**b**) TD of the battery according to time; (**c**) Pressure drop for different ambient temperatures; (**d**) Cooling capacity for different ambient temperatures.

Figure 10 illustrates the temperature distribution of a battery module depending on a change in flow rate when applying summer outdoor conditions in a model where the number of ports of the battery cooling block is three and the header width is 5 mm. In the red zone of all flow bands, the battery showed a high temperature distribution, and the blue zone showed a relatively low temperature distribution compared to the red zone. The maximum velocity of R134a in the blue zone of the battery cooling block was 37.8 mm/s, 121 mm/s, and 220 mm/s when the flow rate was 1 lpm, 3 lpm, and 5 lpm, respectively, and the maximum velocity in the red zone was 25.9 mm/s, 75.4 mm/s, and 139 mm/s, respectively. When the R134a flow rate of the battery cooling block was 1 lpm, 3 lpm, 5 lpm, the blue zone showed a relatively low temperature distribution because the maximum velocity of the blue zone was 11.9 mm/s, 45.6 mm/s, and 81 mm/s faster than the maximum velocity of the red zone. In addition, as the flow rate of R134a increases from 1 lpm to 3 lpm and 5 lpm, the MXT of the battery decreases to 7.5 °C and 9.2 °C, respectively, confirming that an increase in the flow rate of R134a could significantly reduce the MXT of the battery. As a result of the analysis, according to the change in ambient temperature, the TD between batteries was 5 °C or higher under the 1 lpm condition of the flow rate of R134a, which was not suitable for the battery thermal management condition. Under the standard conditions of 25 °C, the MXT of the battery, which is a safe operation management temperature of the battery, and the difference between batteries were maintained at 40 °C and 5 °C or less, but as the pressure drop increases by about 2.74 times as the pressure increases from 3 lpm to 5 lpm, it is verified that 3 lpm is the most suitable condition for battery heat management under the standard conditions. However, in summer, unlike standard conditions, the MXT of the battery and the TD between the batteries were 40.5 °C and 4.91 °C, so the difference between batteries was maintained at 5 °C or lower, but the MXT was 40 °C or higher, exceeding the safe operation management temperature of the battery. Therefore, in the summer condition of 35 °C, the condition where the flow rate of R134a is 5 lpm was confirmed as a suitable condition for battery thermal management.

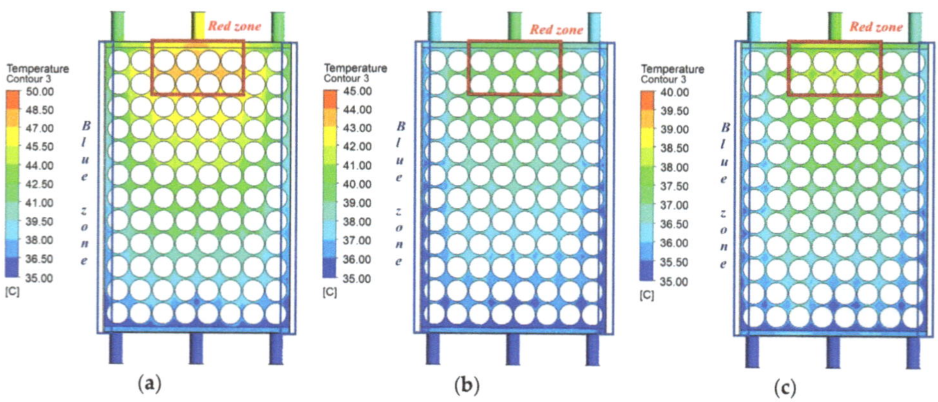

Figure 10. Temperature variation of battery module according to the flow rate under 35 °C outdoor air condition: (**a**) 1 lpm; (**b**) 3 lpm; (**c**) 5 lpm.

4. Conclusions

In this study, to examine the applicability of direct contact cooling, the cooling performance characteristics of the working fluids (silicone oil, transformer oil, HFE-6120, R134a) that can be recommended as direct contact cooling for the system with 96 batteries of 21700 type are analyzed and optimized. In addition, the effect on the number and arrangement of the inlet and outlet ports was investigated. Finally, the influence of outside air temperature on direct contact cooling was also studied. The main conclusions obtained through this study are as follows.

(1) The analysis result of the different working fluids in direct contact cooling with a single port verified that silicone oil and TO have a pressure drop and required power of about ten times more than R134a and HFE-6120. In addition, among the studied working fluids, HFE-6120 was found to have a cooling capacity of 4.47 W higher than that of R134a, but the required power was 0.8 mW smaller than that of HFE-6120, showing a small

maintenance ratio. In addition, among the four fluids, R134a, which has advantages in terms of power requirement and price, has also been identified as a fluid for proper cooling because the HFE-6120 has a price 5.29 times higher than that of R134a.

(2) As a result of the analysis of cooling performance and pressure drop according to the number of ports and header width of the direct contact battery cooling block applied, R134a, the cooling performance was lower than that of one and three ports. Analysis showed that a model with excellent performance had a header width of 0 mm for one port and 5 mm for three ports. A model with a header width of 5 mm for three ports had an MXT of the battery of 2.08 °C. Still, the pressure drop was 4.38 times smaller, so it was verified that the battery cooling block had an excellent cooling performance.

(3) Analysis results were obtained according to the flow rate change of R134a applying the optimal model with respect to the number of ports and header width under standard and summer conditions. The MXT and the TD between the batteries were 30.5 °C, 4.91 °C, 28.7 °C, and 3.28 °C, respectively, confirming that it is an appropriate refrigerant flow condition for battery safety. However, as the battery pressure drop increases from 3 lpm to 5 lpm, it increases 2.47 times, so 3 lpm is more effective in terms of maintenance costs. In addition, the MXT of the battery at 1 lpm and 3 lpm, excluding 5 lpm, was 48.0 °C and 40.5 °C, respectively, in the summer condition of 35 °C in outdoor air, which was not suitable for battery thermal management. Therefore, it was verified that 5 lpm or more is a flow condition suitable for battery thermal management.

Author Contributions: Conceptualization, H.C. and D.S.; methodology J.H.; validation, D.S. and H.C.; formal analysis, M.K. and J.H.; investigation, M.K. and J.H.; resources, M.K. and D.S.; writing—original draft preparation, M.K. and H.C.; writing—review and editing, H.C.; supervision, H.C. All authors have read and agreed to the published version of the manuscript.

Funding: This work was supported by "Eco-friendly Car Sector in Development Technology" of the Korea Institute of Industrial Technology Evaluation and Management (KEIT), granted financial resource from the Ministry of Trade, Industry & Energy, Korea. (No. 20011906).

Data Availability Statement: Not applicable.

Conflicts of Interest: The authors declare no conflict of interest.

Nomenclature

$BTMS$	Battery thermal management system
c	Specific heat capacity (J/kg·°C)
D	Diameter (m)
EES	Engineering equation solver
g	Gravitational acceleration (m/s^2)
H	Height (m)
$HVAC$	Heating, ventilation, and air conditioning
I	Current (A)
L	Length, (m)
MXT	Maximum temperature of the battery (°C)
$\dot{m}$	Mass flow rate (kg/s)
P	Pressure (Pa)
PCM	Phase change material
$\dot{Q}$	Flow rate (m^3/s)
Q	Heat generation rate (W)
$\dot{q}$	Volumetric heat generation rate (W/m^3)
R	Equivalent internal resistance, Ω
T	Temperature (°C)
TD	Temperature difference between batteries (°C)
TO	Transformer oil
t	Time (s)
U	Voltage (V)
u	Velocity, (m/s)
W	Power
$1S16P$	1 serial × 16 parallel

Greek symbols

ρ	Density (kg/m^3)
λ	Thermal conductivity (W/m·°C)

Subscript

avg	average
ba	Battery
in	inlet
ir	irreversible heat
max	Maximum
min	Minimum
$pump$	Pump
re	reversible heat
OCV	Open circuit voltage
wf	Working fluid

References

1. Chen, S.; Bao, N.; Garg, A.; Peng, X.; Gao, L. A Fast Charging–Cooling Coupled Scheduling Method for a Liquid Cooling-Based Thermal Management System for Lithium-Ion Batteries. *Engineering* **2021**, *7*, 1165–1176. [CrossRef]
2. Amjad, S.; Neelakrishnan, S.; Rudramoorthy, R. Review of design considerations and technological challenges for successful development and deployment of plug-in hybrid electric vehicles. *Renew. Sustain. Energy Rev.* **2010**, *14*, 1104–1110. [CrossRef]
3. Roe, C.; Feng, X.; White, G.; Li, R.; Wang, H.; Rui, X.; Li, C.; Zhang, F.; Null, V.; Parkes, M.; et al. Immersion cooling for lithium-ion batteries—A review. *J. Power Sources* **2022**, *525*, 231094. [CrossRef]
4. Chitta, S.D.; Akkaldevi, C.; Jaidi, J.; Panchal, S.; Fowler, M.; Fraser, R. Comparison of lumped and 1D electrochemical models for prismatic 20Ah LiFePO4 battery sandwiched between minichannel cold-plates. *Appl. Therm. Eng.* **2021**, *199*, 117586. [CrossRef]
5. Zhang, J.; Shao, D.; Jiang, L.; Zhang, G.; Wu, H.; Day, R.; Jiang, W. Advanced thermal management system driven by phase change materials for power lithium-ion batteries: A review. *Renew. Sustain. Energy Rev.* **2022**, *159*, 112207. [CrossRef]
6. Gan, Y.; He, L.; Liang, J.; Tan, M.; Xiong, T.; Li, Y. A numerical study on the performance of a thermal management system for a battery pack with cylindrical cells based on heat pipes. *Appl. Therm. Eng.* **2020**, *179*, 115740. [CrossRef]
7. Ramadass, P.; Haran, B.; White, R.; Popov, B.N. Capacity fade of Sony 18650 cells cycled at elevated temperatures: Part II. Capacity fade analysis. *J. Power Sources* **2002**, *112*, 614–620. [CrossRef]
8. Huang, P.; Yao, C.; Mao, B.; Wang, Q.; Sun, J.; Bai, Z. The critical characteristics and transition process of lithium-ion battery thermal runaway. *Energy* **2020**, *213*, 119082. [CrossRef]
9. Wang, S.; Takyi-Aninakwa, P.; Jin, S.; Yu, C.; Fernandez, C.; Stroe, D.I. An improved feedforward-long short-term memory modeling method for the whole-life-cycle state of charge prediction of lithium-ion batteries considering current-voltage-temperature variation. *Energy* **2022**, *254*, 124224. [CrossRef]
10. Li, L.; Chen, X.; Yuan, Q.; Wang, T.; Ji, H.; Papović, S.; Raleva, K.; Pan, F.; Yang, T.; Li, J. Effects of Minor Mechanical Deformation on the Lifetime and Performance of Commercial 21700 lithium-Ion Battery. *J. Electrochem. Soc.* **2022**, *169*, 060544. [CrossRef]
11. Sun, Y.; Lu, H.; Jin, Y. Experimental and Numerical Study on Mechanical Deformation Characteristics of Lithium Iron Phosphate Pouch Battery Modules under Overcharge Conditions. *Energy Fuels* **2021**, *35*, 15172–15184. [CrossRef]
12. Akbarzadeh, M.; Kalogiannis, T.; Jaguemont, J.; Jin, L.; Behi, H.; Karimi, D.; Beheshti, H.; Van Mierlo, J.; Berecibar, M. A comparative study between air cooling and liquid cooling thermal management systems for a high-energy lithium-ion battery module. *Appl. Therm. Eng.* **2021**, *198*, 117503. [CrossRef]
13. Egab, K.; Oudah, S.K. Thermal management analysis of li-ion battery-based on cooling system using dimples with air fins and perforated fins. *Int. J. Therm. Sci.* **2022**, *171*, 107200. [CrossRef]
14. Xu, H.; Zhang, X.; Xiang, G.; Li, H. Optimization of liquid cooling and heat dissipation system of lithium-ion battery packs of automobile. *Case Stud. Therm. Eng.* **2021**, *26*, 101012. [CrossRef]
15. Wang, N.; Li, C.; Li, W.; Chen, X.; Li, Y.; Qi, D. Heat dissipation optimization for a serpentine liquid cooling battery thermal management system: An application of surrogate assisted approach. *J. Energy Storage* **2021**, *40*, 102771. [CrossRef]
16. Liu, H.; Ahmad, S.; Shi, Y.; Zhao, J. A parametric study of a hybrid battery thermal management system that couples PCM/copper foam composite with helical liquid channel cooling. *Energy* **2021**, *231*, 120869. [CrossRef]
17. Xin, Q.; Xiao, J.; Yang, T.; Zhang, H.; Long, X. Thermal management of lithium-ion batteries under high ambient temperature and rapid discharging using composite PCM and liquid cooling. *Appl. Therm. Eng.* **2022**, *210*, 118230. [CrossRef]
18. Jang, D.S.; Yun, S.; Hong, S.H.; Cho, W.; Kim, Y. Performance characteristics of a novel heat pipe-assisted liquid cooling system for the thermal management of lithium-ion batteries. *Energy Convers. Manag.* **2022**, *251*, 115001. [CrossRef]
19. E, J.; Yi, F.; Li, W.; Zhang, B.; Zuo, H.; Wei, K.; Chen, J.; Zhu, H.; Zhu, H.; Deng, Y. Effect analysis on heat dissipation performance enhancement of a lithium-ion-battery pack with heat pipe for central and southern regions in China. *Energy* **2021**, *226*, 120336. [CrossRef]
20. Zhao, G.; Wang, X.; Negnevitsky, M.; Zhang, H. A review of air-cooling battery thermal management systems for electric and hybrid electric vehicles. *J. Power Sources* **2021**, *501*, 230001. [CrossRef]
21. Jouhara, H.; Delpech, B.; Bennett, R.; Chauhan, A.; Khordehgah, N.; Serey, N.; Lester, S.P. Heat pipe based battery thermal management: Evaluating the potential of two novel battery pack integrations. *Int. J. Thermofluids* **2021**, *12*, 100115. [CrossRef]
22. Wang, X.; Liu, S.; Zhang, Y.; Lv, S.; Ni, H.; Deng, Y.; Yuan, Y. A Review of the Power Battery Thermal Management System with Different Cooling, Heating and Coupling System. *Energies* **2022**, *15*, 1963. [CrossRef]
23. Shen, M.; Gao, Q. Structure design and effect analysis on refrigerant cooling enhancement of battery thermal management system for electric vehicles. *J. Energy Storage* **2020**, *32*, 101940. [CrossRef]
24. Hong, S.H.; Jang, D.S.; Park, S.; Yun, S.; Kim, Y. Thermal performance of direct two-phase refrigerant cooling for lithium-ion batteries in electric vehicles. *Appl. Therm. Eng.* **2020**, *173*, 115213. [CrossRef]
25. Tan, X.; Lyu, P.; Fan, Y.; Rao, J.; Ouyang, K. Numerical investigation of the direct liquid cooling of a fast-charging lithium-ion battery pack in hydrofluoroether. *Appl. Therm. Eng.* **2021**, *196*, 117279. [CrossRef]
26. Xie, J.; Wang, Y.; He, S.; Zhang, G.; Liu, X.; Yang, X. A simple cooling structure with precisely-tailored liquid cooling plate for thermal management of large battery module. *Appl. Therm. Eng.* **2022**, *212*, 118575. [CrossRef]
27. Chen, S.; Zhang, G.; Zhu, J.; Feng, X.; Wei, X.; Ouyang, M.; Dai, H. Multi-objective optimization design and experimental investigation for a parallel liquid cooling-based Lithium-ion battery module under fast charging. *Appl. Therm. Eng.* **2022**, *211*, 118503. [CrossRef]

28. Lu, Y.; Wang, J.; Liu, F.; Liu, Y.; Wang, F.; Yang, N.; Lu, D.; Jia, Y. Performance optimization of Tesla valve-type channel for cooling lithium-ion batteries. *Appl. Therm. Eng.* **2022**, *212*, 118583. [CrossRef]
29. Wu, S.; Lao, L.; Wu, L.; Liu, L.; Lin, C.; Zhang, Q. Effect analysis on integration efficiency and safety performance of a battery thermal management system based on direct contact liquid cooling. *Appl. Therm. Eng.* **2022**, *201*, 117788. [CrossRef]
30. Liu, J.; Fan, Y.; Xie, Q. Feasibility study of a novel oil-immersed battery cooling system: Experiments and theoretical analysis. *Appl. Therm. Eng.* **2022**, *208*, 118251. [CrossRef]
31. Al-Zareer, M.; Dincer, I.; Rosen, M.A. Heat and mass transfer modeling and assessment of a new battery cooling system. *Int. J. Heat Mass Transf.* **2018**, *126*, 765–778. [CrossRef]
32. Fluent, I. Fluent 2021 user's guide. *Fluent Solut.* **2021**.
33. Sheng, L.; Zhang, Z.; Su, L.; Zhang, H.; Zhang, H.; Li, K.; Fang, Y.; Ye, W. A calibration calorimetry method to investigate the thermal characteristics of a cylindrical lithium-ion battery. *Int. J. Therm. Sci.* **2021**, *165*, 106891. [CrossRef]
34. Yataganbaba, A.; Kilicarslan, A.; Kurtbaş, İ. Exergy analysis of R1234yf and R1234ze as R134a replacements in a two evaporator vapour compression refrigeration system. *Int. J. Refrig.* **2015**, *60*, 26–37. [CrossRef]
35. A/C Pro Certified Auto Air Conditioner R-134a Refrigerant, 12 oz, 301CA. Available online: Walmart.com (accessed on 17 August 2022).
36. Silicone oil Dow Corning Corporation 200 fluid, 60,000cSt 25 °C 63148-62-9. Available online: Sigmaaldrich.com (accessed on 20 September 2022).
37. Sigma Aldrich Fine Chemicals Biosciences Novec 7000 Engineered Fluid, 375-03-1. Fisher Scientific: Waltham, MA, USA. Available online: Fishersci.com (accessed on 17 August 2022).
38. HD TO4 NO.10 18L | Transmission Oil. Sinopec: Sydney, NSW, Australia. Available online: Sinopeconline.com (accessed on 20 September 2022).
39. Sheng, L.; Zhang, H.; Zhang, H.; Su, L.; Zhang, Z. Lightweight liquid cooling based thermal management to a prismatic hard-cased lithium-ion battery. *Int. J. Heat Mass Transf.* **2021**, *170*, 120998. [CrossRef]
40. Lai, Y.; Wu, W.; Chen, K.; Wang, S.; Xin, C. A compact and lightweight liquid-cooled thermal management solution for cylindrical lithium-ion power battery pack. *Int. J. Heat Mass Transf.* **2019**, *144*, 118581. [CrossRef]
41. Bernardi, D.; Pawlikowski, E.; Newman, J. A general energy balance for battery systems. *J. Electrochem. Soc.* **1985**, *132*, 5–12. [CrossRef]
42. Ghahfarokhi, P.S.; Kallaste, A.; Vaimann, T.; Rassolkin, A.; Belahcen, A. Determination of natural convection heat transfer coefficient over the fin side of a coil system. *Int. J. Heat Mass Transf.* **2018**, *126*, 677–682. [CrossRef]
43. Shin, Y.; Im, G.; Yu, K.; Cho, H. Experimental study on the change in driver's physiological signals in automobile HVAC system under Full load condition. *Appl. Therm. Eng.* **2017**, *112*, 1213–1222. [CrossRef]
44. Dong, J.; Lan, H.; Liu, Y.; Wang, X.; Yu, C. Indoor environment of nearly zero energy residential buildings with conventional air conditioning in hot-summer and cold-winter zone. *Energy Built Environ.* **2022**, *3*, 129–138. [CrossRef]
45. Huang, Z.; Li, G.; Tian, S.; Song, X.; Sheng, M.; Shah, S. Operation Parameters Calculation. *Abras. Water Jet Perforation Multi-Stage Fract.* **2018**, 191–203. [CrossRef]

batteries

MDPI

Article

A Critical Analysis of Helical and Linear Channel Liquid Cooling Designs for Lithium-Ion Battery Packs

Rob Lloyd and Mohammad Akrami *

Department of Engineering, University of Exeter, Exeter EX4 4QF, UK
* Correspondence: m.akrami@exeter.ac.uk

Abstract: Thermal management systems are integral to electric and hybrid vehicle battery packs for maximising safety and performance since high and irregular battery temperatures can be detrimental to these criteria. Lithium-ion batteries are the most commonly used in the electric vehicle (EV) industry because of their high energy and power density and long life cycle. Liquid cooling provides superior performance with low power draw and high heat transfer coefficient. Two liquid cooling designs-the Linear Channel Design (LCD) and Helical Channel Design (HCD)-underwent multiple numerical and geometrical optimisations, where inlet mass flow rate, channel diameter, and inlet and outlet locations were analysed using CFD (computational fluid dynamics). The primary objectives were to maintain maximum temperatures and thermal uniformity within the operational limits derived from the literature. These were both achieved with the LCD using a mass flow rate of 7.50E-05 kgs^{-1}. The T_{max} goal was met for the HCD but not the thermal uniformity goal. The LCD achieved 1.796 K lower in maximum temperature and 8.740 K lower in temperature difference compared to the HCD, proving itself superior in both metrics. The HCD required a higher mass flow rate than the LCD to regulate temperatures, resulting in an undesirably high power consumption.

Keywords: lithium-ion battery; thermal management; liquid cooling; linear channel; helical channel; design optimisation; electric vehicle

Citation: Lloyd, R.; Akrami, M. A Critical Analysis of Helical and Linear Channel Liquid Cooling Designs for Lithium-Ion Battery Packs. *Batteries* **2022**, *8*, 236. https://doi.org/10.3390/batteries8110236

Academic Editors: Jinsheng Xiao, Hengyun Zhang, Sousso Kelouwani and Torsten Brezesinski

Received: 12 September 2022
Accepted: 9 November 2022
Published: 12 November 2022

1. Introduction

1.1. Background

With ever-increasing concern for pollution due to the use of fossil fuels, the demand for electric and hybrid vehicles has noticeably been on the incline in recent years. In 2015, 65% of global oil consumption was due to the transport sector [1], and in 2021, a declaration was made at COP26 (the UN climate change conference) entailing the sales of all new cars and vans to be zero emission vehicles globally by 2040 and by no later than 2035 in leading markets [2], meaning that developments for sustainable EV (Electric Vehicle) technology are in high demand by automotive companies.

Lithium-ion batteries have been a popular choice within the industry for EV powertrains due to their high energy density, high power density and long life cycle [3]. When lithium-ion batteries undergo the discharge cycle during operation, they generate a considerable amount of heat that needs to be expelled, because high and irregular temperatures can severely impact the battery pack's performance, life cycle and safety [4]. The main aim of this research is, therefore, to provide a suitable and effective thermal management system for the use of EV battery packs. The criteria needed to reach this aim include maintaining the maximum temperature of the cells within the optimal operating range of 298 K < T_{max} < 313 K [5] and the thermal uniformity of the cells in the form of temperature difference in the range of $\Delta T < 5$ K [5]. Thermal uniformity is an important metric for analysing thermal performance due to the application of the Arrhenius law [6], where areas of high cell temperature increase the current and accelerate electrochemical reactions, causing a local peak SOC (State of Charge). Uneven SOCs within electrodes shorten the operating lifespan of the cell by accelerating the ageing of the electrodes at high SOCs [6]. Furthermore, the thermal management solution should be realistic for use in an EV and not consume unnecessarily large quantities of power in order to operate that would otherwise be more usefully diverted to other systems in the vehicle. Additionally, the cost and manufacturing complexity will be minor factors to consider towards the end of the design optimisation process. These criteria will be achieved through a series of numerical and geometrical optimisations to the given designs.

1.2. Literature Review

Over the past decade, extensive research has been conducted on EV lithium-ion battery packs' thermal management systems. The demand for new innovative ways to improve battery performance from various electric car manufacturers, has always been prevalent due to cooling being a major factor in the vehicle's overall performance and efficiency. This literature review will be categorised in order to cover the three key areas of EV battery pack cooling research, comprising PCM cooling, air cooling and liquid cooling.

1.2.1. PCM Cooling

One of the fundamental areas for research within the thermal management of EV battery packs involves the use of a PCM enclosure for the Li-ion cells. PCM cooling relies on the absorption and release of thermal energy when the coolant material changes phase. PCMs have been explored in recent research such as that of Li et al. [7] where porous cellulose nanofibril (CNF)/silver nanowire (AgNW) hybrid supporting materials were used to increase the phonon propagation of the cellulosic material lattice and improve the thermal transmission ability of the hybrid carriers.

Multiple structural arrangements of PCM enclosures have been proposed in the literature, such as Choudhari et al. [8] comparing the use of rectangular, triangular, trapezoidal, I-shape and T-shape fins within the PCM enclosure in order to counteract the heat accumulation in PCMs due its low heat conduction [8]. It was concluded that the I-shape and rectangular fins had the strongest performance in terms of maximum temperature reduction and thermal uniformity. Although it should be said that each of these fin types had very similar results, leading to the conclusion that fin shape is not a major determining factor in the performance of a battery thermal management system. Additionally, this test was conducted on a single Li-ion battery and not a pack, which may limit the usefulness of this data for car manufacturers, so a battery pack should be tested for future research. Landini et al. [9] studied the use of direct-metal-laser-sintered aluminium heat exchangers containing PCMs, finding that the system showed promising results in comparison to natural convection and maintained an even thermal uniformity.

An interesting development in the research of PCM cooling involves the combination with liquid cooling, as researched by Cao et al. [10], who proposed a structure containing EG/RT44HC acting as a PCM alongside a cold plate situated in amongst the battery pack. It was found that in order to best control the battery pack's thermal performance, the coolant water temperature should be limited to less than 313K and as close to the ambient temperature as possible [10]. It was found that these low inlet temperatures reduce the general temperature of the battery pack but increase the temperature in the axial and radial directions, worsening the thermal uniformity of the Li-ion cells. Hekmat et al. [11] investigated a hybrid PCM and liquid-cooled structure for prismatic cells, finding that the PCM was especially beneficial in creating a strong thermal uniformity. The addition of coolant pipes reduced the maximum temperature significantly compared to singularly using PCM, indicating that liquid cooling systems will majorly influence the thermal performance of an EV battery pack. Perhaps the liquid cooling structures should be researched heavily to optimise cooling performance instead of purely relying on passive cooling.

1.2.2. Air Cooling

The use of natural convection (air cooling) is commonly used for effectively managing the thermal performance of an EV battery pack. Fans are generally used in practice [12] as opposed to purely relying on the exterior winds around the moving vehicle which would be unreliable and not keep temperatures steady enough over an extended period of time. Air cooling systems for EV battery packs are generally cheaper and simpler in design when compared to liquid cooling loops [12]. However, some issues can arise in terms of the thermal uniformity of Li-ion cells since predicting the exact flow path of the air and making sure it evenly cools each Li-ion cell can be troublesome to model, which is why much literature has focused on this, along with minimising maximum temperatures.

Forced air cooling systems have been developed by scholars such as Wang et al. [13], where two cooling channels were used and tested at multiple discharge rates. Results showed that the cooling efficiency reached 73%, and the maximum temperature decreased from 343 K to 328 K at a discharge rate of 1C. The main appeal of this design is the alternating airflow system, which has a noticeably beneficial impact on the thermal uniformity of the battery pack. However, it had a negligible effect in reducing maximum temperatures. The idea of alternating inlet airflow arrangements was also researched by Sharma et al. [14], with the conclusion that dual-direction airflow provided great benefits to thermal uniformity and required a lower power consumption compared to unidirectional airflow. The airflow velocity was also optimised to be 2 ms^{-1} for 1, 2 and 3C discharge rates and 4 ms^{-1} for 4C.

The battery pack structure and layout of the cells are important considerations and determining factors when it comes to the thermal performance of an EV battery pack. Xi et al. [15] proposed a novel Z-type structure, which produced promising results for thermal performance when an optimisation process was conducted. Furthermore, the addition of deflectors and rounding-off chamfers reduced the maximum temperature by 2.52 K and enhanced the thermal uniformity due to the more balanced airflow within the battery pack. A J-type airflow structure was studied by Liu et al. [16] with a comparison to U-type and Z-type structures. It was found that after the J-type system was optimised using surrogate-based optimisation, it showed a 4.12 K reduction in maximum temperature in relation to the U-type system and a 15.42 K reduction compared to the Z-type structure since the optimised system benefits from a highly balanced and uniform airflow distribution.

1.2.3. Liquid Cooling

Liquid cooling is the method of EV battery thermal management employed by car manufacturers such as Tesla, BMW and Ford [17] and is the most widely researched in the literature, with many different types of designs. It should be noted that liquid cooling can be split into direct and indirect cooling, where the former involves the submersion of the Li-ion cells directly into the coolant fluid. While direct cooling comes with the advantages of maximal thermal contact area for the cells and greatly improved thermal uniformity [12], this method is uncommon in the real-world manufacturing of battery thermal management systems. This is due to the troublesome requirement of the water-resistant performance of the battery pack, where electrical short circuits and electrochemical reactions may occur [12]. This is why indirect liquid cooling (designs involving exterior channels for coolant flow) is a well-researched area in the literature and is held at a higher degree of importance by both scholars and EV manufacturers.

Car manufacturers such as Tesla make use of a serpentine wavy channel cooling structure, which a fair level of academic research has been conducted on, such as Zhao et al. [18], that used a simulation of a serpentine channel cooling arrangement battery pack. The pack comprises 71 18,650 Li-ion cells, where the heat generation is based on experimental data of a single cell. Zhao et al. [18] found that increasing the discharge/charge rate of the cells increases the battery pack temperature and worsens its thermal uniformity and increasing the coolant flow rate significantly lowers the temperature and improves thermal uniformity. However, this was only tested with two inlet velocity rates, 0.1 ms^{-1}

and 0.5 ms^{-1}, providing very little sample space, so there is room to research and find an optimal flow rate. Zhao et al. [18] also found that modelling a contact area between each cell had a negligible effect on the thermal performance and that increasing the contact area between each cell and the coolant loop resulted in lower battery pack temperatures (due to the increase in contact surface area resulting in a higher rate of thermal dissipation). Although, again this was tested arbitrarily with contact angles of 10°, 20°, 30° and 40°, so an optimal or maximum can still be researched.

The following year, this research was built upon again by Zhao et al. [19], with the main focus being to optimise thermal uniformity with the battery pack. The removal of the coolant loop's U-turn resulting in multiple single wavy channels was tested, with the findings that this improved the thermal uniformity and reduced the maximum temperature of the battery pack. This time, the contact area of the cell and coolant loop was tested much more in-depth. It was found that increasing the contact area improved thermal uniformity. The serpentine wavy cooling channel structure was also researched by Xu et al. [20], comparing it with a U-shaped coolant loop with grooves. It was found that the serpentine wavy structure had better performance in terms of maximum temperature reduction, that being 0.37 K lower than the U-shaped structure, and demonstrated a better thermal uniformity. It would be potentially useful to combine the research of Zhao et al. [18] and Xu [19] to test a single-channel system with grooves.

In terms of alternative liquid cooling structures, many novel designs have been proposed in the literature that are not similar to those currently adopted by car manufacturers, such as the use of a helical duct structure encasing each Li-ion cell. Multiple parameters within this design were tested by Zhao et al. [21], and it was found that the battery pack's maximum temperature was lowered, and the temperature uniformity was improved as the inlet mass flow rate increased, concurring with the findings of Zhao et al. [18] even for a vastly different coolant structure. It was also discovered that when the pitch and number of helical ducts were varied on the optimal flow rate model of 3×10^{-4} kgs^{-1}, this had a negligible effect on the thermal performance of the battery pack. Various cases of alternating fluid flow were tested in an attempt to regulate temperature (since outlet temperature will inevitably be lower than inlet temperature). The findings highlight that having three inlets and two outlets at the positive electrode of the battery gave a T_{max} of 306.95 K and ΔT of 7.4 K at 40% DOD compared to a T_{max} of 307.85 K and ΔT of 8.7 K for a one inlet, four outlet case. This technique improved cooling effectiveness and thermal uniformity when the fluid flow is alternated. Finally, the helical duct diameters were tested with the conclusion that smaller diameter ducts result in slightly lower maximum temperatures but a worsened thermal uniformity.

The use of helical liquid ducts was also researched by Dong et al. [22], who analysed a novel double helix structure. It was also found that an increase in coolant mass flow rate resulted in lower maximum temperatures. Although in this paper, a wider sample was used in order to derive a converging relationship, where the cooling benefits of higher mass flow rates began to diminish as the flow rate increased beyond 5×10^{-4} kgs^{-1}. The pitch and diameter parameters were also tested, with the conclusion concurring with the findings of Zhao et al. [21] that the smaller diameters result in lower maximum temperatures but a worsened thermal uniformity.

The use of linear mini-channel cooling structures has been a key area of EV battery liquid cooling research over recent years, with many design variations being proposed and evaluated in the literature. Some designs have linear channels flowing perpendicular to the length of the cells, such as the structure proposed by Rao et al. [23], comprising aluminium blocks between the Li-ion cells with mini channels for coolant flow. The particular design shown in this paper provides adequate cooling performance for the battery pack since the temperatures are kept well within operating conditions. However, it is obvious from the schematics of this design that since the straight mini-channels flow perpendicular to the lengths of the Li-ion cells, the thermal uniformity of the cells is weakened. This is mainly due to the distance between a given coolant channel and the cells is constantly changing along the length of the channel. This is evident in the results of Rao et al. [23], where the temperature contours demonstrate large variation within the cell volumes. This can be a major issue as denoted by Zeng et al. [6], where the volumetric heat generation rate increases with the energy density and charge rate. There is more battery heat to be dissipated, requiring more cooling power. Zeng et al. [6] mentions the application of the Arrhenius law, where areas of high cell temperature increase the current and accelerate electrochemical reactions, causing a local peak SOC. Uneven SOCs within electrodes shorten the operating lifespan of the cell by accelerating the ageing of the electrodes at high SOCs. From this, the thermal uniformity of the Li-ion cells shall be an important consideration for the proposed designs in this paper.

Most of the proposed linear cooling channel designs within the literature comprise channels parallel to the lengths of the cells. Lai et al. [24] researched designs in this area and identified a gap in the literature research, being the lack of focus on the weight and compactness of the battery pack, which is an important factor for the performance of electric vehicles. A lightweight battery pack was constructed comprising multiple aluminium thermal conductive structures containing coolant channels, allowing the cooling water to flow from an upper reservoir and down the lengths of each cell to the lower reservoir. It was also stated that liquid cooling, in general, provided far superior heat transfer coefficient and power consumption levels in relation to PCM and air cooling. A similar concept was presented by Sheng et al. [25], utilising a cellular liquid cooling jacket. Sheng et al. [24] discovered that interlaced flow directions provide a lower temperature standard deviation and improved thermal uniformity, concurring with the findings associated with the helical channel designs of Zhao et al. [21]. A glycol aqueous solution was also tested, with the conclusion that this is a more effective coolant than regular liquid water. Although, there is still room to potentially test more coolants to fully optimise this design.

Another such design would be the mini channel cylinder cooler proposed by Zhao et al. [26]. Similar to the design proposed by Lai et al. [24], the battery pack comprises an upper and lower reservoir connected via vertical ducts for the coolant to flow along the length of the cells. However, this design makes use of smaller channels (0.8 mm diameter) compared to that of the design proposed by Lai et al. [24] (2 mm diameter), and utilises up to sixteen channels surrounding each cell as opposed to three per cell from Lai et al. [24]. Zhao et al. [26] and Lai et al. [24] made use of different form factors of Li-ion batteries, 42,110 and 18,650, respectively. Comparing cooling performance among these designs provide little analytical benefit, so perhaps a comparison of these designs under the same boundary conditions and using the same battery form factor should be studied in future studies.

With liquid cooling showing very promising results within this field of research, the aim of this study shall be to further develop the specific types of cooling loops used within the literature. A comprehensive evaluation shall be conducted to determine the applicability of each design, to encourage further research for the superior model.

2. Materials and Methods

2.1. Battery Chemistry

On a basic level, the purpose of a battery is to convert chemical energy to electrical energy via an electrochemical oxidation-reduction reaction, for which the inverse of this reaction is used to recharge a rechargeable battery. This type of reaction involves the transfer of electrons from one material to another through an electric circuit [27]. The batteries store and discharge energy in the form of charged atoms or ions, and the charge and discharge rate of the battery is restricted by the speed that ions can move through the cell [28].

A typical cell comprises three primary components:

1. The anode—gives up electrons to the external circuit and is oxidized during the electrochemical reaction [26].
2. The cathode—accepts electrons from the external circuit and is reduced during the electrochemical reaction [26].

3. The electrolyte—is usually a liquid medium to sustain ionic conductivity and transfer ions between the cathode and anode.

In order to maximise the efficiency of the battery, the materials for the anode and cathode should be light and provide high cell voltage and high capacity [27]. The metal with the lowest density on the periodic table is lithium, which also has a high electrochemical potential, making this a logical choice for the production of batteries. Research into intercalation electrodes has led to the use of lithiated carbons and lithium alloys in lithium-ion technology [27]. There are many different types of lithium-ion batteries, such as $LiCoO_2$, $LiFePO_4$, $LiMn_2O_4$, $LiNiMnCoO_2$ and Li_2TiO_3. Each of these types comes with its own benefits and drawbacks. $LiCoO_2$ is the most commonly used form of lithium-ion battery, and is under consideration for use in innovative solid-state batteries [29]. The $LiFePO_4$ battery uses lithium iron phosphate as the cathode material and in comparison to $LiCoO_2$, has a lower energy density, longer lifetime, and higher power density [27]. A study performed by Tang et al. [30] showed that under external heating conditions, the $LiFePO_4$ battery was able to withstand a higher heat until failure compared to both the Li_2TiO_3 and $LiNiMnCoO_2$ batteries, proving itself superior in thermal stability. Developments have recently been made using a silicon anode for solid state lithium-ion batteries such as Gu et al. [31], where a novel SiO_2@Li_3PO_4@carbon shell coated on micron-sized silicon was shown to maintain mechanical integrity and stable cycling performance with high capacities. Han et al. [32] also demonstrated the benefits of a novel silicon MXene with interfacial nitrogen engineering, which also achieved high mechanical stability as well as stable and fast lithium-ion transportation pathways. Both of these silicon-based anodes could potentially be incorporated into the proposed thermal management systems in this study, although requiring a moderate reconstruction of geometries to account for dimensional changes. Table 1 shows data for the lithium iron phosphate battery modelled in this study [27,28,33].

Table 1. Lithium Iron Phosphate battery properties.

Battery	Cell Voltage (V)	Specific Energy ($MJkg^{-1}$)	Cycle Life (Cycles)	Charge Rate (Hours)
Lithium Iron Phosphate ($LiFePO_4$)	3.3	>0.32	1000+	<1

Under extreme conditions, the safety of a lithium-ion battery can be compromised, for example during a short circuit, when temperatures can increase rapidly [34]. The use of hydrogel to rapidly absorb thermal energy and mitigate thermal runaway when was explored to a high degree of success by Zhao et al. [34]. Collisions are also a major risk to the battery pack integrity and safety, for which shock-absorbing "sacrifice" tubes and crushable foam were tested by Kukreja et al. [35] and were successful in reducing the severity of collisions.

When a battery is under operational use, it is undergoing its discharge cycle. For practical purposes, the performance efficiency of batteries is often rated not on a full discharge cycle, but instead using a percentage of DOD (Depth of Discharge), since batteries are seldom fully discharged due to long term damage this could cause to the batteries.

Figure 1 shows a typical representation of the discharge process of a cell. It can be seen that the electron flow takes place from the oxidised anode to the cathode via the external load. As the electrons are received by the cathode, the cathode material is gradually depleted. Anions (negative) flow from the cathode to the anode, and cations flow vice versa through the electrolyte to complete the circuit. Over the charging cycle of a rechargeable battery, the flow of current is reversed, and the anode and cathode switch places to become the positive and negative electrodes, respectively. The electrochemical reactions during both of these processes are represented by Equations (1) and (2) [26]:

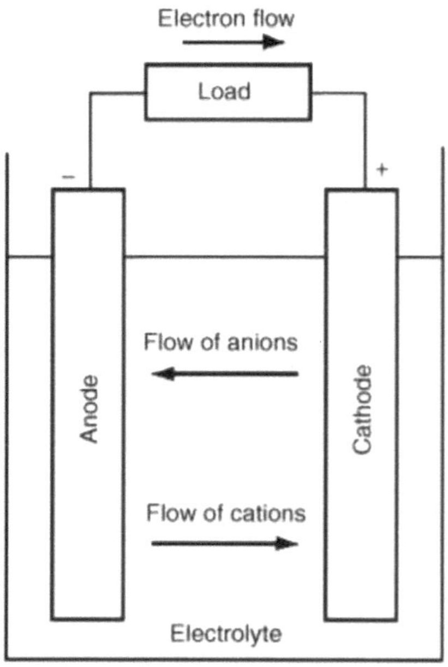

Figure 1. Discharge process of a cell [26].

Anode:

$$LiC_6 \xleftrightarrow{\text{discharge/charge}} Li_{1-x}C_6 + xLi^+ + e^- \quad (1)$$

Cathode:

$$xLi + xe^- + Li_{1-x}FePO_4 \xleftrightarrow{\text{discharge/charge}} LiFePO_4 \quad (2)$$

2.2. Numerical Model

In this investigation of liquid-cooled battery pack thermal management systems, the computational fluid dynamics (CFD) method is introduced, where it is important to understand the governing equations behind the simulation of lithium-ion batteries and the energy transfer due to the liquid cooling channels. First of all, the heat generation of a single lithium-ion cell was modelled according to the following Equation (3) [36]:

$$\frac{\partial}{\partial t}\left(\rho_b \cdot c_{p,b} \cdot T_b\right) = \nabla \cdot (k_b \cdot \nabla T_b) + q_{gen} \quad (3)$$

where ρ_b is the generalised density of the battery, $c_{p,b}$ is the specific heat capacity of the battery, k_b is the generalised thermal conductivity of the battery, finally q_{gen} is the heat generation of the battery and can be determined by [21]:

$$q_{gen} = \frac{Q_b}{V_b} \quad (4)$$

where Q_b is the rate of heat flow across the battery and V_b is the volume of the battery. Much work relating to lithium-ion battery heat generation within recent literature has been based upon the work of Bernardi et al. [37], for which a simplified Bernardi equation is used within this research and is as follows [38]:

$$q_{gen} = \frac{1}{V_b}\left[I_b{}^2 R_b + I_b T_b \frac{dU_b}{dT}\right] \quad (5)$$

where dU_b/dT refers to the temperature coefficient, a parameter related to the electrochemical reaction within a lithium-ion cell derived from the open-circuit voltage of the battery, U_b. I_b is the current through the battery and R_b is the internal resistance of the battery. The coolant used within this research was liquid water, for which the fluid flow and heat transfer model can be expressed using the three following equations [21], firstly the continuity equation [21]:

$$\frac{\partial \rho_w}{\partial t} + \nabla \cdot \left(\rho_w \vec{v}_w\right) = 0 \tag{6}$$

where ρ_w is the density of water, t is time and $\vec{v}_w$ is the velocity of the water. The momentum conservation equation [21]:

$$\frac{\partial \vec{v}_w}{\partial t} + \left(\vec{v}_w \cdot \nabla\right) \vec{v}_w = -\frac{1}{\rho_w} \nabla \cdot P + \mu_w \nabla^2 \left(\vec{v}_w\right) \tag{7}$$

where P is pressure, μ_w is the dynamic viscosity of water Lastly, the energy conservation equation [21]:

$$\frac{\partial}{\partial t}\left(\rho_w \cdot c_{p,w} \cdot T_w\right) + \nabla \cdot \left(\rho_w \cdot c_{p,w} \cdot \vec{v}_w \cdot T_w\right) = \nabla \cdot \left(k_w \nabla T_w\right) \tag{8}$$

where the aforementioned variables are associated with water denoted by subscript "w". The heat transfer between the coolant, aluminium channels and the battery can be modelled through the following heat flow equations:

$$Q = \dot{m} \cdot c_p \cdot \Delta T \tag{9}$$

$$Q = h \cdot A \cdot \Delta T \tag{10}$$

where $\dot{m}$ is the mass flow rate, h is the heat transfer coefficient and A is the contact area for heat transfer. When dealing with fluid flow, the distinction between laminar, transitional and turbulent flow is important to consider and is characterised by the Reynolds number:

$$Re = \frac{\rho v L}{\mu} = \frac{4\dot{m}}{\pi L \mu} \tag{11}$$

The properties of the materials used in this investigation are summarised in Table 2, for which the properties of aluminium and liquid water were sourced directly from the materials database in ANSYS Fluent. The properties of the battery are sourced from the work of Zhao et al. [4] and are assumed to be uniform throughout the geometry of the battery.

Table 2. Material properties.

Materials	**ρ (kg m^{-3})**	**c_p (Jkg^{-1} K^{-1})**	**k (Wm^{-1} K^{-1})**	**μ (kgm^{-1} s^{-1})**
Liquid water	998.2	4128	0.6	1.003×10^{-3}
Aluminium	2719	871	202.4	-
Battery	1760	1108	3.91	-

2.3. *Modelling a Single Lithium-Ion Cell*

There are many different types of lithium-ion batteries to consider for study such as cylindrical, prismatic and pouch. The $LiFePO_4$ cylindrical lithium-ion battery was selected for modelling, given this has shown promising electrochemical and thermal characteristics as described earlier, and there has been a recent wave in the EV manufacturing industry to mass adopt this particular battery chemistry [39]. Many form factors for cylindrical batteries exist, such as 18,650, 21,700 and 4680. The numbering system is characterised by the dimensions of the battery. For example, the 18,650 battery is 18 mm in diameter and 65 mm in height. The new "tabless" 4680 batteries, which are still in development by Tesla, were originally considered for study. Although this, unfortunately, means that there is limited technical information available to the general public and academic scholars, which would make the heat generation model of the battery impossible to validate. This is why the 18,650 form factor was eventually chosen for study, given there is plenty of technical information available about this battery in the literature as summarised in the literature review.

Firstly, the heat generation of a single cell undergoing its discharge cycle must be accurately modelled to produce meaningful results. The geometry for a simple 18,650 cylindrical lithium-ion cell was created in SolidWorks 2021 (Dassault Systèmes, SolidWorks Corp., Waltham, MA, USA) with the relevant dimensions of 18 mm diameter and 65 mm height and was imported into ANSYS Fluent (ANSYS Inc., Canonsburg, PA, USA) as a step file. The applied material properties are shown in Table 2 and are assumed to be uniform throughout the geometry of the battery. This would provide a limitation to the accuracy of the battery model since different components of the battery, such as the can, separator, cathode, and anode foil, would all comprise unique material properties, but for the purpose of this study, a generalised material composition provides accurate enough data without compromising on highly complex geometry and computational power requirements.

Papers covered in the literature review, as well as other research in this field, use the experimental results of Wang et al. [40], for which a cell operated at a 5C discharge rate achieved a maximum temperature of 351 K after 720 s, and where the heat generation and temperature of the cell increased approximately linearly with respect to time throughout the 720 s discharge cycle. Furthermore, from the results of Wang et al. [34], it was shown that as the discharge rate was increased, the heat generated in the battery was greater. For this study, the single cell model was given a heat generation rate of qgen = 138,000 Wm^{-3} within the cell zone conditions and set to run transiently for 720 s. At the end of the discharge cycle, the cell reached a maximum temperature of 351 K through a linear increase in temperature with respect to discharge time (shown in Figure 2), validating this thermal model against the literature. Given this was a very simple calculation for Fluent with simple geometry, different mesh densities provided infinitesimal differences in results, so a mesh dependency study provides little analytical benefit here. Mesh dependency will become significant for the full simulated battery packs with cooling, given the higher complexity of the problem.

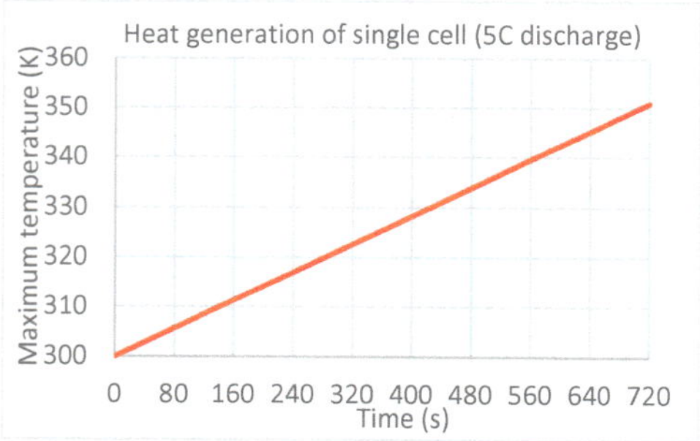

Figure 2. Heat generation of a single cell (5C).

2.4. Construction of the Linear Channel Design (LCD) Battery Pack

The first proposed liquid cooling design is one based upon the work of Lai et al. [24], where SolidWorks 2021 was used in order to plan the design for and build the CAD model, shown in Figure 3 (all dimensions shown are in mm). The LCD considers a pack of four lithium-ion cells to alleviate computational power requirements, with nine coolant flow channels, as presented in Figure 3c. The channels each have four curved edges that interlock with the cylindrical cell surfaces, and the upper reservoir carries fluid from the inlet through the channels to the outlet on the lower reservoir. As opposed to the staggered battery arrangement with three contact surfaces per channel proposed by Lai et al. [24], the LCD uses a uniform structured battery arrangement with four contact surfaces per channel to provide a more useful comparison to the HCD. This is covered in the next section. It should be noted that this arrangement of batteries is slightly less compact, providing a minor limitation to the usefulness of this design within an EV. The channel diameter and the inlet and outlet locations were chosen arbitrarily for now, as they are to be optimised later in this study, as well as the inlet mass flow rate.

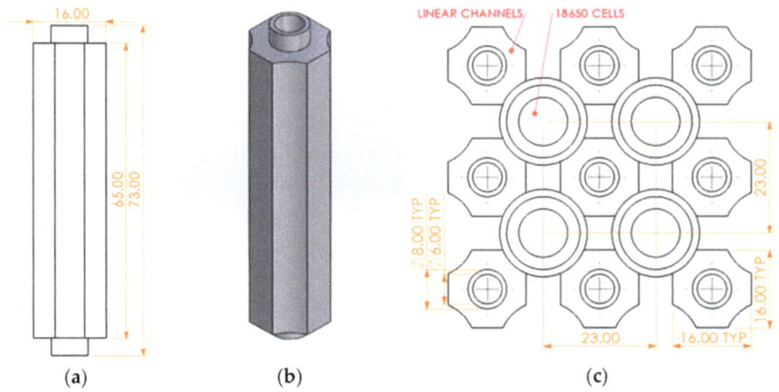

Figure 3. LCD views and dimensions, all in millimetres: (**a**) Single channel side view; (**b**) Single channel ISO view; (**c**) Battery pack top view with fluid domain removed.

2.5. Construction of the Helical Channel Design (HCD) Battery Pack

The second proposed liquid cooling design is one based upon the work of Zhou et al. [21] and Dong et al. [22], involving helical coolant channel structures weaving around each cell. Ideally, the main benefit of a helical channel structure should be the allowance for a more even distribution of coolant around the circumference of the cell, improving thermal uniformity. The CAD model created in Solidworks 2021 is presented in Figure 4 (all dimensions in mm). Illustrated in Figure 4, a helix was swept, and an extruded cylindrical cut of radius 9 mm was taken through the centroid, creating a half helix structure similar to the work of Zhou et al. [21]. This allows the lithium-ion cell to have a maximal contact area with the helical channel via a continuous flat edge, allowing for the greatest quantity of heat transfer rate, as described in Equation (10). The HCD also comprises four lithium-ion cells which are spaced in the same arrangement as the LCD, to provide the most relevant comparison. Each cylindrical cell has its own helical channel, for which the coolant flows from the inlet into the upper reservoir and through the helical channels to the outlet on the lower reservoir.

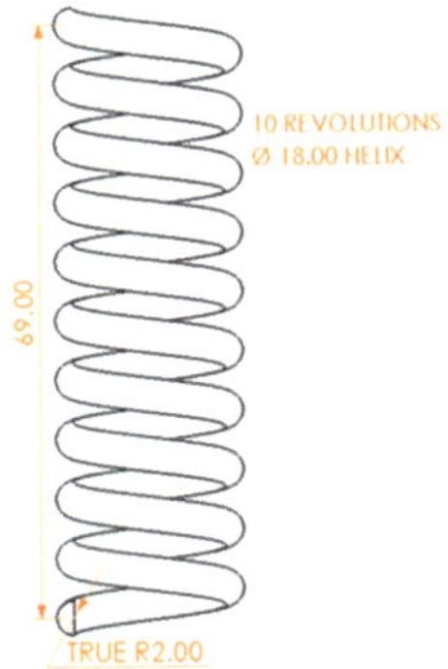

Figure 4. Helical channel–side (mm).

2.6. CFD Configuration and Boundary Conditions

Once each battery pack assembly had been created in SolidWorks 2021, they were imported as a step file into ANSYS 2021 R1. A hexahedral mesh was generated for each battery pack within ANSYS Mechanical. This provided low numerical diffusion, resulting in a solution of higher accuracy [41] compared to meshing geometries such as tetrahedral. Body and face sizings were applied to enhance the mesh density of critical areas of heat transfer, such as the channels and sections of the fluid domain, as well as inflation layers between the boundaries of the fluid, channel and batteries. Inflation layers are vital for resolving the boundary layer between two regions where a significant temperature gradient is occurring due to energy imbalance. The first layer thickness was set to 0.1 mm based on the y+ value of 1 [42], with a growth rate of 1.2. It is generally good practice with inflation layers to include between 10 and 15 layers to accurately resolve the boundary layer and predict separation or reattachment points [43], which is why 10 layers have been used for these studies. According to the Fluent User's Guide [44], the orthogonal quality should be kept close to 1 and skewness close to 0 (perfectly equiangular cells), which was confirmed for every case and geometrical optimisation using the mesh statistics in ANSYS Mechanical. Lastly, the named selections are applied to the relevant geometries for ease of access within Fluent.

In Fluent, the double precision mode was activated to provide more accurate simulations, and the transient model was selected. The energy model was activated to allow the equations of heat transfer Equation (8), along with the 2-equation k-ε turbulence model. Liquid water was used as the coolant in this study, which is commonly used in the literature, and a custom material was created for the battery using the properties described in Table 2. The appropriate materials were assigned in the cell zone conditions: aluminium for the channels, liquid water for the fluid domain and the custom battery material for the lithium-ion cells. The heat generation model described earlier is applied here for the cells. The fluid inlet boundary condition is set to 300 K and as a mass flow inlet, initially as 2.5×10^{-5} kgs^{-1}, well within the laminar flow region as described by Equation (12) (derived from Equation (11), using 8 mm as the inlet diameter and the dynamic viscosity of water from Table 2.

$$\dot{m}_{max} = \frac{Re \cdot \pi \cdot L \cdot \mu}{4} = \frac{2300 \times \pi \times 8 \times 10^{-3} \times 0.001003}{4} = 0.0145\ kgs^{-1} \quad (12)$$

Any walls separating fluid or solid domains were set as coupled, allowing natural convection for energy transfer, and the interior faces of the channels were set to a no-slip condition and as stationary. The ambient conditions are 300 K, and each case was assumed to be in an adiabatic environment via the battery pack exterior surfaces set to thermally insulating boundaries, due to the constricted airflow in real EV battery pack environments where the thermal impact is negligible. As each optimisation is completed in this study, the newfound boundary conditions are applied for the consecutive stages.

Monitors were set to record the maximum and minimum temperatures for the volumes of the lithium-ion cells over time, and a hybrid initialisation was utilised. For the calculation, each time step was set to a magnitude of 1 s for 720 time steps in order to recreate the 5C discharge cycle [40], as described earlier. The maximum iterations per time step was set to 300 in order to significantly increase the likelihood of convergence, although it was found that most time steps converged between 1 and 10 iterations.

2.7. Mesh Dependency

A mesh refinement study was performed on the LCD and HCD to find convergence in the solution where an optimal mesh can be applied to provide accurate solution data without extending the computational time and power requirements unreasonably. This is achieved by gradually applying denser element meshing via altering the global mesh size, the body and face sizing, and inflation layers. The results were recorded in each case (in this case, the maximum temperature of the cells), and the graphs presented in Figure 5 show the convergences. The data was plotted on separate graphs for clarity. Since the data range for T_{max} in each case is small, the graphs would appear as almost horizontal lines otherwise.

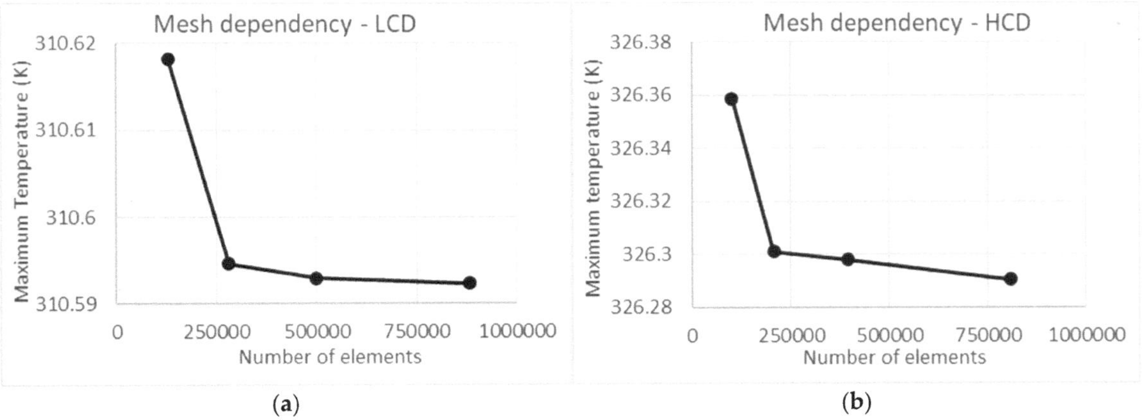

Figure 5. Mesh dependency study for the: (**a**) LCD; (**b**) HCD.

From Figure 5a, it can be seen that the maximum temperature of the lithium-ion cells decreases in magnitude rapidly as a denser mesh is used, up until roughly 300,000 elements, where the maximum temperature thereafter levels off and converges tightly toward an asymptote as the element number increases. For the HCD in Figure 5b, the same relationship applies, except with roughly 200,000 elements as the convergence point. This is why a mesh of 279,995 elements is to be used for the LCD model and 207,943 elements for the HCD since these accurately predict the thermal models for the LCD and HCD to a reasonable degree of scrutiny without the need for excessive computation time, especially in these cases where Fluent calculations can take upwards of 5 h to execute. However, it should be noted that the maximum temperatures at each of the mesh qualities tested in Figure 5a are extremely similar, with a total variation of only 0.0083% from the minimum density mesh to the maximum. The temperature reached the selected HCD mesh density (Figure 5b) at T_{max} = 326.301 K with 0.0032% error from the temperature recorded at 808,411 elements, which is also considered a precise enough result for this study.

3. Results

3.1. Linear Channel (LCD) Results and Analysis

3.1.1. Influence of Mass Flow Rate on the LCD

The first step of the design optimisation process for the LCD was to vary the inlet mass flow rate and examine the effect this has on the thermal performance of the battery pack. Six mass flow rates were tested, within the range of 2.5E-05 to 7.5E-04 kgs^{-1}, based on the range used by Yates et al. [45], and it should be noted that the three largest mass flow rates are an entire order of magnitude higher than the three lowest. This was done to achieve a wide range of results. The thermal performance was measured through the maximum temperature the lithium-ion cells would reach at each time step and the thermal uniformity of the cells (difference from minimum to maximum cell temperature at each time step) over the full 5C discharge cycle of 720 s. Table 3 shows a snapshot of the results taken at the 720th time step, after the discharge cycle concluded, and Figure 6a,b show the maximum temperature of the cells and the temperature difference in the cells, respectively.

Table 3. Influence of mass flow rate at 720 s–LCD.

Mass Flow Rate (kgs^{-1})	T_{max} (K) at 720 s	ΔT (K) at 720 s
2.5E-05	310.595	1.105
5.0E-05	310.134	1.285
7.5E-05	309.695	1.431
2.5E-04	307.180	1.888
5.0E-04	305.004	1.881
7.5E-04	303.851	1.749

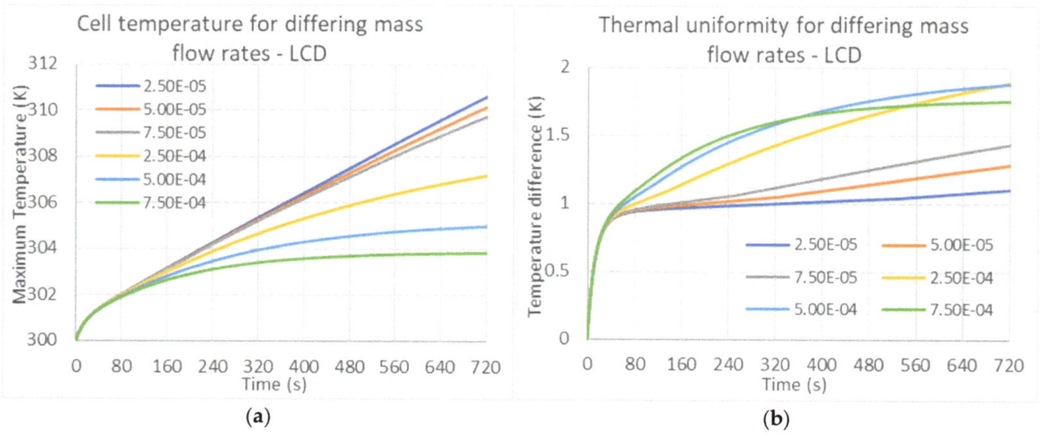

Figure 6. LCD-effects of mass flow rate on: (**a**) Maximum cell temperature; (**b**) Cell temperature difference.

A relationship can be seen from Figure 6a that as the inlet mass flow rate increases, the maximum temperature of the cells decreases. Within the first 80 s of the battery discharge, the temperatures recorded for each mass flow rate are very similar, and after this point, the temperatures deviate increasingly. This is due to the relationship of mass flow rate to the heat transfer coefficient and rate of heat flow, demonstrated in Equations (9) and (10), since the rate of heat flow varies linearly with mass flow rate. Figure 6b illustrates the thermal uniformity of the cells at each mass flow rate, wherein the first 40 s ΔT increases rapidly. This is due to the surface of cells rapidly undergoing a change in temperature from the external liquid channel compared to the centre of the cells, creating a significant increase in temperature differential, for which the effects are gradually mitigated after 40 s as the core of the lithium-ion cells begin to cool at a similar rate to the surface. A temperature differential will be inevitably caused due to this being an indirect cooling method, where there are four liquid channels split evenly around each cell rather than each cell being evenly submerged in coolant liquid. Furthermore, with the LCD model, the central channel coolant will rise in temperature more rapidly than the other channels due to this channel being in closest proximity and equidistant from each cell. The channels on the corners will increase less in temperature due to being in proximity to only one cell. This is a limitation mainly induced by the size of the considered model, given it analyses only a sample of four cells from a full battery pack in order to not require unreasonable amounts of computational power to execute the Fluent calculation.

Interestingly in Figure 6b, it would seem that increasing the mass flow rate actually worsened the thermal uniformity of the battery pack, especially evident in the lowest three mass flow rates. This phenomenon was also documented in different cooling designs by Bai et al. [46] and Chen et al. [47]. The temperature difference for 7.5E-04 kgs^{-1} is the largest from 0–320 s, thereafter, it becomes lower than that of 5.00E-4 kgs^{-1} and then lower than that of 2.50E-4 kgs^{-1} after 530 s. This is likely due to the aforementioned phenomena where the surface of the cell is cooling rapidly compared to the core when the heat transfer coefficient is higher due to the higher mass flow rate, and as the temperature regulates through the discharge cycle, this effect lessens since the rate of increase in temperature in the core begins to match that of the surface. Considering the optimal operating temperature for lithium-ion batteries lies between 298 K and 313 K [5] with a temperature difference of no greater than 5 K [5], the LCD battery pack immediately performed strongly in both regards, in that none of the mass flow rates tested resulted in results greater than these values (evident in Table 3). It is important to consider that using a high inlet mass flow rate in the cooling system inevitably draws more power, lowering the overall efficiency of the EV and potentially reducing its driving range. This is why although the concept of using the highest mass flow rate may be attractive, a compromise must be made to maintain higher efficiency in the EV. This is why the mass flow rate of 7.50E-05 kgs^{-1} has been chosen for further study, as it provides solid maximum temperature regulation (298 K < T_{max} < 313 K) and thermal uniformity (ΔT < 5 K) whilst not requiring excessive power to run in an EV.

3.1.2. Influence of Linear Channel Diameter, d_L, on the LCD

The next step of the design optimisation process for the LCD involved varying the diameter of the coolant channel, which would control the volume flow of fluid down the aluminium channels, and examine the effect this has on the thermal performance of the battery pack. Three values for channel diameter were tested, being d_L = 3 mm, 6 mm and 9 mm, illustrated in Figure 7a–c, respectively. In order to keep the simulations consistent, the total number of mesh elements for each simulation was kept as close as possible, despite the change in geometry. This was achieved through altering the body sizing of the aluminium channels, where the total mesh density was kept roughly between 277,000 and 281,000. The thermal performance was again measured through the maximum temperature reached by the cells for each time step and the temperature difference across the cells for each time step over the full 5C 720 s battery discharge cycle. Figure 8a,b show the maximum temperature of the cells and the temperature difference in the cells, respectively.

Figure 8a shows that as the diameter of the coolant channel increased, the maximum temperature of the lithium-ion cells decreased. This, however, was only over a small temperature range, taking the values at 720 s to be T_{max} = 309.97 K when d_L = 3 mm, and T_{max} = 309.31 K at d_L = 9 mm, indicating that the diameter of the coolant channels perhaps does not hold a large influence of the maximum temperatures, but still a noticeable one that is worth considering for optimisation. The study by Lai et al. [24], of a similar design to the LCD, only altered the diameter from 1–3 mm, providing a small range of results where a distinct relationship could not be found. The study performed in this paper, however, has used a significantly larger range of input parameters for diameter, where the relationship is well defined.

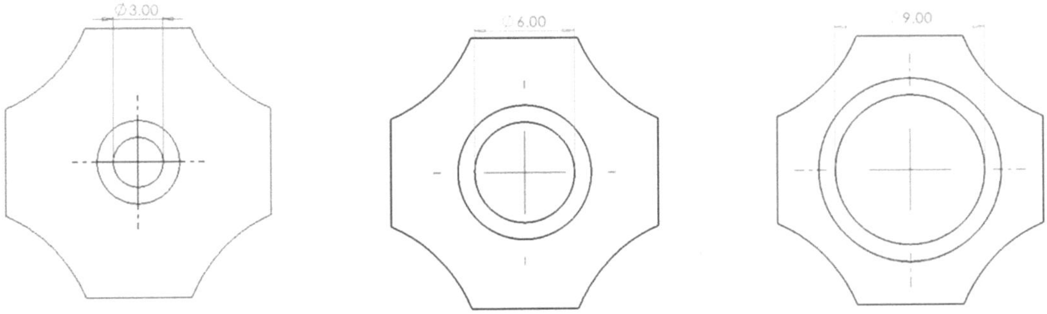

(a) (b) (c)

Figure 7. Top view of LCD channels for varying diameters: (**a**) d_L = 3 mm; (**b**) d_L = 6 mm; (**c**) d_L = 9 mm.

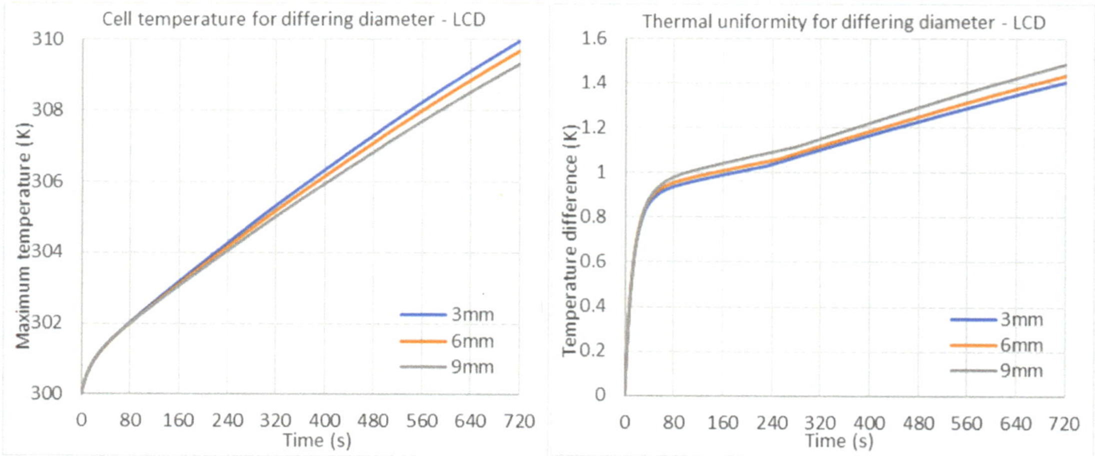

(a) (b)

Figure 8. LCD—effects of channel diameter on: (**a**) Maximum cell temperature; (**b**) Cell temperature difference.

The relationship presented in Figure 8a is due to a combination of factors. On the one hand, since the inlet mass flow rate remains the same in each case but the fluid channel diameter is decreased (note the inlet in question is the inlet to the upper reservoir, for which the parameters are not changed), the water pressure and hence the fluid velocity is increased, which increases the heat transfer coefficient (see Equations (9) and (10)), providing increased levels of cooling to the lithium-ion cells. On the other hand, this effect is cancelled out by the fact that the heat transfer area between the fluid and the inside of the aluminium channel is decreased when the diameter is decreased. This results in a lower heat transfer (see Equation (10)) which leads to reducing the cooling capability. From these results, evidently, the latter effect has a slightly larger influence over the maximum cell temperature compared to the former effect, but the cancellation is also still evident in that T_{max} only has a maximum variance of 0.66 K across the smallest and largest diameter tested.

Figure 8b shows that as the channel diameter increases, the thermal uniformity worsens, but only marginally. At the end of the 5 C discharge cycle (t = 720 s), $\Delta T = 1.404$ K when $d_L = 3$ mm and $\Delta T = 1.484$ K when $d_L = 9$ mm, which is a very small range of temperature difference and is due to the same aforementioned reasoning for the results in Figure 8a. Both the maximum cell temperature and thermal uniformity are kept within the optimal range of 298 K < T_{max} < 313 K and $\Delta T < 5$ K for all diameters, although since the maximum temperature reduction is more apparent for the $d_L = 9$ mm case in relation to the worsening of thermal uniformity, and also since T_{max} values are closer to the upper boundary of 313 K, the 9 mm diameter channel geometry is selected as most optimal and will be considered for further study.

3.1.3. Influence of Inlet and Outlet Locations on the LCD

Based on the methods of Yates et al. [45], the final optimisation step for the LCD involved varying the locations of the fluid domain inlets and outlets and examining the effect this has on the thermal performance of the battery pack. Five different combinations of locations were tested, for which the locations are presented in diagram form in Figure 9 and the combinations of these locations are presented in Table 4.

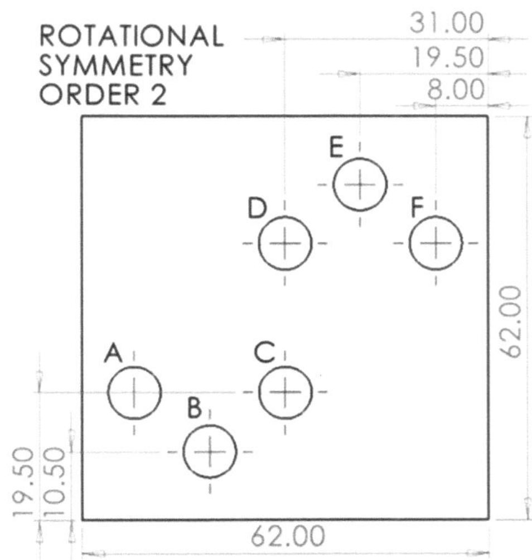

Figure 9. Inlets (A–C) and outlets (D–F) locations (dimensions in mm)–LCD.

Table 4. Thermal performance of location combinations at t = 720 s–LCD.

Location Combination	T_{max} (K) at 720 s	ΔT (K) at 720 s
Case 1 = A, F	309.366	1.796
Case 2 = B, E	309.362	1.779
Case 3 = C, D	309.308	1.484
Case 4 = A, D	309.364	1.799
Case 5 = C, F	309.321	1.492

When selecting the locations, it was important to make sure that the inlet was not directly above one of the aluminium channels since, in reality, the coolant would not be able to spread evenly in the upper reservoir to flow through each channel. The metrics used for assessing the thermal performance of the battery pack were again the maximum temperature reached in the lithium-ion cells and the temperature difference across the cells over the full 5C 720 s battery discharge cycle. Table 4 shows the maximum temperature and temperature difference measured at the 720th time step (end of the discharge cycle), and Figure 10 shows the temperature difference in the cells. The main objective of this stage was to optimise thermal uniformity since only location conditions are being altered, which will not have a significant impact on the maximum temperature reached. This is why the graph for maximum temperature is not plotted, as the curves for each combination are more or less on top of each other, providing little analytical benefit.

Figure 10 shows the results of the five combination cases applied in this study, where line markers were applied for clarity. Cases 3 and 5 performed the strongest in terms of thermal uniformity, whereas cases 1, 2 and 4 were lacking in comparison. Cases 3 and 5 both used location C as an inlet (see Table 4 and Figure 9), and as this is the common factor, it is most likely the reason for higher performance in these cases. This is because location C provided the most central location out of all the inlets in relation to the LCD battery pack, providing the most even distribution of coolant to all of the lithium-ion cells. This data also shows that the location of the inlet holds greater significance to the thermal performance than the location of the outlet since using outlets D and F in cases 3 and 5, respectively, gave very similar results, as shown in Figure 10 and Table 4.

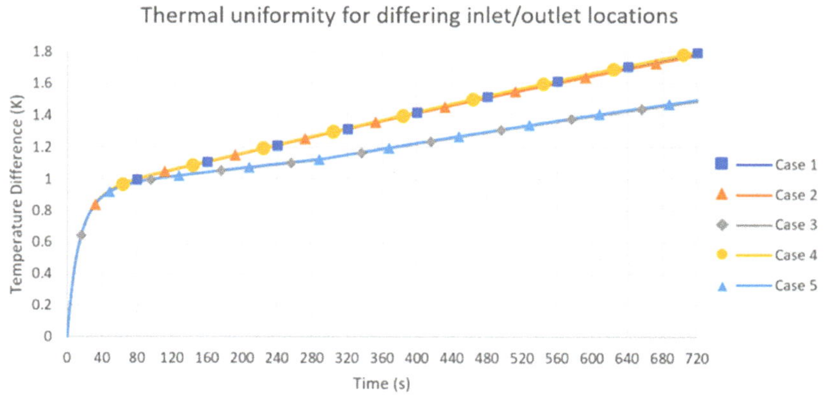

Figure 10. Cell temperature difference over discharge cycle for differing inlet/outlet locations–LCD.

All of the inlet and outlet combinations tested fell well within the optimal temperature regions of 298 K < T_{max} < 313 K and ΔT < 5 K. Although the results in Table 4 show that case 3 provided the best results in both regards, with T_{max} = 309.308 K and ΔT = 1.484 K at the conclusion of the discharge cycle, just slightly ahead of case 5 in performance. The LCD battery pack is now considered optimised within the bounds of this research, using inlet location C and outlet location D (Figure 10). The final temperature contours for the LCD are shown in Figure 11, whereas in Figure 11a, the effects of the aluminium channel contact area are evident in the lower temperature regions.

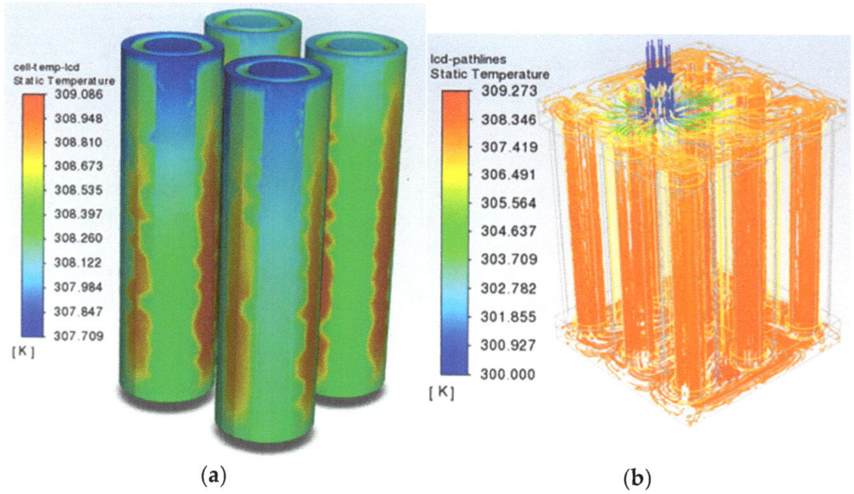

Figure 11. LCD contours: (**a**) Cell temperature (K); (**b**) Fluid pathlines.

An experimental case was conducted by Sheng et al. [25] with the cellular liquid cooling jacket containing liquid channels running parallel to the battery length, similar in design to this study. Using cylindrical 21,700 Li-ion cells operating at 2.5C discharge rate, the maximum cell temperature reached was 312 K, just falling under the upper boundary in the optimal region of 298 K < T_{max} < 313 K, in similar fashion to the LCD, having a T_{max} value just below at 309.308 K. The maximum temperature difference in this case found by Sheng et al. was approximately ΔT = 1.4 K, extremely close to the ΔT = 1.484 K achieved by the LCD, validating this study's results.

3.2. Helical Channel Design (HCD) Results and Analysis

3.2.1. Influence of Mass Flow Rate on the HCD

The inlet mass flow rate of coolant must be optimised for the HCD in order to contribute to the objectives of maximum temperature regulation of 298 K < T_{max} < 313 K and thermal uniformity of ΔT < 5 K whilst remaining realistic for the use in an electric vehicle. Six mass flow rates were tested in the range of 2.50E-05 to 7.50E-04 kgs^{-1}. Table 5 shows a snapshot of the results taken at the 720th time step (after the discharge cycle has concluded). Figure 12a,b show the maximum temperature of the cells and the temperature difference in the cells across the entire discharge cycle, respectively.

Table 5. Influence of mass flow rate at 720 s–HCD.

Mass Flow Rate (kg s^{-1})	T_{max} (K) at 720 s	ΔT (K) at 720 s
2.50E-05	326.301	16.462
5.00E-05	324.562	19.332
7.50E-05	322.081	18.964
1.00E-04	319.264	17.197
2.00E-04	311.201	10.396
7.50E-04	303.702	3.439

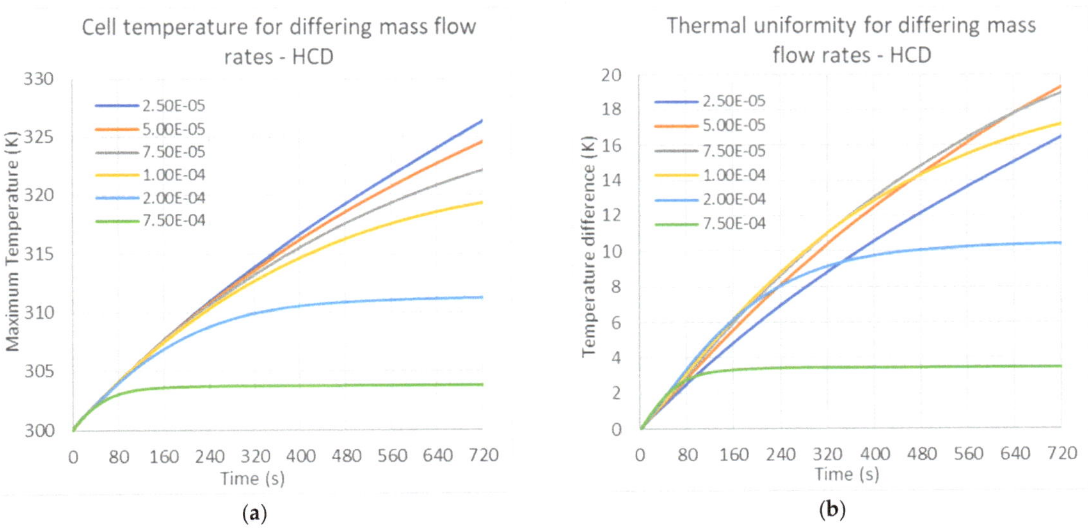

Figure 12. HCD–effects of mass flow rate on: (**a**) Maximum cell temperature; (**b**) Cell temperature difference.

As can be seen in Figure 12a, the same relationship as in the LCD (Figure 6b) can be found where as the mass flow rate increases, the maximum cell temperature decreases. The same explanation of this relationship as given for the LCD can also be applied here, where the link of mass flow rate to heat transfer coefficient given in Equations (9) and (10) provides increased cooling capability with a higher mass flow rate. Figure 12b shows the thermal uniformity in the form of temperature difference in the cells, where the relationship with mass flow rate is unclear. In general, thermal uniformity is improving with higher mass flow rates. Although the test with 2.50E-05 kgs^{-1} seems to not align with this trend, as it exhibits a lower temperature difference than with the 5.00E-05 kgs^{-1}, 7.50E-05 kgs^{-1} and 1.00E-04 kgs^{-1} mass flow rates. Additionally, the temperature difference over the full 720 s discharge cycle for the 5.00E-05 kgs^{-1}, 7.50E-05 kgs^{-1} and 1.00E-04 kgs^{-1} mass flow rates seem to overlap each other at varying points in the cycle. These occurrences were also reported by Zhou et al. [21] in a similar study and can be accounted for simply by a computational error.

The results of the HCD immediately seem to give the impression of a lack of performance relative to the LCD. The maximum temperatures reached for the HCD (Table 5) are much higher for the same mass flow rates used as the LCD (Table 3). For example, where both designs used m = 2.50E-05 kgs^{-1}, the LCD gave T_{max} = 310.595 K, but the HCD gives T_{max} = 326.301 K, an increase in temperature by 15.706 K, and well beyond the desired value within 298 K < T_{max} < 313 K. The thermal uniformity of the lithium-ion cells in the HCD has also worsened in comparison to the LCD, wherein the LCD ΔT remained between 1.105 K and 1.888 K (Table 3) for all tested mass flow rates. However, the HCD reached its highest temperature difference of 19.332 K at m = 5.00E-05 kgs^{-1}. This can be due to a multitude of reasons, one being the lessened contact area with aluminium acting as a heat sink for the HCD, meaning the rate of heat dissipation through the aluminium channel will be lessened. Furthermore, the coolant within the helical channels is inevitably in closer proximity to the cells due to the channels needing to weave in complete circumference to the batteries. This includes the points where the distance between each cylindrical battery is a minimum, rather than with the LCD where the centroid of a coolant channel is always at the point of maximum distance in-between each cell. This means that the top of the cells (positive electrode) will be cooled at a higher rate, with the coolant fluid increasing in temperature more rapidly as it reaches the bottom (negative electrode), creating a more significant imbalance in temperature throughout each cell.

Since the mass flow rates of 2.50E-05 kgs^{-1} to 1.00E-04 kgs^{-1} resulted in a higher temperature than the objective of 298 K < T_{max} < 313 K evident in Table 5, they have been automatically rejected, and the temperature difference also lies above the goal of ΔT < 5 K. A mass flow rate of 2.00E-04 kgs^{-1} was selected for further study, given this is the lowest tested mass flow rate for T_{max} to fall below 313 K, providing T_{max} = 311.201 K and ΔT = 10.396 K. Although ΔT remains above the desired 5 K, further optimisation to the geometry of the HCD shall be carried out to reduce this. Additionally, the selected mass flow rate should not be too high as this requires more power to operate within an EV, reducing the overall efficiency and driving range. This means the HCD is already at a disadvantage compared to the LCD since a higher mass flow rate must be used in order to keep T_{max} within acceptable levels.

3.2.2. Influence of Helical Channel Diameter, d_H, on the HCD

The next chosen design optimisation for the HCD involved altering the diameter of the helical coolant channels in order to examine and evaluate the effect this has on the thermal performance of the battery pack. Four values for diameter were tested, comprising d_H = 2 mm, 3 mm, 4 mm, and 5 mm, illustrated in Figure 13, where the total height of each helix and the dimensions of the reservoirs were altered slightly each time to keep the structure consistent. Due to the nature of the helical geometry, the diameters, unfortunately, cannot be increased beyond 5 mm, since the channels would be in contact with each other at the point of closest proximity between the batteries, which provides another limitation in comparison to the LCD. The drawings shown in Figure 13 are cross-section views of the helical channel, where the channel would be in contact with the battery along the flat edge. Once again, to keep the simulations consistent, the total number of mesh elements for each case was kept between 207,212 and 207,943 by altering the body sizing of the channel since this mesh density was considered accurate from the HCD mesh dependency study. Figure 14a,b show the maximum temperature and temperature difference, respectively, achieved in the full 5C 720 s discharge cycle.

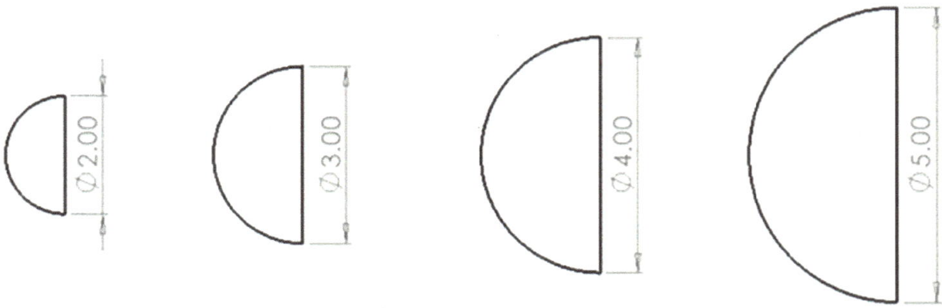

Figure 13. Varying diameter for HCD channels–cross-section view (dimensions in mm).

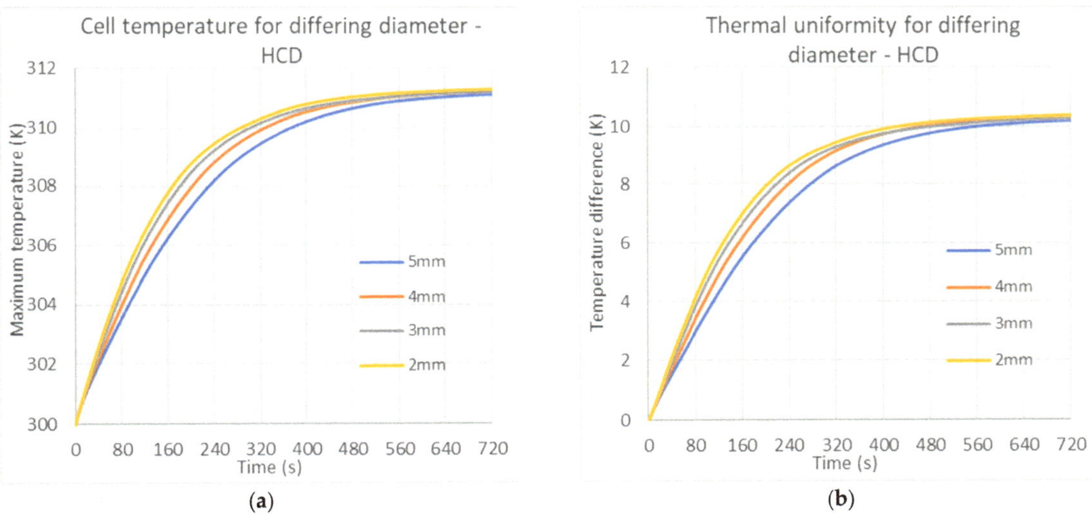

Figure 14. HCD–effects of channel diameter on: (**a**) Maximum cell temperature; (**b**) Cell temperature difference.

Figure 14a,b both show a very similar relationship in the characteristics of each tested diameter over the 720 s discharge cycle. From roughly 5% DOD (t = 40 s) to 60% DOD (t = 440 s), the difference in performance for each diameter is most noticeable, where the larger diameter channel performs better in terms of maximum temperature reduction and thermal uniformity. Beyond this point, up until the conclusion of the discharge cycle, the performance difference of each diameter is less apparent but still follows the aforementioned relationship. These results follow a similar relationship to the findings of the double helix structure proposed by Dong et al. [22], although a larger range of diameters has been tested within this research compared to Dong et al. [22] and the half helical design proposed by Zhou et al. [21], in order to fully define the relationship between channel diameter and thermal performance. In a similar fashion to the LCD channel diameter test, the larger helical diameters for HCD provide stronger overall cooling due to the increased contact surface area with the lithium-ion cells, which in turn increases the heat transfer rate (described in Equation (10)).

At t = 720 s, when d_H = 5 mm the value of T_{max} = 311.104 K and ΔT = 10.224 K, whereas when d_H = 2 mm the value of T_{max} = 311.282 K and ΔT = 10.384 K. This shows that the increase in diameter reduced maximum temperatures by 0.178 K and improved thermal uniformity by 0.16 K. Although this is not a significant amount, it is still worth considering, especially when the values of T_{max} and ΔT were improved by up to approximately 1 K at the earlier DOD, which in the long term will mitigate the de-ageing of the lithium-ion cells [5]. This is why the 5 mm diameter for the helical channels is to be selected as optimal. At this stage, the maximum temperature is within the objective region of 298 K < T_{max} < 313 K, but unfortunately, the thermal uniformity could not be reduced to the desired ΔT < 5 K, which will be attempted during the final stage of optimisation.

3.2.3. Influence of Inlet and Outlet Locations on the LCD

The final stage of optimisation for the HCD involved altering the inlet and outlet locations for the fluid domain and examining the effect this has on the thermal performance of the battery pack. Three combinations of inlet/outlet locations were tested, where the locations are mapped in Figure 15, and the combinations are noted in Table 6. The central hole shown in Figure 15 is named by both I (inlet) and J (outlet) since they overlap in the top view, the inlet is on the upper reservoir, and the outlet is on the lower reservoir.

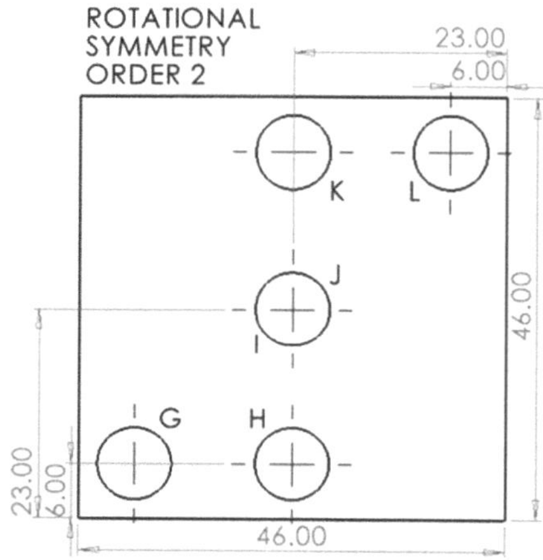

Figure 15. Inlets (G–I) and outlets (J–L) locations (dimensions in mm)–HCD.

Table 6. Thermal performance of location combinations at t = 720 s–HCD.

Location Combination	T_{max} (K) at 720 s	ΔT (K) at 720 s
Case 1 = G, L	311.279	10.794
Case 2 = I, J	311.104	10.224
Case 3 = H, K	311.328	10.733

It was established for the LCD that the combinations that were not symmetrical in both axes performed weakest and that the inlet location is the dominant factor compared to the outlet location, so in this study, only the three cases need to be analysed. For each tested combination, the values for T_{max} and ΔT at 720 s are presented in Table 6, along with the temperature difference graph shown in Figure 16 representing thermal uniformity.

Figure 16 shows the results of the three combinations, where it is conclusive that Case 2 has performed the strongest in terms of improving thermal uniformity, making use of inlet I and outlet J (see Figure 15). This is due to this inlet and outlet providing the most central location of all the possibilities, ensuring an even distribution of coolant for the lithium-ion cells. All the combinations tested provided maximum temperatures within the goal of 298 K < T_{max} < 313 K, however, the thermal uniformity still could not be reduced down to ΔT < 5 K for the HCD, although Case 2 did get the closest to reaching this goal, with a value for T_{max} = 311.104 K and ΔT = 10.224 K. Figure 17 shows the final temperature contours for the HCD.

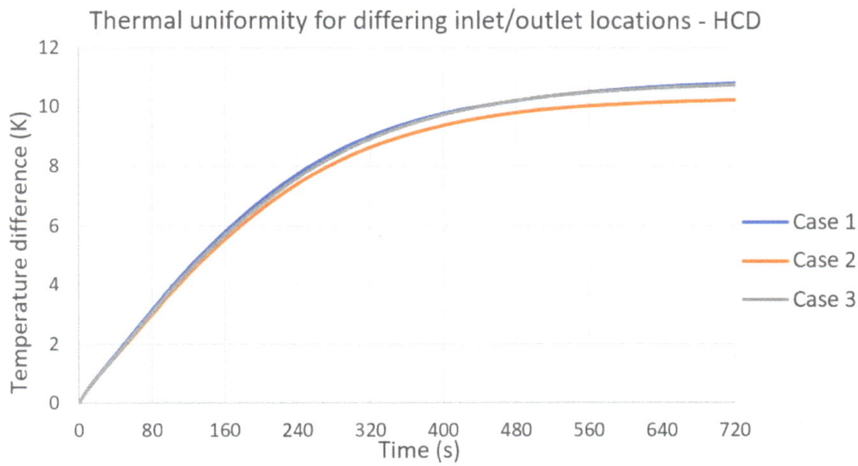

Figure 16. Cell temperature difference over discharge cycle for differing inlet/outlet locations–HCD.

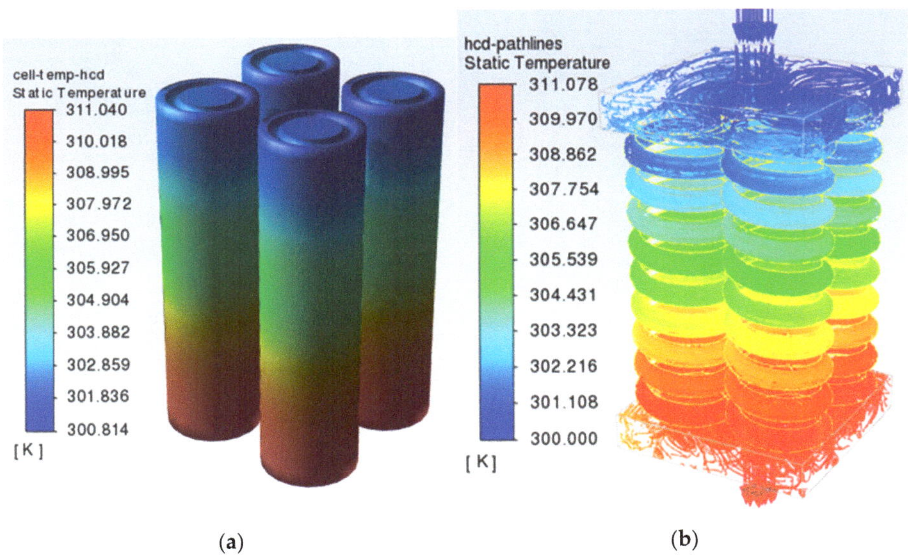

Figure 17. HCD contours: (**a**) Cell temperature (K); (**b**) Fluid pathlines.

In terms of experimental validation, a double helix cooling plate with PCM cooling for 18,650 cylindrical cells was investigated by Yu et al. [7], achieving maximum temperatures varying between 310 K < T_{max} < 313 K, where the HCD in this study achieves a similar result of 311.104 K. The CFD simulations conducted in the literature on helical duct liquid cooling structures similar to the HCD also seem to have issues with regulating temperature difference. Dong et al. [22] achieved ΔT = 7.01 K after the use of optimisation algorithms for the double helix structure, and Zhou et al. [34] achieved ΔT = 6.7 K for the half-helical duct structure, both falling outside the optimal range of ΔT < 5 K.

3.3. Comparison of the LCD and HCD

The fully optimised LCD and HCD structures illustrated in Figure 18a,b, respectively, shall now be compared and evaluated. Figures 19 and 20 show the maximum temperatures reached and the temperature difference in the lithium-ion cells, respectively, over the same

5C 720 s discharge. The LCD clearly provides a superior cooling performance compared to the HCD in both metrics, where T_{max} = 309.308 K and ΔT = 1.484 K at 720 s, compared to that of T_{max} = 311.104 K and ΔT = 10.224 K reached by the cells in the HCD, a reduction of 1.796 K in maximum temperature and 8.740 K in temperature difference. This is on top of the fact that the LCD could make use of a lower value for inlet mass flow rate, providing a lower power consumption requirement to run the thermal management system, maximising the efficiency of the given EV.

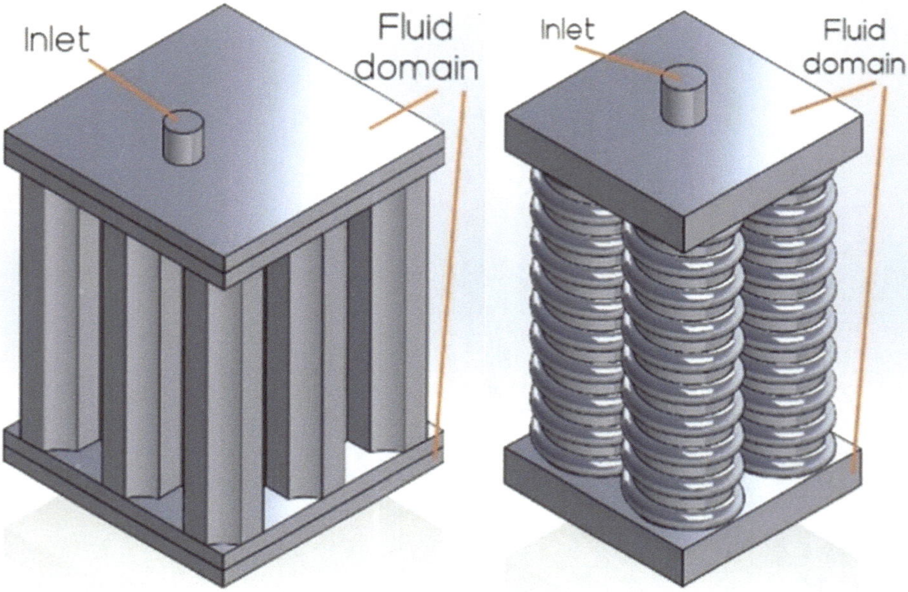

(a) (b)

Figure 18. ISO views for the final optimised battery packs: (**a**) LCD; (**b**) HCD.

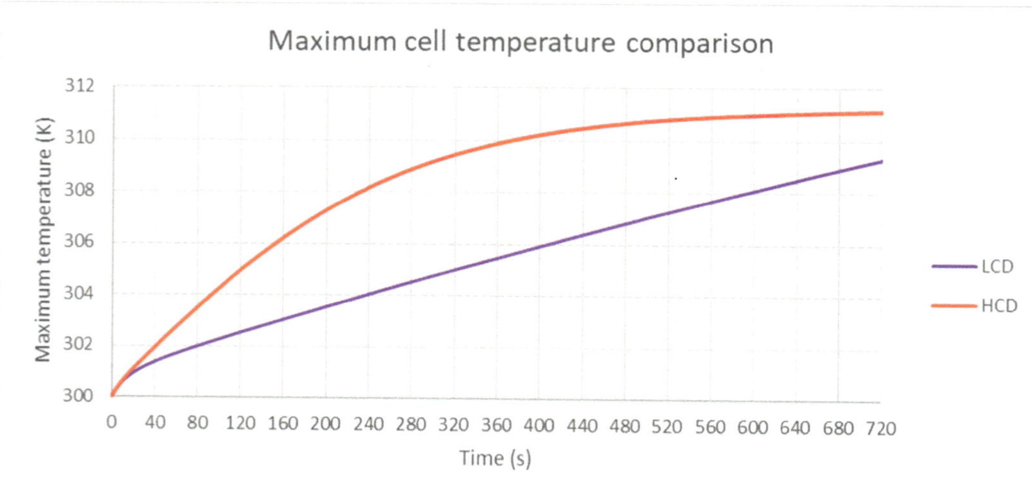

Figure 19. Maximum cell temperature comparison–LCD & HCD.

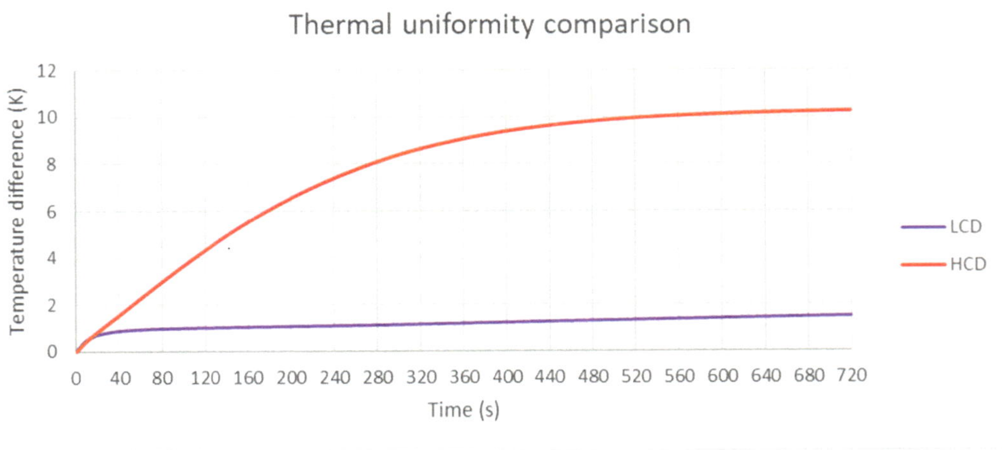

Figure 20. Cell temperature difference comparison–LCD & HCD.

Both objectives of 298 K < T_{max} < 313 K and ΔT < 5 K were achieved for the LCD comfortably, and the maximum temperature objective for the HCD was also achieved, albeit slightly higher in magnitude compared to the LCD. However, the objective to control temperature difference under 5 K was not achieved by the final HCD structure. Theoretically, the objective could be achieved by using a higher mass flow rate, as shown in Table 5. Although this would have to compromise even further on the overall efficiency of the EV, given its requirement of much higher power consumption, preventing this power from being more usefully allocated to other systems in the EV.

Excluding the masses of the lithium-ion cells, the mass of the HCD structure comes to 50.2 g, and the mass of the LCD comes to 285 g, given the volumes computed from SolidWorks 2021 and the mass density of aluminium to be 2719 kgm^{-3} from Table 2. This at least gives the HCD the advantage of being much more lightweight, which is a quality that vehicle manufacturers will strive for to achieve maximal efficiency. Given that the cells are arranged in the same spacing for both designs and both designs output the same power (but at different temperatures), the volumetric energy densities are the same for each design when considering a full EV battery pack. The HCD has a 5.7 times higher gravimetric energy density, based on the ratio of weights for each design. From a manufacturing standpoint, the helical channels could be made in a similar fashion to a compression spring using a CNC spring coiler machine, followed by heat treatment and grinding. Although it would need to be heat treated and moulded awkwardly to shell out the pipe providing the medium for fluid flow. This would be a significantly more complicated manufacturing process than that of the linear channels, whereby, for example, a drill press could be used for the extruded cut. This also would lead to the LCD being cheaper to manufacture and simpler to maintain and clean, given there is less need for specialised equipment to clean a straight pipe compared to a helical pipe.

The LCD would prove to be a more applicable thermal management system for use in electric vehicles given its lower power consumption, and overall superior cooling effectiveness compared to the HCD. The LCD could also be adapted to large scale energy storage applications due to the simplicity of its manufacturing, meaning high volumes for a large battery pack would be more economical to produce.

The shortfalls of the HCD are mainly due to the lack of significant contact area for a heatsink and the lack of aluminium. This is corroborated by the designs tested by Yates et al. [45], where the use of a heat sink in one design provided better thermal uniformity compared to one without. Additionally, due to the closer proximity of the coolant to the cells in the HCD, an imbalance in temperature is created from the top of the cells (positive electrode) to the bottom (negative electrode).

4. Discussion and Conclusions

In this research, two liquid cooling structures for EV lithium-ion battery packs, denoted as the LCD and HCD, were designed and evaluated. The heat generation of a 18,650 cell undergoing a 5C discharge cycle was modelled with the standardised 300 K inlet temperature for the coolant water. The thermal performance of the cells were analysed throughout the discharge cycle. Each design underwent a series of numerical and geometrical optimisations, comprising the variance of inlet mass flow rate, channel diameter and inlet and outlet locations. The metrics used for measuring the thermal performance were the maximum temperature and temperature difference (thermal uniformity). The important findings are as follows:

- The performance of each design, especially the HCD, was most significantly dependent on the inlet mass flow rate. It was found that as mass flow rate increases, the maximum cell temperature decreases due to the relationship between flow velocity and the heat transfer coefficient. The thermal uniformity improved for the HCD, and thermal uniformity worsened for the LCD, with a few discrepancies.
- As the channel diameter increased for each design, the maximum temperature decreased due to the relationship between contact area and heat transfer rate. The thermal uniformity for the HCD improved, whereas it worsened slightly for the LCD. The improvement in temperature regulation was more apparent towards the end of the discharge cycle for the LCD, but for the HCD, it was more apparent during the 5% to 60% DOD range.
- The thermal uniformity was optimal when the inlet and outlet were centralised, providing the evenest distribution of coolant among the four cells. Additionally, the location of the inlet was more significant in influencing the thermal performance than the location of the outlet.
- The optimal LCD provided a cell T_{max} of 309.308 K and ΔT of 1.484 K at t = 720s, well within the objectives of 298 K < T_{max} < 313 K and ΔT < 5 K. The final design used a mass flow rate of 7.50E-05 kgs^{-1}, channel diameter of 9 mm and manifold locations C & D (see Figure 10).
- The optimal HCD provided a cell T_{max} of 311.104 K and ΔT of 10.224 K at t = 720 s, meeting the objective of 298 K < T_{max} < 313 K but failing to satisfy ΔT < 5 K. The HCD could have theoretically reached the ΔT objective by using a higher inlet mass flow rate, but in the interest of keeping the design realistic for use in an EV, the mass flow rate was limited. The final design used a mass flow rate of 2.00E-04 kgs^{-1}, channel diameter of 5 mm and manifold locations I & J (see Figure 15).
- The LCD provided much stronger results than the HCD, even while making use of a lower mass flow rate, reducing the power draw for the cooling system to operate. The LCD is also simpler to manufacture, although it does weigh more, potentially harming the overall efficiency and the driving range of the EV in question.

Some limitations of this study include that the material properties for the battery were generalised and that the heat generation model was assumed to be uniform throughout the volume of the battery, along with not being a function of time or temperature. Due to the nature of the cooling system geometry, the cylindrical cells could not be structured in a staggered layout, which would be more compact and applicable in the design of EVs. Furthermore, only four lithium-ion cells were modelled in this battery pack simulation in order to mitigate unnecessarily high computational requirements. Future work could therefore evaluate these models on a larger scale more accurate to a real EV battery pack. Additionally, analysing different materials for the cooling structure other than aluminium as well as different coolants such as various glycol aqueous solutions would be useful to optimise these designs further. It is also recommended that experimental studies are conducted in future research for helical type channel designs similar to the HCD, as currently very little data exists for structures similar to this. Finally, the MSMD battery model within Fluent could be used to develop a more accurate heat generation model providing more applicable results, although requiring a slight reconstruction of some geometries.

Author Contributions: Conceptualization, R.L. and M.A.; methodology, R.L. and M.A.; software, R.L. and M.A.; validation, R.L. and M.A.; formal analysis, R.L. and M.A.; investigation, R.L. and M.A.; resources, M.A.; data curation, R.L.; writing—original draft preparation, R.L.; writing—review and editing, M.A.; visualization, R.L. and M.A.; supervision, M.A.; project administration, M.A. All authors have read and agreed to the published version of the manuscript.

Funding: This research received no external funding.

Institutional Review Board Statement: Not applicable.

Informed Consent Statement: Not applicable.

Data Availability Statement: Not applicable.

Conflicts of Interest: The authors declare no conflict of interest.

Nomenclature

Variables	
A	contact area (m^2)
c	specific heat capacity (Jkg^{-1} K^{-1})
d	channel diameter (mm)
h	heat transfer coefficient (Wm^{-2} K^{-1})
I	current (A)
k	thermal conductivity (Wm^{-1} K^{-1})
L	characteristic length (m)
m	mass (kg)
$\dot{m}$	mass flow rate (kgs^{-1})
P	pressure (Pa)
q	heat generation (Wm^{-3})
Q	rate of heat flow (W)
R	resistance (Ω)
Re	Reynolds number
T	temperature (K)
t	time (s)
U	open-circuit voltage (V)
V	volume (m^3)
v	velocity (ms^{-1})
Acronyms	
CAD	computer-aided design
CFD	computational fluid dynamics
DOD	depth of discharge
EV	electric vehicle
HCD	helical channel design
LCD	linear channel design
MSMD	multi-scale multi-domain
PCM	phase change material
SOC	state of charge
Greek letters	
Δ	difference
μ	dynamic viscosity (kgm^{-1} s^{-1})
ρ	mass density (kgm^{-3})
Subscripts	
b	battery
gen	generation
H	helical
L	linear
max	maximum
p	constant pressure process
w	water

References

1. Navas-Anguita, Z.; García-Gusano, D.; Iribarren, D. A review of techno-economic data for road transportation fuels. *Renew. Sustain. Energy Rev.* **2019**, *112*, 11–26. [CrossRef]
2. COP26 Declaration on Accelerating the Transition to 100% Zero Emission Cars and Vans. Available online: https://www.gov.uk/government/publications/cop26-declaration-zero-emission-cars-and-vans/cop26-declaration-on-accelerating-the-transition-to-100-zero-emission-cars-and-vans (accessed on 4 August 2022).
3. Liu, Y.; Zhang, R.; Wang, J.; Wang, Y. Current and future lithium-ion battery manufacturing. *iScience* **2021**, *24*, 102332. [CrossRef] [PubMed]
4. Zhao, J.; Rao, Z.; Huo, Y.; Liu, X.; Li, Y. Thermal management of cylindrical power battery module for extending the life of new energy electric vehicles. *Appl. Therm. Eng.* **2015**, *85*, 33–43. [CrossRef]
5. Pesaran, A. Battery thermal models for hybrid vehicle simulations. *J. Power Sources* **2002**, *110*, 377–382. [CrossRef]
6. Zeng, Y.; Chalise, D.; Lubner, S.; Kaur, S.; Prasher, R. A review of thermal physics and management inside lithium-ion batteries for high energy density and fast charging. *Energy Storage Mater.* **2021**, *41*, 264–288. [CrossRef]

7. Yu, Y.; Tang, A.; Chen, W.; Jia, H.; Jin, Y. Experimental Investigation of a Battery Thermal Management Device Based on a Composite Phase Change Material Coupled with a Double Helix Liquid Cooling Plate. *Energy Fuels* **2022**, *36*, 13370–13381. [CrossRef]
8. Choudhari, V.; Dhoble, A.; Panchal, S. Numerical analysis of different fin structures in phase change material module for battery thermal management system and its optimization. *Int. J. Heat Mass Transf.* **2020**, *163*, 120434. [CrossRef]
9. Landini, S.; Waser, R.; Stamatiou, A.; Ravotti, R.; Worlitschek, J.; O'Donovan, T. Passive cooling of Li-Ion cells with direct-metal-laser-sintered aluminium heat exchangers filled with phase change materials. *Appl. Therm. Eng.* **2020**, *173*, 115238. [CrossRef]
10. Cao, J.; Luo, M.; Fang, X.; Ling, Z.; Zhang, Z. Liquid cooling with phase change materials for cylindrical Li-ion batteries: An experimental and numerical study. *Energy* **2020**, *191*, 116565. [CrossRef]
11. Hekmat, S.; Molaeimanesh, G. Hybrid thermal management of a Li-ion battery module with phase change material and cooling water pipes: An experimental investigation. *Appl. Therm. Eng.* **2020**, *166*, 114759. [CrossRef]
12. Zhang, X.; Li, Z.; Luo, L.; Fan, Y.; Du, Z. A review on thermal management of lithium-ion batteries for electric vehicles. *Energy* **2022**, *238*, 121652. [CrossRef]
13. Wang, Y.; Jiang, J.; Chung, Y.; Chen, W.; Shu, C. Forced-air cooling system for large-scale lithium-ion battery modules during charge and discharge processes. *J. Therm. Anal. Calorim.* **2018**, *135*, 2891–2901. [CrossRef]
14. Sharma, A.; Patil, Y.; Krishnaiah, R.; Ashok, B.; Garg, A.; Gao, L. Study on effect of diverse air inlet arrangement on thermal management of cylindrical lithium-ion cells. *Heat Transf.* **2020**, *49*, 4626–4656. [CrossRef]
15. Xi, Y.; Feng, Y.; Xiao, Y.; He, G. Novel Z-Shaped Structure of Lithium-Ion Battery Packs and Optimization for Thermal Management. *J. Energy Eng.* **2020**, *146*, 04019035. [CrossRef]
16. Liu, Y.; Zhang, J. Design a J-type air-based battery thermal management system through surrogate-based optimization. *Appl. Energy* **2019**, *252*, 113426. [CrossRef]
17. Mohammed, A.; Elfeky, K.; Wang, Q. Recent advancement and enhanced battery performance using phase change materials based hybrid battery thermal management for electric vehicles. *Renew. Sustain. Energy Rev.* **2022**, *154*, 111759. [CrossRef]
18. Zhao, C.; Cao, W.; Dong, T.; Jiang, F. Thermal behavior study of discharging/charging cylindrical lithium-ion battery module cooled by channeled liquid flow. *Int. J. Heat Mass Transf.* **2018**, *120*, 751–762. [CrossRef]
19. Zhao, C.; Sousa, A.; Jiang, F. Minimization of thermal non-uniformity in lithium-ion battery pack cooled by channeled liquid flow. *Int. J. Heat Mass Transf.* **2019**, *129*, 660–670. [CrossRef]
20. Xu, H.; Zhang, X.; Xiang, G.; Li, H. Optimization of liquid cooling and heat dissipation system of lithium-ion battery packs of automobile. *Case Stud. Therm. Eng.* **2021**, *26*, 101012. [CrossRef]
21. Zhou, H.; Zhou, F.; Zhang, Q.; Wang, Q.; Song, Z. Thermal management of cylindrical lithium-ion battery based on a liquid cooling method with half-helical duct. *Appl. Therm. Eng.* **2019**, *162*, 114257. [CrossRef]
22. Dong, F.; Cheng, Z.; Zhu, J.; Song, D.; Ni, J. Investigation and optimization on cooling performance of a novel double helix structure for cylindrical lithium-ion batteries. *Appl. Therm. Eng.* **2021**, *189*, 116758. [CrossRef]
23. Rao, Z.; Qian, Z.; Kuang, Y.; Li, Y. Thermal performance of liquid cooling based thermal management system for cylindrical lithium-ion battery module with variable contact surface. *Appl. Therm. Eng.* **2017**, *123*, 1514–1522. [CrossRef]
24. Lai, Y.; Wu, W.; Chen, K.; Wang, S.; Xin, C. A compact and lightweight liquid-cooled thermal management solution for cylindrical lithium-ion power battery pack. *Int. J. Heat Mass Transf.* **2019**, *144*, 118581. [CrossRef]
25. Sheng, L.; Zhang, H.; Su, L.; Zhang, Z.; Zhang, H.; Li, K.; Fang, Y.; Ye, W. Effect analysis on thermal profile management of a cylindrical lithium-ion battery utilizing a cellular liquid cooling jacket. *Energy* **2021**, *220*, 119725. [CrossRef]
26. Zhao, J.; Rao, Z.; Li, Y. Thermal performance of mini-channel liquid cooled cylinder based battery thermal management for cylindrical lithium-ion power battery. *Energy Convers. Manag.* **2015**, *103*, 157–165. [CrossRef]
27. Reddy, T.; Linden, D. *Linden's Handbook of Batteries*, 4th ed.; McGraw-Hill: New York, NY, USA, 2010; pp. 23–26, 967.
28. Parsons, K.; Mackin, T. Design and Simulation of Passive Thermal Management System for Lithium-Ion Battery Packs on an Unmanned Ground Vehicle. *J. Therm. Sci. Eng. Appl.* **2016**, *9*, 011012. [CrossRef]
29. Cheng, E.; Taylor, N.; Wolfenstine, J.; Sakamoto, J. Elastic properties of lithium cobalt oxide (LiCoO2). *J. Asian Ceram. Soc.* **2017**, *5*, 113–117. [CrossRef]
30. Tang, W.; Tam, W.C.; Yuan, L.; Dubaniewicz, T.; Thomas, R.; Soles, J. Estimation of the critical external heat leading to the failure of lithium-ion batteries. *Appl. Therm. Eng.* **2020**, *179*, 115665. [CrossRef]
31. Gu, L.; Han, J.; Chen, M.; Zhou, W.; Wang, X.; Xu, M.; Lin, H.; Liu, H.; Chen, H.; Chen, J.; et al. Enabling robust structural and interfacial stability of micron-si anode toward high-performance liquid and solid-state lithium-ion batteries. *Energy Storage Mater.* **2022**, *52*, 547–561. [CrossRef]
32. Han, X.; Zhou, W.; Chen, M.; Chen, J.; Wang, G.; Liu, B.; Luo, L.; Chen, S.; Zhang, Q.; Shi, S.; et al. Interfacial nitrogen engineering of robust silicon/mxene anode toward high energy solid-state lithium-ion batteries. *J. Energy Chem.* **2022**, *67*, 727–735. [CrossRef]
33. Primary and Rechargeable Battery Chemistries with Energy Density. Available online: https://www.epectec.com/batteries/chemistry/ (accessed on 13 July 2022).
34. Zhao, R.; Liu, J.; Gu, J. Simulation and experimental study on Lithium Ion Battery Short Circuit. *Appl. Energy* **2016**, *173*, 29–39. [CrossRef]

35. Kukreja, J.; Nguyen, T.; Siegmund, T.; Chen, W.; Tsutsui, W.; Balakrishnan, K.; Liao, H.; Parab, N. Crash analysis of a conceptual electric vehicle with a damage tolerant battery pack. *Extrem. Mech. Lett.* **2016**, *9*, 371–378. [CrossRef]
36. Qian, Z.; Li, Y.; Rao, Z. Thermal performance of lithium-ion battery thermal management system by using mini-channel cooling. *Energy Convers. Manag.* **2016**, *126*, 622–631. [CrossRef]
37. Bernardi, D.; Pawlikowski, E.; Newman, J. A General Energy Balance for Battery Systems. *J. Electrochem. Soc.* **1985**, *132*, 5–12. [CrossRef]
38. Huang, Y.; Lu, Y.; Huang, R.; Chen, J.; Chen, F.; Liu, Z.; Yu, X.; Roskilly, A.P. Study on the thermal interaction and heat dissipation of cylindrical Lithium-Ion Battery cells. *Energy Procedia* **2017**, *142*, 4029–4036. [CrossRef]
39. Walvekar, H.; Beltran, H.; Sripad, S.; Pecht, M. Implications of the Electric Vehicle Manufacturers' Decision to Mass Adopt Lithium-Iron Phosphate Batteries. *IEEE Access* **2022**, *10*, 63834–63843. [CrossRef]
40. Wang, Z.; Li, X.; Zhang, G.; Lv, Y.; Wang, C.; He, F.; Yang, C.; Yang, C. Thermal management investigation for lithium-ion battery module with different phase change materials. *RSC Adv.* **2017**, *7*, 42909–42918. [CrossRef]
41. Sosnowski, M.; Krzywanski, J.; Grabowska, K.; Gnatowska, R. Polyhedral meshing in numerical analysis of conjugate heat transfer. *EPJ Web Conf.* **2018**, *180*, 02096. [CrossRef]
42. Shahid, S.; Agelin-Chaab, M. Development and analysis of a technique to improve air-cooling and temperature uniformity in a battery pack for cylindrical batteries. *Therm. Sci. Eng. Prog.* **2018**, *5*, 351–363. [CrossRef]
43. Inflation Layer Meshing in ANSYS. Available online: https://www.computationalfluiddynamics.com.au/tips-tricks-inflation-layer-meshing-in-ansys/ (accessed on 2 July 2022).
44. ANSYS, Inc. *ANSYS Fluent User's Guide*; ANSYS, Inc.: Canonsburg, PA, USA, 2021; pp. 972–973.
45. Yates, M.; Akrami, M.; Javadi, A. Analysing the performance of liquid cooling designs in cylindrical lithium-ion batteries. *J. Energy Storage* **2021**, *33*, 100913. [CrossRef]
46. Bai, F.; Chen, M.; Song, W.; Feng, Z.; Li, Y.; Ding, Y. Thermal management performances of PCM/water cooling-plate using for lithium-ion battery module based on non-uniform internal heat source. *Appl. Therm. Eng.* **2017**, *126*, 17–27. [CrossRef]
47. Chen, D.; Jiang, J.; Kim, G.; Yang, C.; Pesaran, A. Comparison of different cooling methods for lithium ion battery cells. *Appl. Therm. Eng.* **2016**, *94*, 846–854. [CrossRef]

batteries

MDPI

Article

Research on Bionic Fish Scale Channel for Optimizingt Thermal Performance of Liquid Cooling Battery Thermal Management System

Yutao Mu [1], Kai Gao [2,3,*], Pan Luo [2], Deng Ma [1], Haoran Chang [1] and Ronghua Du [2,3]

1 International College of Engineering, Changsha University of Science & Technology, Changsha 410114, China
2 College of Automotive and Mechanical Engineering, Changsha University of Science & Technology, Changsha 410114, China
3 Hunan Key Laboratory of Smart Roadway and Cooperative Vehicle-Infrastructure Systems, Changsha 410114, China
* Correspondence: kai_g@csust.edu.cn

Abstract: Liquid cooling battery thermal management systems (BTMSs) are prevalently used in electric vehicles (EVs). With the use of fast charging and high-power cells, there is an increasing demand on thermal performance. In this context, a bionic fish scale (BFS) channel structure optimization design method is proposed to optimize the thermal performance. The effects of different structural parameters of the liquid cooling plate in BTMS on its cooling performance, including BFS notch diameter (D), BFS notch depth (H), and BFS notch spacing (S), are investigated. To minimize the maximum temperature (T_{max}) and the maximum temperature difference (ΔT_{max}) as optimization indicators, experimental tests and numerical calculations are performed for a battery pack consisting of 36 square cells. Sixteen sets of thermal performance are discussed for different structural parameters in the transient thermal fluid simulation by using orthogonal tests. Under the optimal structural parameters, T_{max} decreases by 1.61 °C (10.8%) and ΔT_{max} decreases by 0.43 °C (16.7%). In addition, the maximum increase in outlet flow velocity is 2.72% and the pressure is reduced by 4.98%. Therefore, the proposed BTMS will have effective cooling performance in high-power dissipation.

Keywords: battery thermal management system; liquid cooling; structure optimization; thermal performance; high power dissipation

Citation: Mu, Y.; Gao, K.; Luo, P.; Ma, D.; Chang, H.; Du, R. Research on Bionic Fish Scale Channel for Optimizing Thermal Performance of Liquid Cooling Battery Thermal Management System. *Batteries* **2023**, *9*, 134. https://doi.org/10.3390/batteries9020134

Academic Editors: Hengyun Zhang, Jinsheng Xiao, Sousso Kelouwani and Carlos Ziebert

Received: 17 December 2022
Revised: 7 February 2023
Accepted: 11 February 2023
Published: 14 February 2023

1. Introduction

In recent years, there is an increasing emphasis on the development and use of clean energy because of the rising CO_2 emissions. As the transportation industry with high emissions, EV is rapidly developing and rising due to their clean, emission-reducing and energy-efficient characteristics [1]. As the heart of EV, the working condition and performance of power battery largely determines the performance of EV. Compared with other batteries, lithium-ion batteries are widely used in EV today due to their high energy density [2], and the battery's performance is closely associated with its temperature and temperature difference [3]. If battery temperature can be controlled at 20–50 °C and the temperature difference can be controlled below 5 °C [4], the battery will show good working performance and long service life. However, its internal violent chemical reaction makes the temperature difficult to control. Too high a temperature will trigger the side reactions in it, which will result in thermal runaway and explosion. Therefore, there is a need to develop an efficient BTMS for effective heat dissipation of the battery to ensure that it is maintained at an optimal operating condition temperature.

Nowadays, BTMS based on different cooling media have been developed, which are divided into four main categories: air cooling [5,6], liquid cooling [7,8], phase change material (PCM) cooling [9,10] and heat pipe cooling [11,12]. However, PCM and heat pipe technologies are not yet mature enough to be applied to BTMS due to thermal saturation [13] and heat transfer limit [14]. Air cooling has been developed earlier as well as applied to BTMS due to its simple structure and lower cost. Zhang et al. [15] proposed a Z-parallel air-cooling model with an infinite number of secondary outlets, and the effects of three parameters, namely the location, number and width of outlets, on the cooling performance of BTMS were investigated. However, due to the increasing thermal power density demand of the battery pack, the cooling efficiency of air cooling has difficulty in meeting its cooling requirements.

In contrast to the other cooling methods, liquid cooling has become a preferred option [16] and is widely used in industrial manufacturing. Pack thermal performance can be improved by optimizing cooling strategies and designing different liquid cooling plate configurations [17]. Kong et al. [18] proposed a method to control the flow channel inlet temperature and liquid flow velocity by using factors such as phase change material temperature. This method improved the ability of the battery pack to adapt to different ambient temperatures and reduced the power loss. Gao et al. [19] proposed a gradient channel flow channel design that can change the monotonic rise characteristics of temperature in the flow channel. The temperature difference in the gradient channel was reduced by 79.2% compared to the uniform large channel and by 60.2% compared to the uniform small channel. Wang et al. [20] improved the microchannel parameters and flow velocity. The T_{max} of the optimized pack was reduced by 20%. Tang et al. [21] proposed a lightweight liquid cooling structure and investigated the cooling performance at different inlet velocities, discharge rates and flow channel inner diameters. The structure design was optimized based on the results of the study to reduce T_{max} and ΔT_{max} to 34.97 and 4.04 °C, respectively. Although these research results are excellent, they are not of high reference value for industrial manufacturing applications because they are models constructed under ideal parameters.

When optimizing liquid cooling in industrial manufacturing, people are often inspired by the clever construction of living organisms in nature to develop relevant bionic structures [22]. Liu et al. [23] proposed a flow channel structure with bionic leaf vein branching. The T_{max} of BLVB channel with optimal parameters can be reduced by 0.23 °C compared to the unoptimized BLVB channel and 1.12 °C compared to the linear channel. Wen et al. [24] proposed a honeycomb cell BTMS, which included bionic microchannels and phase change materials. The T_{max} of the battery pack was kept stable between 39.0 °C and 3.5 °C using the estimated values from the BP model. Wang et al. [25] proposed a bionic spider web channel. The structure utilizes the heat dissipation structure of a spider web and exhibits excellent thermal performance. An et al. [26], inspired by the growth pattern of kidney glomeruli, investigated the effects of outer diameter, number of layers of bionic microchannels, liquid flow velocity, and channel arrangement on the maximum cell discharge rate of 4C by simulations and physical experiments. The T_{max} can be adjusted at 37.28 °C and 36.99 °C. The intercellular temperature differences are 4.06 °C and 4.21 °C for discharge ratios of 4C and 5C, respectively. These bionic designs can produce good thermal performance optimization while taking into account the possibility of practical industrial manufacturing applications. However, they are often difficult to apply as well due to the complexity of their structural processing.

Although there have been many previous studies and good results around liquid cooling [27,28], the actual engineering has put forward higher demands on the cooling performance of BTMS with the increasing energy density of the cores. The contemporary industrial manufacturing not only needs research methods to reduce the T_{max} and ΔT_{max}, but also needs these researches to fit and apply to the industrial manufacturing practice. However, the complex structure and theoretical experiments of many current BTMS studies with excellent results make it difficult to apply these studies in practice. Moreover, the trend in the field of thermal management is structural optimization and improved cooling efficiency [29]. Considering the excellent thermal performance of bionic structures and the possibility of practical industrial manufacturing processing, it is necessary to develop a liquid cooling structure optimization with simple structure and excellent results.

In this context, this study innovatively proposes an improvement of the liquid cooling plate structure based on the BFS structure. The coolant generates vortex flow in this structure, which increases the coolant flow velocity and reduces the flow pressure inside the liquid cooling plate. Unlike the complex theoretical model mentioned above, it can be processed directly by stamping method. In the study, orthogonal experiments are designed for three parameters of D, H and S of the BFS structure. The thermal performance of BTMS in orthogonal experiments with various structural parameters and the correlation between the parameters are analyzed using transient thermal fluid simulation to provide a reference for the design of structural parameters of liquid cooling plate with BFS channel. At last, the mechanism of action of BFS flow channel is studied and some validation is done.

2. Model Description

The optimized battery pack model based on the BFS structure is constructed, as shown in Figure 1a. The battery pack is composed of 36 lithium batteries with a capacity of 50 Ah in series. The electrodes on both sides of each cell are connected by series connection. The optimized liquid cooling plate is connected to the bottom of the cell by a heat-conducting pad. This liquid cooling plate consists of an inlet and an outlet. The flow channel is a serpentine flow channel [30], which is commonly used in industrial manufacturing today, and a BFS notch is added to it. The coolant in the liquid cooling plate is water, and its good specific heat capacity can effectively achieve the heat dissipation effect [31]. The coolant entering the liquid cooling plate is heat exchanged with the battery through the internal serpentine flow channel, and the heated coolant is discharged through the outlet. It is essential to note that the coolant flow shall enter from the blunt tip of the BFS notch as a way to ensure the effectiveness of this optimized structure, as shown in Figure 1b. Due to this structure, this liquid cooling system can reduce both the maximum pack temperature and the pack temperature inhomogeneity. It is also highly expandable, allowing for rapid modular design for different pack sizes. The entire battery pack is packaged using end caps and insulation pads to ensure the safety of the experiment. The dimensions and thermal properties of the core components of the pack are given in Table 1.

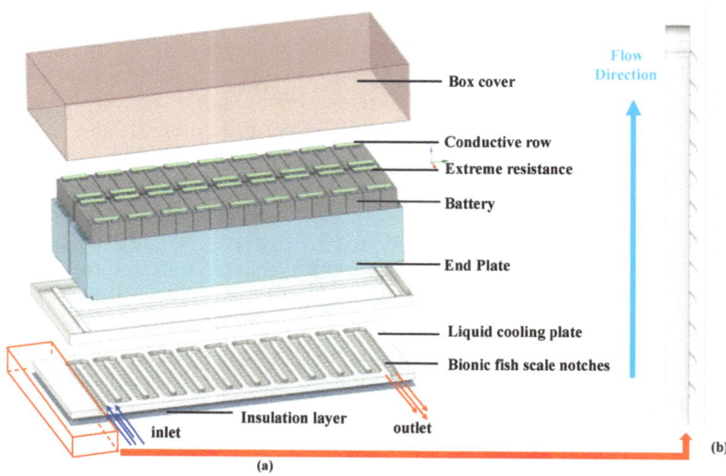

Figure 1. (**a**) Battery pack structure (**b**) Side view of liquid cooling plate.

Table 1. Battery pack structure parameters.

Specifications	Value
Battery pack	864 × 174 × 159 mm
Insulation layer	1 × 174 × 168 mm
Liquid cooling plate	991 × 411 × 51 mm
End plate	912 × 177 × 159 mm
Box cover	991 × 411 × 169 mm
Conductive row	72 × 26 × 2 mm
Nominal voltage	133.2 V
Energy (Cell measurement value calculation)	18 kWh

The channel width of the liquid cooling plate is different from the wide channel design in the previous study. The designed channel is considerably less in width and has some added thickness. The reduction in width reduces the heat dissipation area significantly, resulting in a decrease in heat dissipation efficiency. The increase in thickness resulted in a significant decrease in the flow velocity, leading to a decrease in the heat transfer efficiency. Therefore, it is clearly evident that the use of wide and flat channel design will produce better thermal performance results. However, the reason why the design is not so designed is to consider the actual design requirements of the project. Due to factors such as large flow pressures and vibration in industrial manufacturing, wide and flat channels are extremely susceptible to impact deformation. Both the safety of use and service life will be greatly reduced.

3. Results

To evaluate the effectiveness of the thermal performance optimization of the structure, transient thermal simulations are performed by using ANSYS FLUENT [32]. Detailed structural parameters and computational information are presented in this section.

3.1. Bionic Model Design Parameters

By analyzing the erosion morphology of the fluid on the mechanical surface [33] and the principle of BFS construction [34], a BFS notch structure is established, as shown in Figure 2. The shape of the groove surface is the approximate circle of BFS. After considering the characteristics of erosion morphology, the lowest point of the semicircular groove is shifted backward by 1/2R to constitute the simplified structure of BFS in this paper, in which the important structural parameters are D, H and S. H is calculated by the laminar boundary layer condition in the following equation.

$$\mathrm{H} = \frac{x}{\sqrt{Re_x}} \quad (1)$$

$$Re_x = \frac{vx}{\vartheta} \quad (2)$$

where H denotes the boundary layer thickness, and x denotes the flow direction displacement, and Re_x denotes the corresponding Reynolds number at x, and v denotes the mean velocity of the fluid, and ϑ denotes the fluid kinematic viscosity.

3.2. Boundary Conditions

During the simulation, the ternary Lithium battery is discharged in the 1C state. Due to the large surface area of the cells, the thermal resistance between the cells and the side plates is neglected. The ambient temperature of the battery pack is set to 25 °C. The coolant inlet and outlet temperature is set to 25 °C, and the volume flow rate is set to 5.04×10^{-3} L/min. The end plate and the box are convection heat dissipation with the outside world, and the convection heat transfer coefficient is 3 W/(m^2·K).

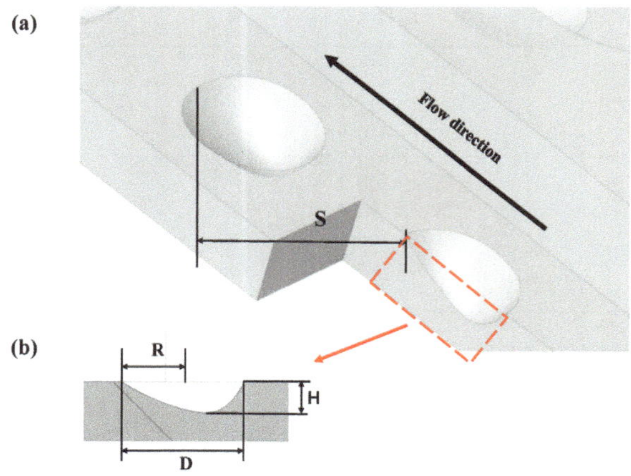

Figure 2. (**a**) BFS notch profile (**b**) Schematic diagram of the flow path of the liquid cooling plate with the BFS structure.

3.3. Control Equations

The simulation model consists of five domains: liquid cooling plate fluid domain, pack internal air domain, battery cell, pack solid components and thermally conductive materials. The simulation time interval is incremented from 0.01 s to 10 s.

The cell heat transfer and heat production equations are defined as follows [35]:

$$(\rho C_p)_{cell}\frac{\partial T}{\partial t} = k_{in}\left(\frac{\partial^2 T}{\partial x^2} + \frac{\partial^2 T}{\partial y^2}\right) + k_{th}\frac{\partial^2 T}{\partial z^2} + q_{gen} \quad (3)$$

where ρ, C_p, k_{in}, k_{th} and T represent the density, specific heat capacity, in-plane thermal conductivity, through-plane thermal conductivity and temperature, respectively.

Based on Bernardi's classical model, the battery heat is represented by the electrochemical reaction during discharge and the internal resistance as [36]:

$$q_{gen} = \frac{1}{V_b}\left(I^2 R + IT_b\frac{dU_{OCV}}{dT}\right)\# \quad (4)$$

where V_b denotes volume, I denotes discharge current, R denotes internal resistance, T_b denotes the thermal conductivity of the cell, and $\frac{dU_{OCV}}{dT}$ denotes the entropy thermal coefficient.

The continuity, momentum and energy equations can be expressed as [37]:

$$\frac{\partial \rho_f}{\partial t} + \frac{\partial \rho_f U_i}{\partial x_i} = 0 \quad (5)$$

$$\frac{\partial \rho_f U_i}{\partial t} + \frac{\partial}{\partial x_j}\left(\rho_f U_i U_j\right) = -\frac{\partial p'}{\partial x_i} + \frac{\partial}{\partial x_j}\left[\mu_{eff}\left(\frac{\partial U_i}{\partial x_j} + \frac{\partial U_j}{\partial x_i}\right)\right] \quad (6)$$

$$\frac{\partial}{\partial t}\left(\rho_f h_{tot}\right) - \frac{\partial p}{\partial t} + \frac{\partial}{\partial x_i}\left(\rho_f U_i h_{tot}\right) = \frac{\partial}{\partial x_i}\left(k_f\frac{\partial T}{\partial x_i}\right) + \frac{\partial}{\partial x_i}(U_i\tau) \quad (7)$$

The transfer equations for the turbulent kinetic energy and energy dissipation rate are as follows, using the k-ε turbulence model.

$$\frac{\partial}{\partial t}\left(\rho_f k\right) + \frac{\partial}{\partial x_i}\left(\rho_f U_i k\right) = \frac{\partial}{\partial x_i}\left[\left(\mu + \frac{\mu_t}{\sigma_k}\right)\frac{\partial k}{\partial x_i}\right] + P_k - \rho_f\varepsilon + P_{kb} \quad (8)$$

$$\frac{\partial}{\partial t}\left(\rho_f \varepsilon\right) + \frac{\partial}{\partial x_i}\left(\rho_f U_i \varepsilon\right) = \frac{\partial}{\partial x_i}\left[\left(\mu + \frac{\mu_t}{\sigma_\varepsilon}\right)\frac{\partial \varepsilon}{\partial x_i}\right] + \frac{\varepsilon}{k}\left(C_{\varepsilon 1}P_k - C_{\varepsilon 2}\rho_f\varepsilon + C_{\varepsilon 1}P_{kb}\right)\# \quad (9)$$

where k and ε denote the turbulent kinetic energy and turbulent dissipation rate, respectively. The model constants $C_{\varepsilon 1} = 1.44$, and $C_{\varepsilon 2} = 1.92$.

Turbulent viscosity μ_t and the production term P_k of the solution equations can be described as:

$$\mu_{eff} = \mu_t + \mu \tag{10}$$

$$\mu_t = C_\mu \rho_f \frac{k^2}{\varepsilon} \tag{11}$$

$$P_k = \mu_t \left(\frac{\partial u_j}{\partial x_i} + \frac{\partial u_i}{\partial x_j} \right) \frac{\partial u_i}{\partial x_j} \tag{12}$$

3.4. Grid Independence Analysis

In order to find the optimal number of meshes for the calculation, a sensitivity analysis is performed on the BTMS. Inappropriate number of meshes in Fluent may lead to poor mesh quality, which can cause the body mesh to fail in the calculation. Therefore, the number of bodies meshes is set between 1 million and 5 million. For a flow velocity of 0.3 m/s, the calculation results are shown in Figure 3a. There is some deviation in data when the grid number is low. When the grid number reaches 3.7 million or more, there is almost no difference in the temperature calculation results at different grid numbers. Therefore, in order to ensure the accuracy of the settlement results and the rapidity of the calculation, a grid number of 4 million is used for the calculation in this study. The face mesh parameters are 2.5 to 24.77 mm and the body mesh parameter is 20 mm. The grid structure of the battery pack is shown in Figure 3b.

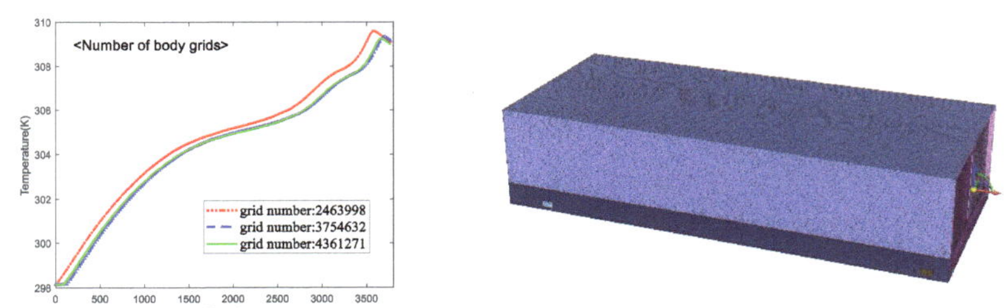

(a) (b)

Figure 3. (**a**) Calculation results using different number of grids in the same time period. (**b**) Schematic of the grid structure of the cell envelope.

4. Experimental Setup and Thermal Model Validation

4.1. Experimental Setup

A total of 36 ternary Lithium batteries with a capacity of 50 Ah are used in series for the experiment. The basic parameters of the Lithium batteries are shown in Table 2. Compared with the batteries used in daily life, the rated capacity of 50 Ah has a discharge current of 50 A at 1C, which involves a lot of heat generation and also places a high demand on the BTMS for heat dissipation. Experiments are conducted using the power battery testing system, thermostatic chamber, as shown in Figure 4. The experiment starts with an initial temperature setting of 25 °C at room temperature and 1C discharge. The coolant uses water, and the pump flow velocity is set to 0.28 m/s. The temperature change of the conductive row of the battery module is recorded every 30 s during the discharge process using a thermal imaging camera. The experimental condition also continues to be used for experimental validation after subsequent replacement of the liquid cooling plate with BFS channel.

Table 2. Lithium battery parameters.

Specifications	Value
Cell Length	148 mm
Cell width	27 mm
Cell height	97 mm
Weight	0.88 kg
AC internal resistance	≤0.1 mΩ
Nominal capacity	50 Ah
Nominal Voltage	3.7 V
Life cycle	≥2000
Maximum allowable charging temperature range	0–55 °C

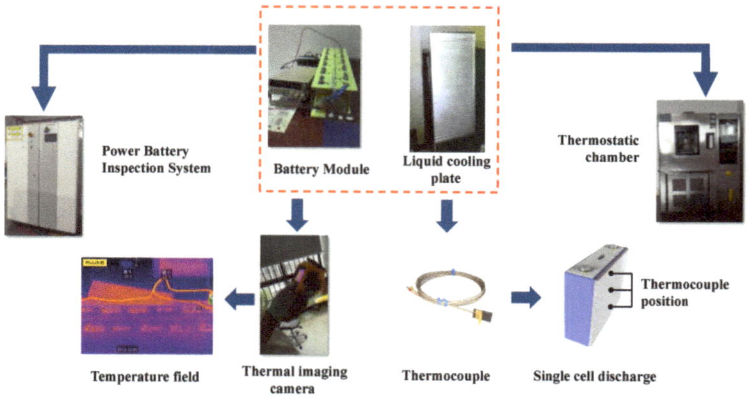

Figure 4. Experimental operation process diagram.

4.2. Thermal Model Validation

To verify the accuracy of the thermal model, CFD calculation is performed according to Table 3. In addition, three thermocouples are placed from top to bottom on the surface of the single cell for temperature acquisition with the experimental test setup described above. After 3600 s of 1C discharge, the average value of the three thermocouples was calculated as the experimental results. The comparison of the simulated temperature with the actual temperature is shown as Figure 5. Compared to the actual temperature, the simulated temperature shows a maximum error of 2.83% at the 3234th second, which is within acceptable limits. This indicates that the accuracy of the established cell thermal model can be used for thermal simulation calculations of liquid cooling BTMS.

Table 3. CFD calculation parameters.

	Specific Heat (J/kg·K)	Density (kg/m^3)	Thermal Conductivity (W/m·K)	Dynamic Density (Pa·s)
Battery pack	1033	2218	17.4, 5.3, 23	
Thermal pad	1800	2000	1.8	
Epoxy plate	1581	1800	0.2	
Insulated board	1260	1150	0.2	
Aerogel	1180	230	0.025	
Conductive row	900	2700	243	
Cabinet	900	2700	209	
	3281	1073	0.38	0.00394
Coolant	3300	1071	0.384	0.00339
(25/30/35/40 °C)	3399	1066	0.391	0.00256
	3358	1063	0.394	0.00226
Insulation layer	1700	65	0.034	
Heating film	1130	1840	1.2	

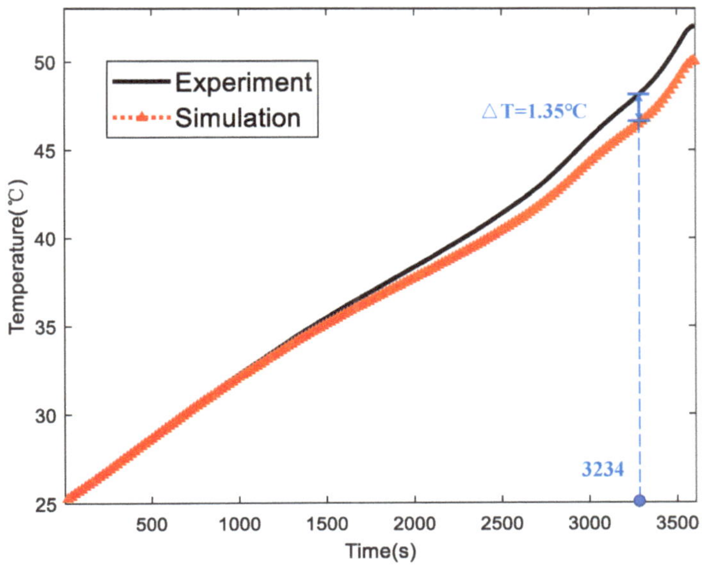

Figure 5. Comparison between simulated temperature data and actual temperature data for 1C discharge.

5. Results and Discussion

5.1. Performance Comparison of BFS Channel and Flat Type Channel

To evaluate the effect of this optimized structure relative to the original structure, the results are first compared with the flat-type channel.

The thermal cloud plots of the flat and BFS battery pack at the same flow velocity (0.28 m/s) are shown as Figure 6a,b. In order to clearly observe the difference between the two thermal clouds, the T_{max} moment is taken for observation and the temperature interval is set from 25 °C to 40 °C. At this point, both structures show a distribution of increasing and then decreasing temperatures in the lateral direction. In contrast, the BFS channel produces a significant cooling effect through structural optimization. The T_{max} in Figure 6a appears at the left of center and reaches close to 40 °C. However, the T_{max} of Figure 6b appears on the right side and is only about 38 °C. In addition, to evaluate the temperature distribution of the 36 battery cells, two lines (x1, x2) are defined on the two battery modules. The measured positions are averaged between the positions of two conductive rows on each module. There are a total of 18 cells along the x-direction, and the conductive rows connect the cells two by two. The eight data points above plus the two data points at each end make a total of ten data points.

The temperature field distribution of the liquid cooling plate for both structures are shown as Figure 6c,d. In comparison, the BFS liquid cooling plate in Figure 6d exhibits a lower temperature field distribution on average. It can be seen that the temperature field of the pack shifts somewhat towards the outlet side and produces better cooling in the second half of the liquid cooling. The channel temperature on the inlet side is closer to the initial temperature, and this effect continues until after the sixth bend. Preliminary analysis suggests that this structure may have had some lifting effect on the flow field flow velocity to produce this temperature field backward shift. In addition, the uniformity of the temperature field of the liquid cooling plate is significantly improved after the structural optimization. Compared to the T_{max} of 30 °C on the outlet side of the flat plate type, the T_{max} on the outlet side is about 29 °C, and the area of the temperature field occupied by the high temperature is reduced by about 1/4.

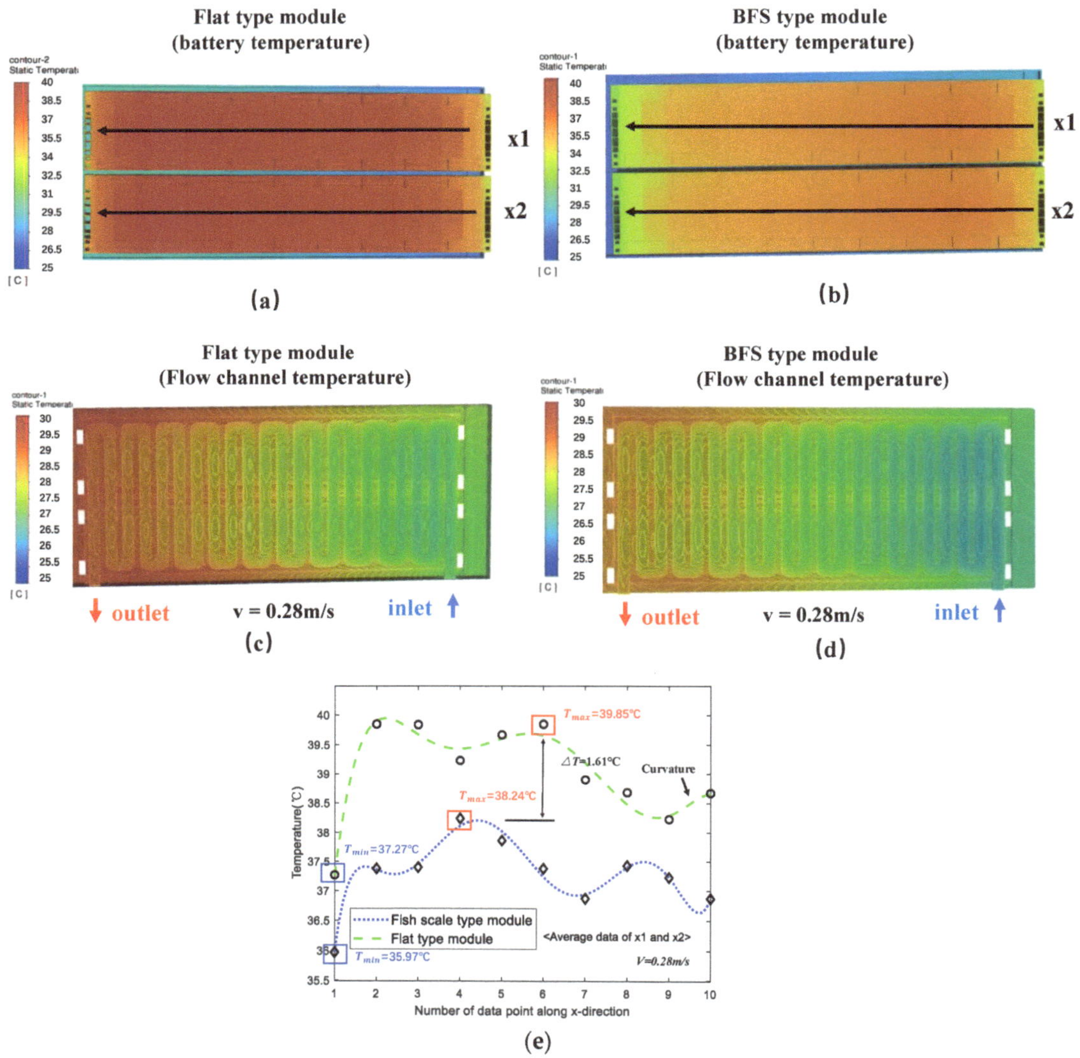

Figure 6. (**a**) Thermal cloud of cell pack at T_{max} of flat type channel (**b**) Thermal cloud of cell pack at T_{max} of BFS type channel (**c**) Thermal cloud of liquid cooling plate at T_{max} of flat type channel (**d**) Thermal cloud of liquid cooling plate at T_{max} of BFS type channel (**e**) Comparison of average values of temperature along x1 and x2 for both structures.

The temperature profile along the x-direction is shown as Figure 6e. For each pack, the average of x1 and x2 in both modules is taken as comparative data. The equation of the fitted curve is as follows.

$$y_{flat} = -4.2 \times 10^{-5}x^7 + 7 \times 10^{-4}x^6 + 0.0085x^5 - 0.28x^4 + 2.58x^3 - 10.76x^2 + 20.84x + 24.88 \quad (13)$$

$$y_{fish} = 5.96 \times 10^{-4}x^7 - 0.0233x^6 + 0.3685x^5 - 3x^4 + 13.7x^3 - 34.13x^2 + 43x + 16 \quad (14)$$

where $R^2_{flat} = 0.9631, R^2_{fish} = 0.9788$.

For the flat panel type channel, the T_{max} is 39.85 °C. There are multiple high temperature fluctuations in the x-direction with a maximum curvature of 2.58 (data points one and two). The ΔT_{max} is 2.58 °C, showing a significant temperature inhomogeneity. In contrast, the optimized model has a maximum curvature of 1.41, and a flat fluctuation in the x-direction, with a ΔT_{max} of only 2.27 °C. Moreover, its maximum temperature point undergoes a certain shift left, which is consistent with the previous findings. This also shows that it has better thermal performance results.

5.2. Orthogonal Experiments

There are three parameters that affect the thermal performance of the battery pack. The orthogonal test method can effectively obtain balanced samples of multiple factors by selecting representative cases. It can also reduce the number of tests for the experiment. In this experiment, four levels are set for each parameter in order to understand the effect of each parameter. The parameters are BFS notch diameter (D), BFS notch depth (H), and BFS notch spacing (S). The study indicators are T_{max} and ΔT_{max}. The values of the different parameters are given in Table 4.

Table 4. Factor level parameters.

Level	1	2	3	4
D	10	12.5	15	17.5
H	1.75	2.5	3.25	4
S	7	9	11	13

In order to analyze the effect of the parameters of this structure on the thermal performance, 16 different parameters of the lower channel structure are designed. Some of the design drawings are shown in Figure 7.

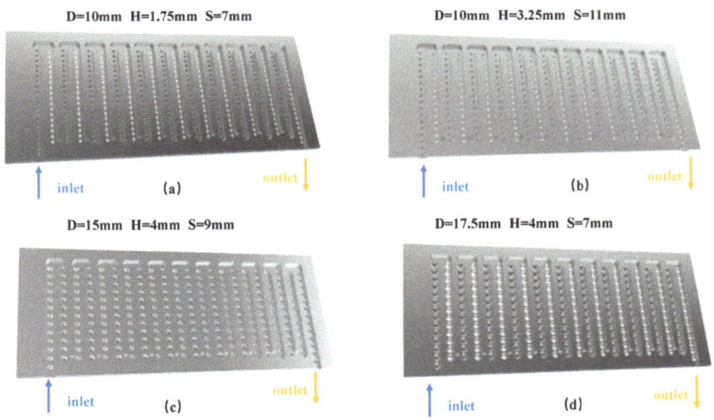

Figure 7. (**a**) Flow channel design drawings for groups I (**b**) Flow channel design drawings for group III (**c**) Flow channel design drawings for group XII (**d**) Flow channel design drawings for group XVI.

Each set of orthogonal experiments is started at an initial temperature of 25 °C, and the parameter settings and results are shown in Table 5. T_{max} represents the maximum temperature of the 36 cells in the pack during discharge in the BFS flow channel. ΔT_{max} represents the maximum value of the difference between the maximum and minimum temperature of the 36 cells during discharge in the BFS flow channel. ΔT represents the difference between the maximum temperature of the 36 cells in the flat-type channel and the BFS channel. The results show that BFS channel produces some thermal performance optimization at different parameters compared to the T_{max} of 39.85 °C and ΔT_{max} of 2.58 °C under the flat-type channel. The T_{max} and ΔT_{max} of battery pack are optimized up to 10.8% and 16.7%, which occur in group 5 and group 2 respectively. The optimization effect of this structure is a result of a multi-parameter combination, and the optimization effect of different parameters for T_{max} and ΔT_{max} have different optimization effects, as shown in Figure 8. Therefore, it is necessary to analyze the effect of the three parameters on the two indicators so that the appropriate combination of parameters can be selected later according to the specific needs, and the best indicator optimization effect has been achieved.

Table 5. Parameter settings and results of 16 sets of orthogonal experiments.

Groups	Caliber (mm)	Breadth (mm)	Density (mm)	ΔT (°C)	T_{max} (°C)	ΔT_{max} (°C)
1	10	1.75	7	0.94	38.91	2.37
2	10	2.5	9	1.1	38.75	2.15
3	10	3.25	11	1.04	38.81	2.28
4	10	4	13	1.05	38.80	2.33
5	12.5	1.75	9	1.61	38.24	2.27
6	12.5	2.5	7	1.07	38.78	2.24
7	12.5	3.25	13	1.09	38.76	2.19
8	12.5	4	11	1.08	38.77	2.24
9	15	1.75	11	1	38.85	2.25
10	15	2.5	13	1.03	38.82	2.23
11	15	3.25	7	1.05	38.80	2.28
12	15	4	9	1.03	38.82	2.27
13	17.5	1.75	13	0.96	38.89	2.27
14	17.5	2.5	11	1.1	38.75	2.18
15	17.5	3.25	9	1.16	38.69	2.29
16	17.5	4	7	1.15	38.70	2.19

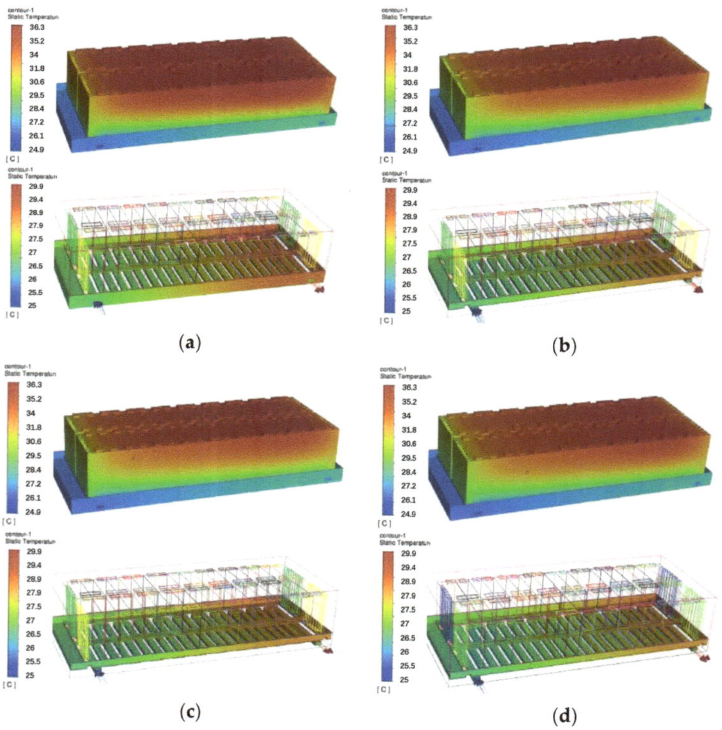

Figure 8. (**a**) Temperature cloud of the battery pack for group I (**b**) Temperature cloud of the battery pack for group III (**c**) Temperature cloud of the battery pack for group XII (**d**) Temperature cloud of the battery pack for group XVI.

5.3. Analysis of Experimental Results

In order to obtain the best combination of parameters, a computational analysis using an orthogonal table is performed to discriminate between primary and secondary factors. Therefore, the influence parameter Q is set, which is calculated as follows.

$$Q = Max\{K_1, K_2, K_3, K_4\} - Min\{K_1, K_2, K_3, K_4\}\# \quad (15)$$

where K_i represents the average of the indicators corresponding to one parameter.

The results of the calculations are shown in Table 6. For the T_{max}, S has the maximum Q, followed by D, and finally H. For ΔT_{max}, the influence size is ranked as H > D > S. The analysis shows that the influence size of the parameters is different for different indicators. Therefore, the combination of parameters needs to be fully considered in the structural design to achieve the best thermal performance optimization.

Table 6. Results of orthogonal experimental analysis.

Indicators		Parameters		
		D	H	S
T_{max}	K_1	38.8175	38.7225	38.7975
	K_2	38.6375	38.775	38.625
	K_3	38.8225	38.765	38.795
	K_4	38.7575	37.7725	38.8175
	Q	0.1850	0.0525	0.1925
ΔT_{max}	K_1	2.2825	2.29	2.27
	K_2	2.235	2.2	2.245
	K_3	2.2575	2.26	2.2375
	K_4	2.2325	2.2575	2.255
	Q	0.05	0.09	0.0325

Further, the influence between the structural parameters under each indicator is analyzed. The correlation between the three parameters is shown as Figure 9. For T_{max}, there is a strong correlation between D and S, with a correlation coefficient of 0.8949, while there is almost no correlation between D and H, with a correlation coefficient of −0.01946. Therefore, the strong correlation between S and D should be fully considered for the optimal structural design of T_{max}, and the parameter designs of S and D are relatively independent compared to H which means the correlation between them can be ignored. For ΔT_{max}, the correlation between the parameters is not strong. The correlation coefficient between D and H is 0.7529, while the correlation coefficients of the other two parameters are about 0.5. It is indicated that the selection and adjustment of the three parameters should be adjusted in a small range to optimize the structural design of ΔT_{max}. Therefore, the parameter design of ΔT_{max} is more difficult than that of T_{max}, which indicates that the temperature difference is more important in BTMS and more difficult to regulate.

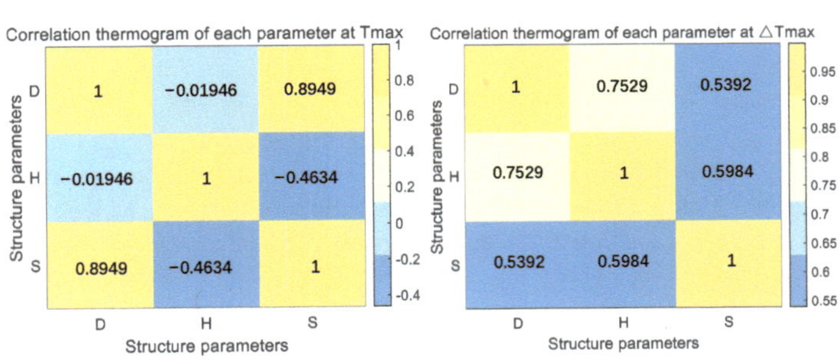

Figure 9. Thermal diagram of the correlation of each parameter at T_{max} and ΔT_{max}.

In addition, the effect of different levels of a single parameter on battery cooling is analyzed. Obviously, the increase in D produces a larger heat exchange area. However, when D increases from 13 mm to 17.5 mm, T_{max} has a slight increase of 0.055 °C rather than a decrease as shown in Figure 10a. Moreover, a larger increase of 0.052 °C in ΔT_{max} occurs during this process. As D increases, T_{max} and ΔT_{max} produce a decrease in the overall trend, but with some degree of upward fluctuation.

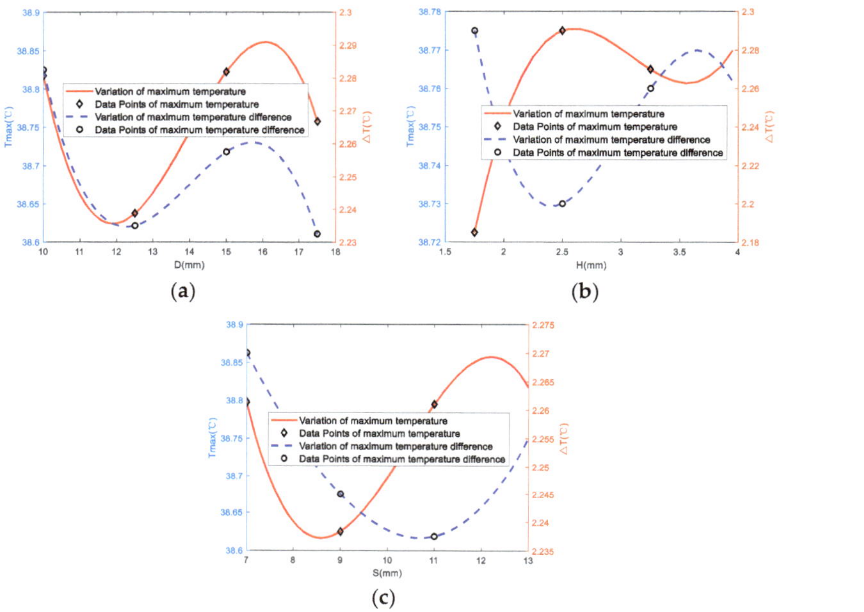

Figure 10. (**a**) Curve of the effect of D on T_{max} and ΔT_{max} (**b**) Curve of the effect of H on T_{max} and ΔT_{max} (**c**) Curve of the effect of S on T_{max} and ΔT_{max}.

T_{max} reaches the best temperature of 38.72 °C at H = 2.5 mm, and subsequently shows an increasing trend again as shown in Figure 10b. ΔT_{max} is positively related to H in the whole, while the worst uniformity of 2.28 °C occurs at H = 2.5 mm. This result shows a clear opposing relationship with T_{max}. Therefore, the H should be appropriately increased to 3.25 mm to balance T_{max} and ΔT_{max} under comprehensive consideration.

Both T_{max} and ΔT_{max} show a trend of decreasing and then increasing with the increase of S as shown in in Figure 10c. ΔT_{max} reaches the peak and valley value of 2.2375 °C one grade before T_{max}. T_{max} reaches the optimum value of 38.625 °C at the third grade S = 11 mm, but the ΔT_{max} at this time has a large increase in value and the thermal uniformity is reduced. Therefore, the value around the second grade S = 9 mm should be selected to achieve the optimal thermal performance.

5.4. Analysis of BFS Structure Design Principles

The original intention of the structure design is to achieve a drag reduction effect using this bionic structure. Therefore, in order to observe the drag reduction effect of the BFS structure, the inlet and outlet flow velocity as well as the pressure are counted in the simulation as shown in Figures 11 and 12.

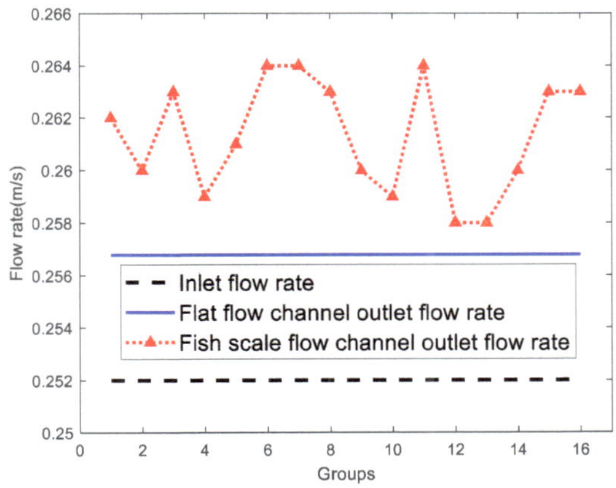

Figure 11. Comparison of the inlet and outlet flow velocity of 16 groups of BFS channels and the inlet and outlet flow velocity of flat flow channels.

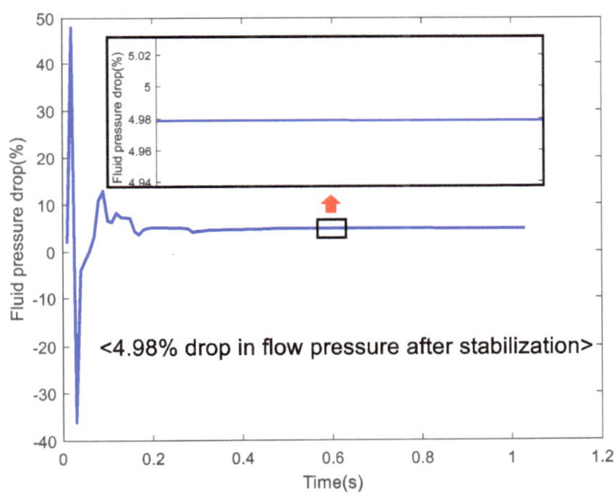

Figure 12. Comparison of the pressure of BFS channel and flat type channel.

Since the set initial flow rate of 0.28 m/s is the flow velocity in the water pump and the coolant is buffered somewhat before it enters the liquid cooling plate. Therefore, the flow velocity in the liquid cooling plate will be somewhat lower than the flow velocity in the water pump. After the steady state, the outlet flow velocities all have different degrees of increase from 0.258 m/s to 0.264 m/s compared to the flow velocity of 0.257 m/s at the outlet of the flat plate type flow channel, as shown in Figure 11. In addition, the flow channel pressure under this structure is reduced by 4.98%, as shown in Figure 12. In summary, this bionic notch structure can enhance the flow velocity by a maximum of 2.72% and reduce the flow pressure by 4.98%.

In order to investigate the principle of drag reduction in this structure, the fluid vector diagram of the liquid cooling plate flow channel is analyzed, as shown in Figure 13. It can be seen that a vortex of fluid is generated in the BFS groove microstructure as shown in Figure 14. The fluid vector diagram shows that this structure acts as a "mechanical gear" to push the coolant forward, greatly reducing the friction of the coolant and allowing the fluid to flow through the flow path efficiently. The tiny vortex formed in this structure creates a convection flow with the main flow direction of the coolant, which avoids direct collision between the coolant and the liquid cooling plate. As a result, the friction between the water and the liquid cooling plate becomes a water-to-water friction, reducing the friction during travel and acting as a drag reduction.

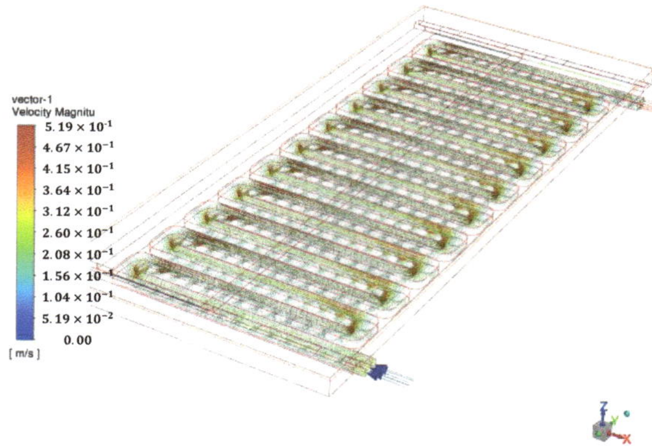

Figure 13. Liquid cooling plate flow channel fluid vector diagram.

Figure 14. Vector diagram of the "gear" return flow direction formed in the BFS structure.

According to the previous statement, the coolant must flow from the blunt tip of the structure, which is essential for the structure to function. Figure 15 shows a vector diagram of the flow in the BFS notch when flowing from the tip. It can be seen that no previous reflux is generated in the BFS notch. Therefore, it also does not produce the previous thermal performance optimization effect and may even be counterproductive.

5.5. Heat Transfer Rate Analysis between Cell and Liquid Cooling Plate

To further verify the effectiveness of the above structure, the maximum heat transfer quantity between the liquid cooling plate and the battery module is counted, as shown in Figure 16. With the different parameters of the BFS structure, the drag reduction effect is different. The resulting flow velocity at different drag reduction effect also leads to a change in the heat transfer quantity. It can be seen that the distribution of the maximum heat transfer quantity between the cell and the liquid cooling plate is approximately the same as the flow velocity distribution in Section 4.2. The maximum heat transfer quantity of 503.41 W is reached at the third set. Its corresponding flow velocity is 0.263 m/s, which is second only to the maximum flow velocity of 0.264 m/s.

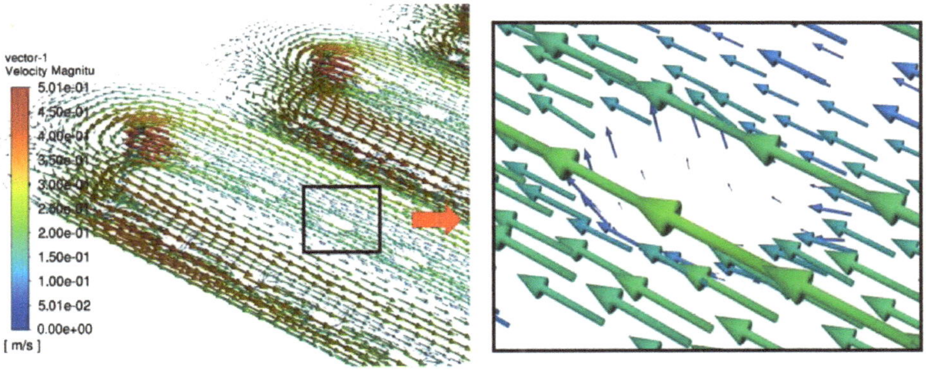

Figure 15. Flow vector diagram for the counterflow case in the BFS structure.

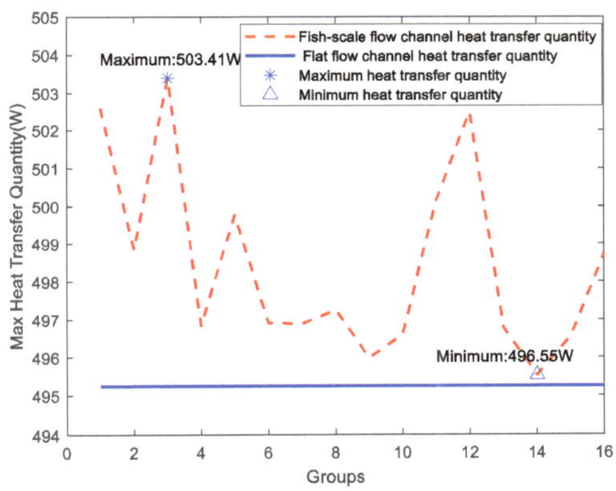

Figure 16. Heat transfer quantity between cell and liquid cooling plate in group 16.

6. Conclusions

(1) In this study, a novel BTMS with a BFS structure is proposed with the aim of providing a solution for optimizing the thermal performance of battery packs with serpentine flow channels in industrial manufacturing.

(2) In this paper, the performance effects of the proposed BTMS are analyzed for different parameters, and the battery pack T_{max} and ΔT_{max} are optimized up to 10.8% and 16.7%, respectively. The effects of three parameters, D, H and S, are analyzed by the transient fluid thermal simulation results under 16 sets of orthogonal experiments, which provide references for the selection of subsequent design.

(3) The mechanism of action of the structure is also deeply explored in this study. It is found that vortex flow is generated in the microstructure, which reduces the friction of coolant and increases the flow velocity. This mechanism of action avoids the collision of coolant with the flow channel and increases the flow velocity by a maximum of 2.72% and reduces the flow pressure by 4.98%. Further research ideas are provided for BTMS using high viscosity with high density coolant.

Author Contributions: Formal analysis, P.L.; resources, D.M.; writing—original draft preparation, Y.M.; writing—review and editing, K.G.; supervision, H.C.; project administration, R.D. All authors have read and agreed to the published version of the manuscript.

Funding: This research received no external funding.

Data Availability Statement: The data presented in this study are available on request from the corresponding author. The data are not publicly available due to privacy.

Conflicts of Interest: The authors declare no conflict of interest.

References

1. Hu, L.; Tian, Q.; Zou, C.; Huang, J.; Ye, Y.; Wu, X. A study on energy distribution strategy of electric vehicle hybrid energy storage system considering driving style based on real urban driving data. *J. Renew. Sustain. Energy Rev.* **2022**, *162*, 112416. [CrossRef]
2. Wu, Y.; Huang, Z.; Zheng, Y.; Liu, Y.; Li, H.; Che, Y.; Peng, J.; Teodorescu, R. Spatial–temporal data-driven full driving cycle prediction for optimal energy management of battery/supercapacitor electric vehicles. *J. Energy Convers. Manag.* **2023**, *277*, 116619. [CrossRef]
3. Fuhao, M.O.; Tian, Y.; Zhao, S.; Xiao, Z.; Zhiling, M.A. Working Temperature Effects on Mechanical Integrity of Cylindrical Lithium-ion Batteries. *Eng. Fail. Anal.* **2022**, *137*, 106399.
4. Mali, V.; Saxena, R.; Kumar, K.; Kalam, A.; Tripathi, B. Review on battery thermal management systems for energy-efficient electric vehicles. *Renew. Sustain. Energy Rev.* **2021**, *151*, 111611. [CrossRef]
5. Zhao, G.; Wang, X.; Negnevitsky, M.; Zhang, H. A review of air-cooling battery thermal management systems for electric and hybrid electric vehicles. *J. Power Sources* **2021**, *501*, 230001. [CrossRef]
6. Lee, J.; Abidi, A.; Sajadi, S.M.; El-Shafay, A.S.; Degani, M.; Sharifpur, M. Study of the effect of the aspect ratio of a cylindrical lithium-ion battery enclosure in an air-cooled thermal management system. *J. Energy Storage* **2022**, *45*, 103684. [CrossRef]
7. Chung, Y.; Kim, M.S. Thermal analysis and pack level design of battery thermal management system with liquid cooling for electric vehicles. *Energy Convers. Manag.* **2019**, *196*, 105–116. [CrossRef]
8. Wu, W.; Wang, S.; Wu, W.; Chen, K.; Hong, S.; Lai, Y. A critical review of battery thermal performance and liquid based battery thermal management. *Energy Convers. Manag.* **2019**, *182*, 262–281. [CrossRef]
9. Wu, W.; Liu, J.; Liu, M.; Rao, Z.; Deng, H.; Wang, Q.; Wang, S. An innovative battery thermal management with thermally induced flexible phase change material. *Energy Convers. Manag.* **2020**, *221*, 113145. [CrossRef]
10. Zhang, J.; Li, X.; Zhang, G.; Wang, Y.; Guo, J.; Wang, Y.; Zhong, Z. Characterization and experimental investigation of aluminum nitride-based composite phase change materials for battery thermal management. *Energy Convers. Manag.* **2020**, *204*, 112319. [CrossRef]
11. Liang, J.; Gan, Y.; Li, Y. Investigation on the thermal performance of a battery thermal management system using heat pipe under different ambient temperatures. *Energy Convers. Manag.* **2018**, *155*, 1–9. [CrossRef]
12. Ren, R.; Zhao, Y.; Diao, Y.; Liang, L. Experimental study on preheating thermal management system for lithium-ion battery based on U-shaped micro heat pipe array. *Energy* **2022**, *253*, 124178. [CrossRef]
13. Murali, G.; Sravya, G.S.N.; Jaya, J.; Vamsi, V.N. A review on hybrid thermal management of battery packs and it's cooling performance by enhanced PCM. *Renew. Sustain. Energy Rev.* **2021**, *150*, 111513. [CrossRef]
14. Wan, Z.; Wang, X.; Feng, C. Heat transfer performances of the capillary loop pulsating heat pipes with spring-loaded check valve. *Appl. Therm. Eng.* **2020**, *167*, 114803. [CrossRef]
15. Zhang, F.; Liu, P.; He, Y.; Li, S. Cooling performance optimization of air cooling lithium-ion battery thermal management system based on multiple secondary outlets and baffle. *J. Energy Storage* **2022**, *52*, 104678. [CrossRef]
16. Akbarzadeh, M.; Kalogiannis, T.; Jaguemont, J.; Jin, L.; Behi, H.; Karimi, D.; Berecibar, M. A comparative study between air cooling and liquid cooling thermal management systems for a high-energy lithium-ion battery module. *Appl. Therm. Eng.* **2021**, *198*, 117503. [CrossRef]
17. Olabi, A.G.; Maghrabie, H.M.; Adhari, O.H.K.; Sayed, E.T.; Yousef, B.A.; Salamah, T.; Abdelkareem, M.A. Battery thermal management systems: Recent progress and challenges. *Int. J. Thermofluids* **2022**, *15*, 100171. [CrossRef]
18. Kong, D.; Peng, R.; Ping, P.; Du, J.; Chen, G.; Wen, J. A novel battery thermal management system coupling with PCM and optimized controllable liquid cooling for different ambient temperatures. *Energy Convers. Manag.* **2020**, *204*, 112280. [CrossRef]
19. Gao, R.; Fan, Z.; Liu, S. A gradient channel-based novel design of liquid cooling battery thermal management system for thermal uniformity improvement. *J. Energy Storage* **2022**, *48*, 104014. [CrossRef]
20. Wang, N.; Li, C.; Li, W.; Chen, X.; Li, Y.; Qi, D. Heat dissipation optimization for a serpentine liquid cooling battery thermal management system: An application of surrogate assisted approach. *J. Energy Storage* **2021**, *40*, 102771. [CrossRef]
21. Tang, Z.; Liu, Z.; Li, J.; Cheng, J. A lightweight liquid cooling thermal management structure for prismatic batteries. *J. Energy Storage* **2021**, *42*, 103078. [CrossRef]
22. Yang, W.; Zhou, F.; Chen, X.; Li, K.; Shen, J. Thermal performance of honeycomb-type cylindrical lithium-ion battery pack with air distribution plate and bionic heat sinks. *Appl. Therm. Eng.* **2023**, *218*, 119299. [CrossRef]

23. Liu, F.; Chen, Y.; Qin, W.; Li, J. Optimal design of liquid cooling structure with bionic leaf vein branch channel for power battery. *Appl. Therm. Eng.* **2023**, *218*, 119283. [CrossRef]
24. Yang, W.; Zhou, F.; Liu, Y.; Xu, S.; Chen, X. Thermal performance of honeycomb-like battery thermal management system with bionic liquid mini-channel and phase change materials for cylindrical lithium-ion battery. *Appl. Therm. Eng.* **2021**, *188*, 116649. [CrossRef]
25. Wang, J.; Liu, X.; Liu, F.; Liu, Y.; Wang, F.; Yang, N. Numerical optimization of the cooling effect of the bionic spider-web channel cold plate on a pouch lithium-ion battery. *Case Stud. Therm. Eng.* **2021**, *26*, 101124. [CrossRef]
26. An, Z.; Zhang, C.; Gao, Z.; Luo, Y.; Dong, Y. Heat dissipation performance of hybrid lithium battery thermal management system using bionic nephrolepis micro-channel. *Appl. Therm. Eng.* **2022**, *217*, 119127. [CrossRef]
27. Liu, H.; Gao, X.; Zhao, J.; Yu, M.; Niu, D.; Ji, Y. Liquid-based battery thermal management system performance improvement with intersected serpentine channels. *Renew. Energy* **2022**, *199*, 640–652. [CrossRef]
28. Yin, B.; Zuo, S.; Xu, Y.; Chen, S. Performance of liquid cooling battery thermal management system in vibration environment. *J. Energy Storage* **2022**, *53*, 105232. [CrossRef]
29. Cabeza, L.F.; Frazzica, A.; Chàfer, M.; Vérez, D.; Palomba, V. Research trends and perspectives of thermal management of electric batteries: Bibliometric analysis. *J. Energy Storage* **2020**, *32*, 101976. [CrossRef]
30. Zichen, W.; Changqing, D. A comprehensive review on thermal management systems for power lithium-ion batteries. *Renew. Sustain. Energy Rev.* **2021**, *139*, 110685. [CrossRef]
31. Jithin, K.V.; Rajesh, P.K. Numerical analysis of single-phase liquid immersion cooling for lithium-ion battery thermal management using different dielectric fluids. *Int. J. Heat Mass Transf.* **2022**, *188*, 122608. [CrossRef]
32. You, Y.; Wang, S.; Lv, W.; Chen, Y.; Gross, U. A CFD model of frost formation based on dynamic meshes technique via secondary development of ANSYS fluent. *Int. J. Heat Fluid Flow* **2021**, *89*, 108807. [CrossRef]
33. Bellwald, B.; Planke, S.; Lebedeva-Ivanova, N.; Piasecka, E.D.; Andreassen, K. High-resolution landform assemblage along a buried glacio-erosive surface in the SW Barents Sea revealed by P-Cable 3D seismic data. *Geomorphology* **2019**, *332*, 33–50. [CrossRef]
34. Dai, W.; Zhang, W.; Zheng, Z.; Li, J. Investigation of micron-sized fish-scale surface structures on tool steel surfaces using laser galvanometer scanning. *Appl. Surf. Sci.* **2019**, *470*, 1111–1121. [CrossRef]
35. Park, S.; Jang, D.S.; Lee, D.; Hong, S.H.; Kim, Y. Simulation on cooling performance characteristics of a refrigerant-cooled active thermal management system for lithium ion batteries. *Int. J. Heat Mass Transf.* **2019**, *135*, 131–141. [CrossRef]
36. Wu, W.; Wu, W.; Wang, S. Thermal management optimization of a prismatic battery with shape-stabilized phase change material. *Int. J. Heat Mass Transf.* **2018**, *121*, 967–977. [CrossRef]
37. Yun, S.; Kwon, J.; Cho, W.; Lee, D.; Kim, Y. Performance improvement of hot stamping die for patchwork blank using mixed cooling channel designs with straight and conformal channels. *Appl. Therm. Eng.* **2020**, *165*, 114562. [CrossRef]

batteries

MDPI

Article

Numerical Study on Cross-Linked Cold Plate Design for Thermal Management of High-Power Lithium-Ion Battery

Huizhu Yang, Zehui Wang, Mingxuan Li, Fengsheng Ren and Binjian Ma *

School of Mechanical Engineering and Automation, Harbin Institute of Technology, Shenzhen 518055, China
* Correspondence: mabinjian@hit.edu.cn; Tel.: +86-137-1386-2671

Abstract: Liquid cooling strategies such as cold plates have been widely employed as an effective approach for battery thermal management systems (BTMS) due to their high cooling capacity and low power consumption. The structural design of the cold plates is the key factor that directly determines the thermal performance of the liquid cooling system. In this study, seven Z-type parallel channel cold plate and two novel cross-linked channel cold plate designs are proposed for the cooling of high-power lithium-ion batteries using two different cooling strategies. The average battery temperature, battery temperature uniformity and energy consumption of all designs are firstly analyzed holistically by three-dimensional conjugated simulation under the scheme of continuous cooling. Two selected designs that demonstrated superior performance (i.e., a Z-type parallel channel cold plate with 8-branches and an improved cross-linked channel design) are further analyzed to explore their integrative performance under different cooling schemes. The results show that within a battery temperature limit of 40 °C, employing the delayed cooling strategy can save 23% energy consumption compared to the continuous cooling strategy. Besides, the cold plate with an improved cross-linked channel configuration requires 13% less pumping power and provides a better temperature uniformity than the Z-type parallel channel cold plate with 8-branches. These results are of great significance to advance the cooling design of BTMS.

Keywords: lithium-ion battery; thermal management; cold plate; continuous and delayed cooling; cross-linked channel

Citation: Yang, H.; Wang, Z.; Li, M.; Ren, F.; Ma, B. Numerical Study on Cross-Linked Cold Plate Design for Thermal Management of High-Power Lithium-Ion Battery. *Batteries* **2023**, *9*, 220. https://doi.org/10.3390/batteries9040220

Academic Editor: Torsten Brezesinski

Received: 7 February 2023
Revised: 21 March 2023
Accepted: 29 March 2023
Published: 5 April 2023

1. Introduction

The last decade has witnessed a fast-growing popularity of electric vehicles (EVs) owing to the increasing concern regarding fossil fuel consumption and tailpipe carbon emissions. Unlike traditional vehicles that run on gasoline, EVs are powered exclusively by rechargeable battery packs such as lead-acid, nickel-cadmium, nickel-metal-hydride, zinc-bromine, sodium sulfur, and lithium batteries. Among them, the lithium-ion battery has been the leading choice for EV manufacturers owing to its high energy density, minimum self-discharging behavior, robust performance, long service lifespan, and relatively small impact on vehicle weight [1–3]. Despite these merits, lithium-ion batteries must be operated at suitable temperature condition to avoid catastrophic failure [4]. Studies have shown that it is desirable to control the temperature of lithium-ion batteries within the range from 25 to 40 °C. Furthermore, the temperature nonuniformity between different cells should be controlled within less than 5 °C [5]. Working in either a low or high temperature environment will lead to poor battery performance, reduced lifespan, or even explosion [6]. Considering these issues, it is a critical task to develop effective battery thermal management systems (BTMS) to ensure that the EVs can run safely over a long period of time under different conditions.

Based on the cooling media, BTMSs can generally be divided into four different types including air cooling [7–9], liquid cooling [10–13], phase change material cooling [14–17], and heat pipe cooling [18–21]. Passive air cooling has a very low cooling capacity, which makes it unsuitable for cooling lithium-ion batteries at high charging or discharging rates. The active air-cooling method can provide a significantly higher heat transfer coefficient, but still suffers from a low cooling capacity resulting from the poor thermophysical property of air. Phase change material (PCM) allows for temporal storage and release of large amounts of heat during the process of melting and freezing. Heat pipes have a very high effective thermal conductivity (>100,000 $W \cdot m^{-1} \cdot K^{-1}$), which allows them to transfer thermal energy with a small temperature gradient. However, both PCM and heat pipes only allow temporary storage or transport of the thermal energy. Therefore, an active cooling device such as air or liquid cooling must be used in combination with PCM or heat pipe to further dissipate the thermal energy from the battery. Liquid cooling has become the most popular for BTMSs in EVs due to its remarkable cooling performance and high compactness [22,23]. So far, liquid-cooled battery systems have already been used in a number of commercial EVs including Tesla, VW, Hyundai, and BYD.

Cold plate is a widely used component in liquid-cooled battery systems for removing the heat generated during the charge–discharge process of battery packs. The cold plates can be installed either between the cells or on the lateral surfaces of the battery pack [24,25]. Many studies have investigated the pattern of fluid flow in cold plates, such as the straight, serpentine, and mini-channel structures. Zhu et al. [26] designed sixteen models using an orthogonal array to quantify and analyze the main and secondary factors of the cooling effect of a liquid-cooled BTMS. It was reported that channel number has the most important effect on the average temperature of batteries, while the channel number and inlet velocity are two major factors controlling the temperature uniformity. Huo et al. [27] also found that the maximum battery temperature increases with increasing channel number in a straight-shaped channel cold plate. Qian et al. [28] showed that the cooling efficiency of a straight channel cold plate can be improved by increasing the channel number. However, the improvement in the cooling performance becomes insignificant when the channel number exceeds five. Li et al. [29] investigated the performance of silicon cold plates based natural air cooling, forced air cooling, and liquid cooling with U-type tubes for BTMSs experimentally. Deng et al. [30] numerically studied the effect of the layout of channels, channel number, and inlet temperature on the performance of a serpentine-channel cooling plate. Kong et al. [31] proposed a divergent-shaped channel cold plate for BTMSs. They found that the divergent-shaped channels can provide enhanced cooling performance, which is characterized by a lower pressure drop and a smaller maximum temperature difference. Huang et al. [32] introduced the streamline concept to design and optimize the performance of an inner mini channel cooling plate. Mo et al. [33] designed a novel cooling plate using the topology optimization method and further analyzed the influences of the flow rate and inlet temperature on the performance of the optimized cooling plate. Kalkan et al. [34] performed experimental studies to explore the thermal performance of different water-cooled cold plates including conventional serpentine tube and novel mini channel designs for thermal management of lithium-ion batteries with a discharging rate from 1C to 5C. Amalesh and Narasimhan [35] proposed seven distinct mini-channel cold plates to analyze the effect of channel profile on the performance of the cold plate. It was found that mini-channels with zigzag and circular slot channels exhibit an exceptional performance.

In addition to the straight and serpentine channel configuration, the Z-type parallel channel cold plate has also been investigated by a significant number of studies. Compared with the traditional serpentine-channel cold plates, the flow path in the Z-type parallel channel cold plate is reduced, which decreases the overall pressure drop in the cooling system. Chen et al. [36] designed a novel cold plate consisting of mini parallel channels (PMCP) to enhance the uniformity of temperature distribution in large battery pack systems. As shown in their results, the proposed Z-type parallel channel design can significantly improve the temperature uniformity and reduce the power consumption of BTMSs. Gungor et al. [37] proposed an efficient liquid cooling system consisting of Z-type parallel channels between different cells in a battery pack by using the constructal canopy-to-canopy architectures. It was found in their study that the cold plate with five branch channels has the best cooling performance while the cold plates with more than seven branch channels require greater pumping power without significant thermal improvement. Sun et al. [38] designed a tapered inlet and outlet channel configuration to further enhance the thermal performance of their cold plate containing Z-shaped parallel minichannels. Guo et al. [39] studied the performance of serpentine channel cooling plates for BTMS. The results showed that the parallel-spiral serpentine channel cooling plate has the best performance. Shen et al. [40] used a tilted Z-shaped air-cooling system for BTMS. Compared to the vertical Z-shaped construction, the maximum battery temperature of the tilted Z-shaped channel was decreased from 38.15 to 34.14 °C. Faizan et al. [41] proposed a cold plate grooved with converging twisted serpentine mini-channels for BTMSs. Guo et al. [42] studied the performance of BTMS with a mini-channel cold plate. While the cold plate with Z-shaped channels has been the most popular for thermal management application, other geometric designs including U-shaped and mixed configuration (i.e., combining Z-shaped and U-shaped) have also been explored by a significant number of studies [43–46]. In addition to the geometric consideration, the controlling strategy can also play an important role in improving the thermal performance of cold plates. For example, Cao et al. [47,48] proposed activating liquid cooling only after the temperature of the battery exceeds certain threshold. Such a method, also known as the delayed cooling scheme, was found to reduce the temperature nonuniformity and power consumption without sacrificing the cooling performance when applied to hybrid cold plates containing phase change materials (PCM).

While cold plates have been extensively studied in past literature studies to address the cooling issue of lithium-ion batteries, most existing cold plate designs only feature long and straight cooling channels. Such a design concept is advantageous in reducing the flow resistance, but fails to maximize the convective heat transfer performance. Furthermore, most existing studies employed continuous flow to cool down the battery, which is not a power efficient solution. On the other hand, the delayed cooling strategy opens a new pathway to simultaneously enhance the uniformity of battery temperature and to reduce energy consumption. In this study, a novel cross-linked channel configuration was proposed for cold plate design. A cross-linked channel cold plate consists of parallel longitudinal and transverse channels which facilitate better fluid mixing and help disrupt the thermal and hydraulic boundary layers. Consequently, such a fluid routing structure has the potential to achieve better temperature uniformity and enhanced heat transfer performance. To the best of the authors knowledge, the concept of cross-linked channel design in cold plates has not been explored in BTMS. The main objective of this study is to investigate the effectiveness of cross-linked channel cold plates coupled with a delayed cooling scheme for thermal management of battery packs at a high discharge rate. We first analyzed seven traditional cold plate designs containing Z-shaped minichannels of different geometries. The thermal performances of these designs were then evaluated and compared with two novel cold plate designs containing cross-linked minichannels. The average temperature of the battery surface, the temperature uniformity, and energy consumption of each cold plate design were studied systematically under the scheme of continuous cooling. Finally, the two best performing cold plate designs (i.e., a Z-shaped parallel channel cold plate with eight-branches and an improved cross-linked channel cold plate) were selected to analyze and compare the thermal performance of a continuous cooling strategy and a delayed cooling strategy.

2. Physical Model

Figure 1 illustrates the geometric model of the BTMS explored in this study. The main components of the BTMS include the batteries (which are also the heat source), the cold plates sandwiched between the neighboring battery cells, and a manifold used for routing the liquid into and out of each single cold plate. As shown in the figure, multiple inlets and outlets were installed on the manifold for connecting the cold plate to an external liquid supply and recirculating system. A commercial prismatic lithium-ion battery with a 45 Ah nominal capacity and 3.2 nominal voltage was selected to construct the battery pack in this study. The positive electrode is made of $LiFeO_4$ and the negative electrode is graphite. The electrolyte consists of 1 M LiPF6–PC/EC/EMC (1/3/5 in weight ratio). The lithium-ion battery has a dimension of 150 mm in length, 200 mm in height, and 30 mm in width. Each cold plate is 3 mm thick and assumed to be manufactured from aluminum. Because of the symmetric geometry of the BTMS, a control volume comprising a half section of the battery and cold plate (shown by the blue dashed line in Figure 1) was selected as the computational domain in this study for analyzing the thermal performance. The cooling medium, selected to be 40% (volume fraction) ethanol solution, is assumed to flow into the cold plate at 25 °C from the top inlet ports and leave via the bottom exit ports. The specifications and thermo-physical properties of the battery, aluminum plate, and the coolant are summarized in Table 1.

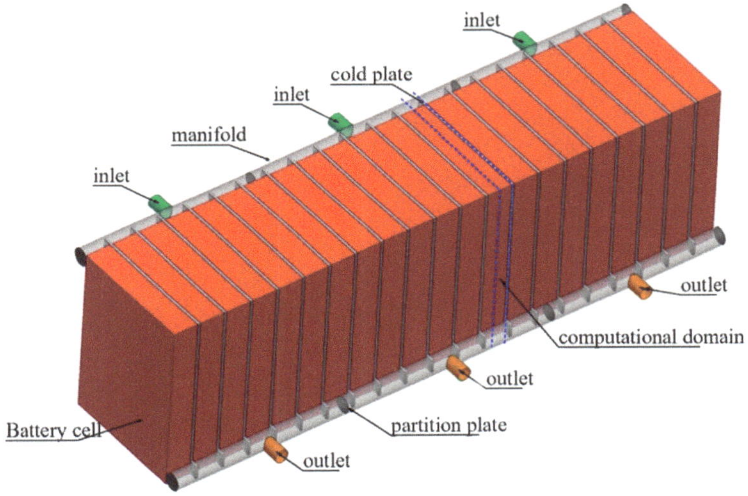

Figure 1. The diagram showing the structure and components of the BTMS for the lithium-ion battery.

Table 1. Main performance parameters and thermo-physical properties of the battery, coolant, and cold plate explored in this study.

Parameters	Battery	Aluminum	Coolant (40% Ethanol Solution)
Nominal Capacity (Ah)	45		
Nominal Voltage (V)	3.2		
Charge Cut-off Voltage (V)	3.6		
Discharge Cut-off Voltage (V)	2.5		
Maximum charging current (A)	≤3		
Maximum discharge current (A)	≤4		
Internal resistance (mΩ)	≤10		
Size (mm)	150 × 200 × 30		
Density ρ ($kg \cdot m^{-3}$)	2090	2719	1055.39
Heat capacity c_p ($J \cdot K^{-1} \cdot kg^{-1}$)	1014.4	871	3502
Thermal conductivity k ($W \cdot m^{-1} \cdot K^{-1}$)	$k_x = 1.696$ $k_y = k_z = 29.94$	202.4	0.412
Dynamic viscosity μ ($kg \cdot m^{-1} \cdot s^{-1}$)			0.00226

Nine cold plate configurations were proposed and analyzed in this work. Based on the geometrical characteristics, these cold plates can be divided into two categories: (1) Z-type parallel channel cold plates (as shown in Figure 2a,b) and (2) cross-linked channel cold plates (as shown in Figure 2c,d). In the Z-type parallel channel cold plate, the coolant flows from the top inlet channel to the bottom outlet channel through a series of long and straight parallel channels without any disruption. Seven different Z-type parallel channel cold plates, each containing 3, 4, 5, 6, 7, 8, and 9 parallel channels, were explored in this work. These designs were referred to as D1 to D7 in the subsequent analysis. In the cross-linked channel cold plate, the coolant is routed through a series of interconnected longitudinal and transverse channels while flowing from the top inlet channel towards the bottom outlet channel. These interconnected channels form a hierarchical flow network which provides greater contact area for convection heat transfer as well as continuous disruption of the boundary layer along the flow path.

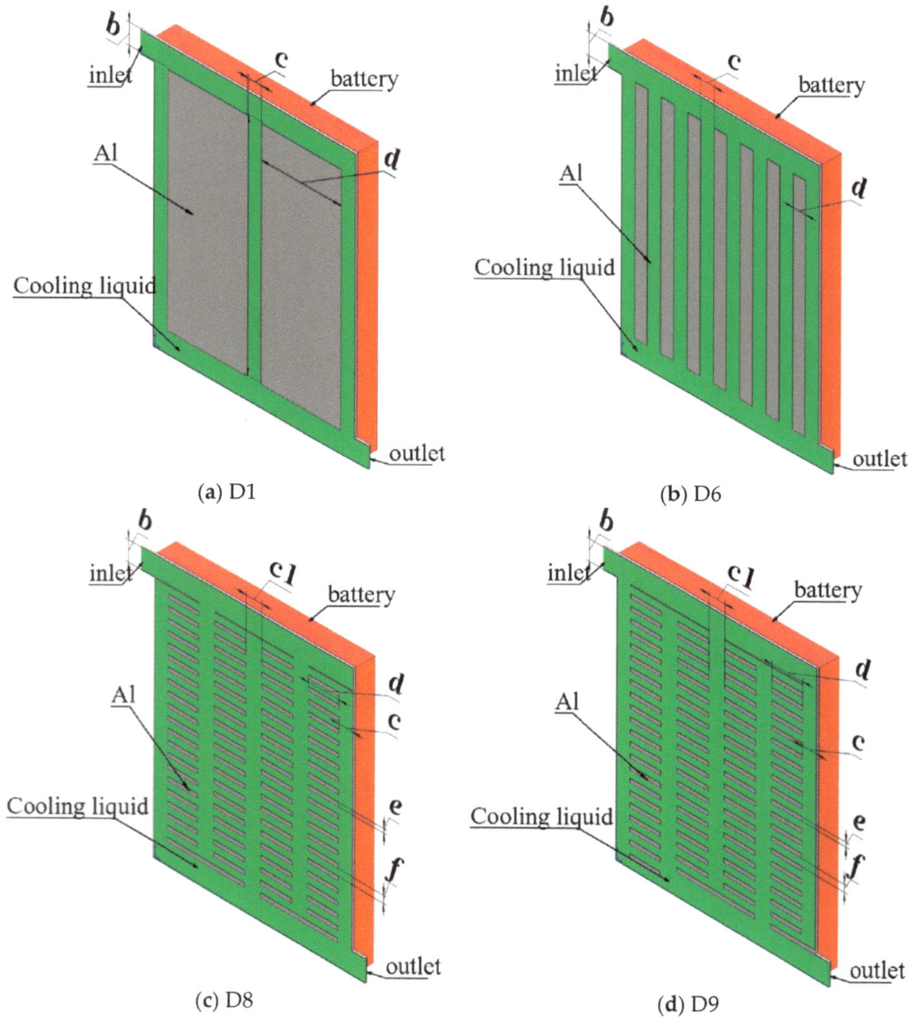

Figure 2. The diagrams showing the geometry of the simulated cold plate. (**a**) Z-type parallel channel design with 3-branches (D1); (**b**) Z-type parallel channel design with 8-branches (D6); (**c**) cross-linked channel design (D8); (**d**) cross-linked channel design with improved topology (D9). The characteristic dimensions associated with each design are marked by geometric parameters a, b, c, $c1$, d, e, and f in each image.

The dimensions of the inlet and outlet channels in all cold plate designs were set as 15 mm × 1 mm. The cross section of the parallel channels in Z-type parallel channel cold plates were designed to be 10 mm × 1 mm. The width of the fin (d) separating two neighboring channels' changes based on the number of parallel channels in different designs. In the cross-linked channel cold plates, the transverse flow channels were designed to be 5 mm wide (labeled as f) and separated by 3 mm wide fins (labeled as e). The neighboring two rows of fins are separated by 12 mm (labeled as $c1$). The spacing between the leftmost/rightmost row of fins and the lateral boundary of the cold plate is 10 mm (labeled as c).

The average temperature and temperature uniformity of the battery pack are two key parameters for assessing its thermal performance. In this study, a temperature maldistribution parameter was introduced to quantify the temperature uniformity of the battery surface. The temperature maldistribution parameter S_T is defined as follows:

$$S_T = \sqrt{\frac{1}{N-1}\sum_{j=1}^{N}\left(T_j - T_{ave}\right)^2} \quad (1)$$

where N stands for the number of temperature detection points, T_j is the temperature measured at the jth detection point, and T_{ave} is the average temperature. Figure 3 shows the locations of the 12 temperature detection points on the battery surface. These points are evenly distributed with a spacing of 50 mm.

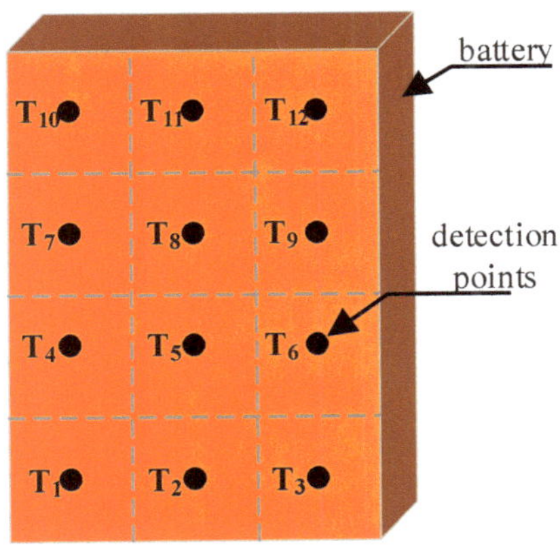

Figure 3. The distribution of the temperature detection points (T_1, T_2, ... T_{12}) on the battery surface.

3. Mathematical Model

3.1. Governing Equations

The computational domain encompasses the battery, aluminum plate, and coolant. The thermal and hydraulic transport behavior in the computational domain are modeled by the continuity, momentum, and energy equations given by

$$\frac{\partial \rho_l}{\partial t} + \nabla\left(\rho_l \vec{v}_l\right) = 0, \quad (2)$$

$$\rho_l\left[\frac{\partial\left(\vec{v}_l\right)}{\partial t} + \vec{v}_l \cdot \nabla \vec{v}_l\right] = -\nabla p + \mu_l \nabla^2 \vec{v}_l, \quad (3)$$

$$\rho_l c_{p,l}\frac{\partial T}{\partial t} + \rho_l c_{p,l}\nabla\left(\vec{v}_l T\right) = \nabla(k_l \nabla T), \quad (4)$$

where ρ_l, μ_l, $c_{p,l}$, and k_l are the density, dynamic viscosity, specific heat capacity, and thermal conductivity of the cooling medium, respectively; p is the liquid pressure; t is the time; $\vec{v}_l$ is the velocity vector; and T is the temperature.

The heat transport in the aluminum plate is governed exclusively by conduction, which can be modeled by the following energy equation:

$$\rho_w c_{pw} \frac{\partial T}{\partial t} = \nabla(k_w \nabla T), \tag{5}$$

where ρ_w, $c_{p,w}$, and k_w are the density, specific heat capacity, and thermal conductivity of the aluminum plate, respectively.

3.2. Battery Model

The heat transport process in the battery is modeled by the following energy equation:

$$\rho_b c_{p,b} \frac{\partial T}{\partial t} = \nabla(k_b \nabla T) + Q_{gen}, \tag{6}$$

where ρ_b, $c_{p,b}$, and k_b are the density, specific heat capacity, and thermal conductivity of the battery. Q_{gen} is the source term, which represents the volume-specific heat generation rate of the battery during operation.

For simplicity, the material properties and heat generation rate are assumed to be constant throughout the battery. Battery heat generation contains both reversible and irreversible components. The irreversible heat contribution is often the dominating heat source resulting from the Joule heating of the electron and ionic flow resistance. The reversible heat contribution is less significant and caused by the electrochemical reactions. Combining these two terms yields the total heat generation rate Q_{gen} as:

$$Q_{gen} = Q_{ir} + Q_{re} = I(E - V) - IT\frac{dE}{dT} = I^2R - IT\frac{dE}{dT}, \tag{7}$$

where I is the current, E is open circuit voltage, V is cell potential, T is the temperature, R is the internal resistance, and dE/dT is the entropy coefficient, which changes with the state of charge (SOC).

Equation (7) indicates that the rate of heat generation inside the battery is dependent on many different parameters, which can vary significantly from one type of battery to another. For a specific type of battery with fixed properties, however, the heat generation rate is often only a strong function of the charging or discharging rate (also known as the C-rate) [49]. The battery C rating is the measurement of current in which a battery is charged and discharged. In general, a charging rate of nC represents that the battery is charged from 0 to 100% within $1/n$ hour's time. A plethora of correlations have been developed to predict the heat generate rate as a function of the charging rate. In this study, the empirical time-dependent model developed by Li [50] was selected to estimate the heat generation rate at different charging rates as:

$$Q_{gen} = A_1t^6 + A_2t^5 + A_3t^4 + A_4t^3 + A_5t^2 + A_6t + A_7, \tag{8}$$

where A_1 to A_7 are the polynomial coefficients as listed in Table 2 [50].

Table 2. The empirical coefficients used to model the heat generation rate of the battery during the discharging cycle [50].

Discharge Rates	A_1	A_2	A_3	A_4	A_5	A_6	A_7
1C	4.9132×10^{-16}	-3.7742×10^{-12}	1.0679×10^{-8}	-1.3417×10^{-5}	0.0076	−2.2208	17,151.7482
2C	1.2578×10^{-13}	-4.8310×10^{-10}	6.8347×10^{-7}	-4.2934×10^{-4}	0.1216	−17.7630	66,623.3365
3C	3.2235×10^{-12}	-8.2542×10^{-9}	7.7851×10^{-6}	-3.2303×10^{-3}	0.6157	−59.9607	148,414.7651

3.3. Initial and Boundary Conditions

The entire BTMS was set at a uniform temperature of 25 °C prior to the simulation as

$$t = 0, \quad T(x, y, z) = T_{ini} = 25\ ^\circ\mathrm{C}. \tag{9}$$

The thermal and hydraulic boundary conditions enforced on the inlet section of the BTMS are given by Equations (10)–(12). When using the continuous cooling strategy, a fixed velocity condition was imposed on the inlet. When the delayed cooling strategy is adopted, a temperature-dependent velocity condition is imposed in the inlet.

$$v = v_{in}, \tag{10}$$

$$\begin{cases} v = v_{in}, & \text{if} \quad T_{ave} \geq 40\ ^\circ\mathrm{C} \\ v = 0, & \text{if} \quad T_{ave} \leq 30\ ^\circ\mathrm{C} \end{cases}, \tag{11}$$

$$T = T_{in} = 25\ ^\circ\mathrm{C}, \tag{12}$$

where T_{ave} is the average temperature of the battery. Different values of the v_{in} were selected in this study to explore the effect of flow rate on the thermal performance. The corresponding Reynolds number varies from 100 to 1000.

A zero-pressure boundary condition was set on the outlet port of the BTMS as:

$$p = 0\ pa. \tag{13}$$

The thermal transport processes in the aluminum and battery were coupled by the energy balance condition at the interface given as:

$$-k_w \frac{\partial T_w}{\partial n} = -k_b \frac{\partial T_b}{\partial n}, \tag{14}$$

where n is the normal vector of interfaces.

The convective heat transfer across the interface between the cooling medium and the aluminum cold plate was modeled on the following heat flux continuity condition and no-slip boundary as:

$$\begin{cases} -k_w \frac{\partial T_w}{\partial n} = -k_l \frac{\partial T_l}{\partial n}, \\ \vec{v} = 0 \end{cases} \tag{15}$$

An adiabatic thermal boundary condition was imposed on the rest of the solid walls of the cold plate and battery pack as:

$$\begin{cases} -k_w \frac{\partial T_w}{\partial n} = 0 \\ -k_b \frac{\partial T_b}{\partial n} = 0 \end{cases}. \tag{16}$$

The central cross-sectional planes of the BTMS (including the cold plate, battery, and cooling medium) were set as symmetric boundary conditions.

3.4. Numerical Method and Validations

The continuity, momentum, and energy equations were solved numerically using Ansys Fluent 2020 R1 commercial software. The differential equations were discretized using the third order MUSCL scheme and solved iteratively with the SIMPLE algorithm. A second order scheme was applied for pressure correction. The iteration was considered to reach convergence when the normalized residues in the momentum and energy equations become smaller than 1×10^{-6} and 1×10^{-12}, respectively.

A structured hexahedral mesh was used in the numerical simulation. For illustration purpose, the front view of the grids is shown in Figure 4. Prior to the numerical simulation, a mesh independent test was conducted on cold plate D1 with six different mesh sizes.

In this test, the Reynolds number at the inlet channel was set as 500 and the discharging rate was set to be 3C. The total number of cells was increased from 3,900,000 to 7,500,000 and the non-dimensional distance y^+ between the first interior node and solid wall was set in the range of 2 to 4.8. Figure 5a shows the average battery temperature of design D1 at time t = 500 and 1080 s as well as the pressure difference between the inlet and outlet at t = 1080 s calculated in the mesh independent test. The results indicate that the deviations between the average battery temperatures obtained with different grid numbers was less than 0.1 °C at t = 500 and 1080 s. Meanwhile, the difference in the pressure drop was less than 0.5 Pa when using different mesh sizes. To balance the computational cost and accuracy, a cell number of 5,690,000 with y^+ = 3.0 was finally selected for the computational analysis of cold plate D1. Similar steps have been followed to determine the mesh size for the analyses of other cold plate designs. Furthermore, a time-step independent study has also been performed to ensure that the time step used in the computational analysis is small enough to capture the transient thermal response of the BTMS accurately. The time step Δt was increased from 0.2 to 2.0 s and the results are shown in Figure 5b. It is evident that changing the time step has negligible impact on the average battery temperature and pressure drop. Therefore, a time step of 1 s was selected for the computational analysis of all cold plate designs in this study.

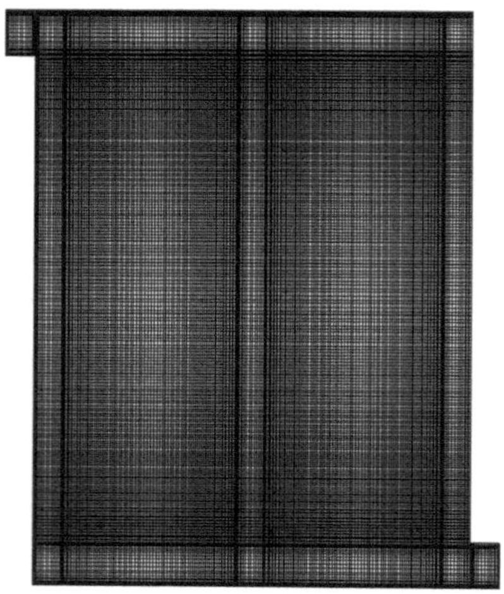

Figure 4. The mesh scheme of design D1 in front view.

To further validate the computational model, we compared the numerical results obtained for cold plate D1 against the experimental results obtained by Li [50] in the literature as shown in Figure 6. The experimental work was conducted using $LiFePO_4$ batteries (3.2 V, 45 Ah) at three different discharging rates. The battery temperature was measured on the surface using five thermocouples and reported as an average value. The comparison shown in Figure 6 reveals a highly consistent change in the average battery temperature between the numerical results and the experimental measurements taken by Li [50]. This agreement confirms the accuracy and reliability of the computational model developed in this study.

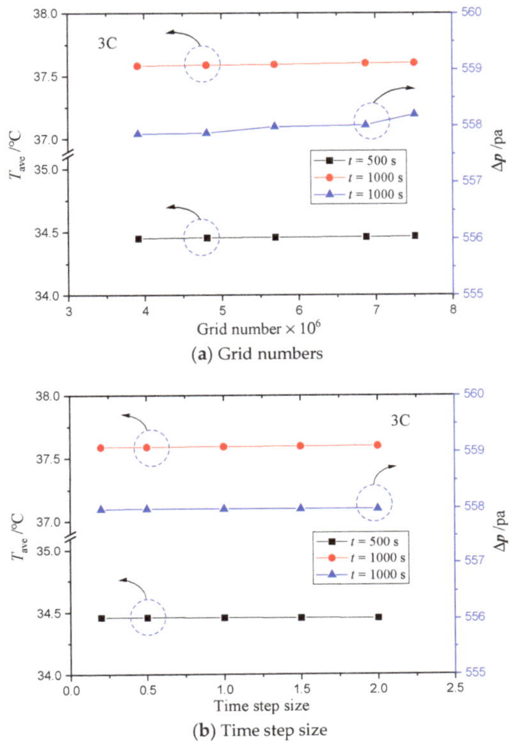

(a) Grid numbers

(b) Time step size

Figure 5. The change in the average battery temperature and pressure difference between inlet and outlet for cold plate with increasing (**a**) grid numbers and (**b**) time step.

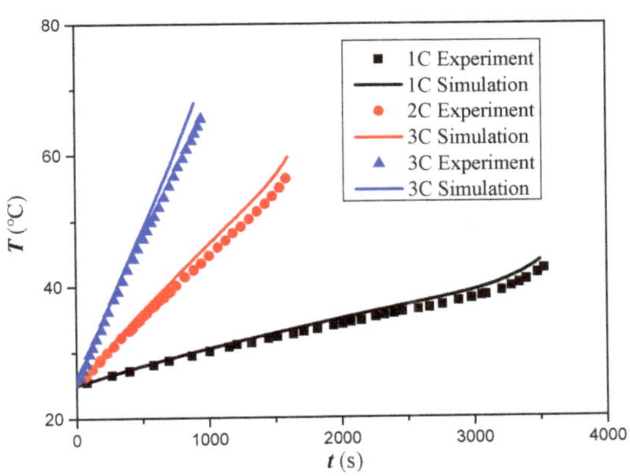

Figure 6. The comparison of the change in the average battery temperature between the simulation results and the experimental measurement taken by Li [50].

4. Results and Discussion

4.1. Temperature Distribution in Different Designs

Figures 7 and 8 separately show the temperature contour with superimposed streamline for Z-type parallel channel cold plates and cross-linked channel cold plates at a discharge rate of 3C when the discharging cycle was completed (t = 1080 s). The Reynolds number was kept constant at 500 and the continuous cooling scheme was employed. As shown in Figure 7a–g, the surface temperatures of the cold plate and battery were found to decrease dramatically with increasing numbers of parallel channels until reaching a total number of 8 (i.e., design D6 shown in Figure 7f). Such a trend was expected since increasing number of cooling channel represents a greater contact area between the coolant and heated surface for convective cooling. However, with a fixed flow rate at the inlet, the velocity and Reynolds number in each parallel channel also becomes smaller with increasing number of channels, which will result in a lower convective heat transfer coefficient in each individual channel. Therefore, when the number of parallel cooling channels is sufficiently large, the reduction in the heat transfer coefficient will start to deprive the merit of greater cooling area and deteriorate the overall cooling performance. As shown by Figure 7f,g, further increasing the number of parallel channels from 8 (i.e., design D7) to 9 leads to a poorer thermal performance represented by a higher surface temperature. Thus, design D6 has the best heat dissipation performance in the Z-type parallel channel cold plates. In addition, the maximum temperature was observed at the middle and lower portions of battery due to the gradual heating of the coolant along the flow path. According to the numerical result, the highest temperatures obtained for cold plate designs D1 to D7 are 50.6 °C, 42.5 °C, 40.9 °C, 40 °C, 39.3 °C, 38.8 °C, and 40.4 °C, respectively.

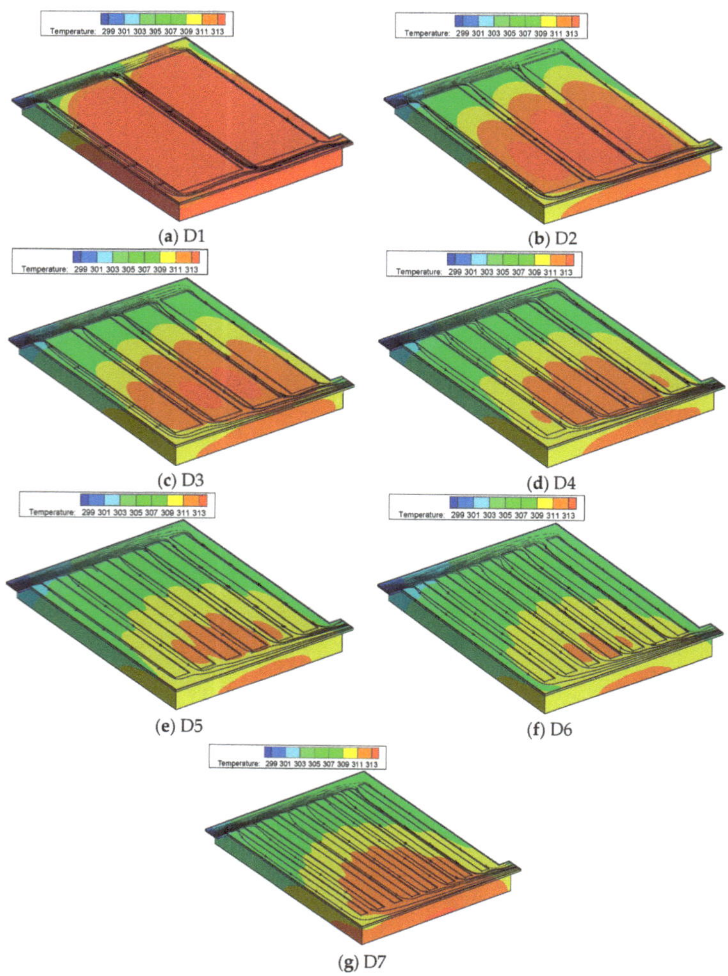

Figure 7. Temperature contours with superimposed streamline at t = 1080 s for Z-type parallel channel cold plates.

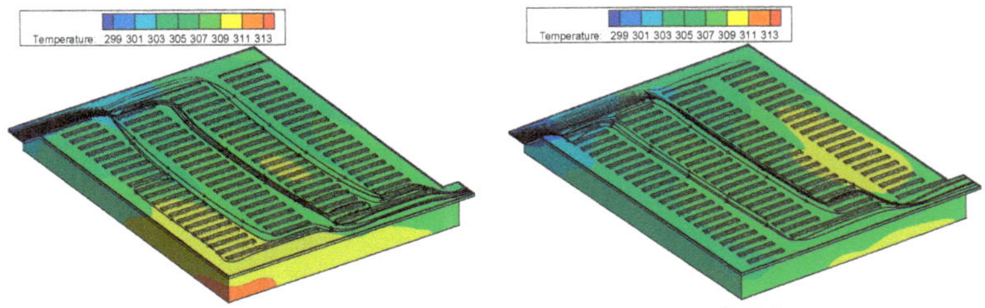

(**a**) D8 (**b**) D9

Figure 8. Temperature contours with superimposed streamline at t = 1080 s for cross-linked channel cold plates.

The dilemma between a greater number of cooling channels and a smaller flow rate in each individual channel can be overcome by introducing the concept of cross-linked channel configuration, which includes two hierarchical levels of interconnected channels. The coolants were first distributed in long and wide parallel longitudinal channels and subsequently routed into short and narrowly spaced transverse channels. Such a design not only allows a greater surface area to be exposed to the convective cooling liquid, but also ensures a sufficiently high heat transfer coefficient in the transverse cooling channels. As shown by Figure 8, the surface temperatures of the battery and aluminum plate obtained for cold plate designs D8 and D9 are significantly lower and more uniform than those obtained for designs D1 to D7. The highest temperature of design D8 was observed in the lower left corner, while that of design D9 was observed in the middle right region. The corresponding maximum surface temperatures of designs D8 and D9 are 38.5 °C and 38.4 °C, respectively. These results demonstrated that the cross-linked channel cold plates have a superior cooling performance than the Z-type parallel channel cold plates.

4.2. *Thermal-Hydraulic Performance in Different Designs*

Figure 9a shows the temporal variation of average battery temperature for all designs. The Reynolds number was still kept at 500 while the discharging rate was set as 3C. As shown in the figure, the average battery temperatures increase rapidly during the initial period of time (about 300 s) followed by a slow and gradual rise for all cold plate designs. Near the end of the discharging cycle (after around 900 s), the average battery temperature started to experience a rapid rise again (about after 900 s). The fast temperature rise at the initial stage is attributed to the slow heat absorption rate by the convective coolant as a result of the small temperature difference between the coolant and battery. At the final stage of the discharge process, the heat generation rate rises sharply which leads to a surge in the battery temperature. As shown in Figure 9a, the average battery temperature calculated for Z type parallel channel cold plates decreases gradually with increasing branch numbers from design D1 to D6 and started rising with increasing branch number from design D6 to D7. In addition, the battery temperatures calculated for designs D8 and D9 are substantially lower than those for the rest of the designs. To better understand such a trend, the effective thermal resistance of all cold plate designs was calculated when the discharging cycle completed (t = 1080 s). The corresponding results are shown in Figure 9b. It is found that the effective thermal resistance decreases monotonically from designs D1 to D9 except for design D7. A smaller thermal resistance represents a smaller temperature difference between the battery surface and the coolant and therefore manifests a better heat transfer performance. It is surprising to notice that the effective thermal resistance of design D7 is even larger than that of design D4, which suggests that convective heat transfer is suppressed severely by the low mass flow rate in each individual channel for design D7. Thus, the average battery temperature becomes significantly higher in design D7. On the other hand, both designs D8 and D9 demonstrated a superior thermal performance than all Z-type cold plates.

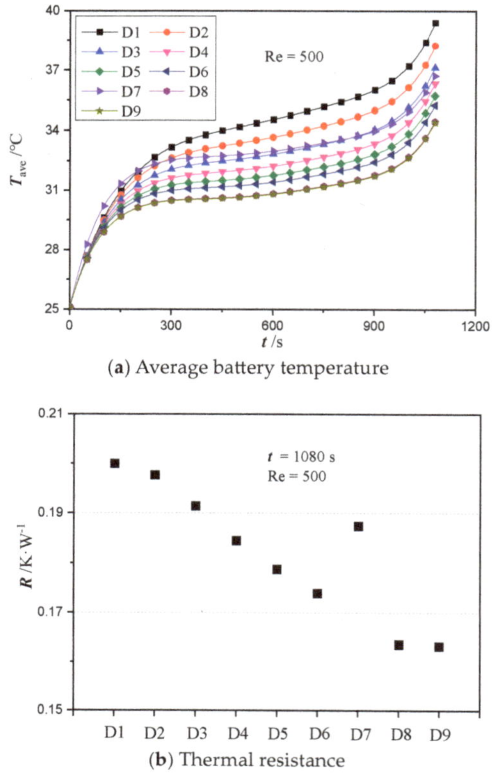

Figure 9. (**a**) The change of average temperature of battery surface during the discharging cycle and (**b**) thermal resistance for different cold plate designs at Re = 500.

Non-uniform temperature distribution has a significant impact on battery performance and lifetime. To analyze the nonuniformity of temperature distribution for different designs, we calculated the variation of the temperature maldistribution parameter over time according to Equation (1) for different cold plate designs as shown in Figure 10. The change of the temperature maldistribution parameter was found to follow a similar trend with that of the average battery temperature. In particular, the temperature maldistribution parameters first increase rapidly, followed by a gradual and slow rise and finally increase quickly again near the end of the discharging process. Among all the designs, D9 has the smallest temperature maldistribution parameter during most of the discharging process. Thus, it can be concluded that design D9 has the best thermal performance than the rest of the designs in terms of both the thermal resistance and temperature uniformity.

The cooling efficiency, which is evaluated based on the energy consumption and heat removal rate, is also another important factor in a practical battery temperature management system. A better cooling performance can always be achieved at a higher flow rate, but at a cost of significantly greater power consumption. The energy consumption can be calculated based on the flow rate and pumping pressure by:

$$Q = \Delta p \cdot V \cdot t, \tag{17}$$

where Δp is the difference between the inlet and outlet pressure, V is the volumetric flow rate of the cooling medium, and t is the operating time of the cold plate. Figure 11 shows the energy consumption and the average temperature of the battery when the discharging cycle completed (t = 1080 s) for comparing the integrative performance of different designs. The inlet Reynolds number was kept constant at Re = 500 and the batteries were operated at a discharge rate of 3C. Because the fluid properties are independent of the temperature, the pressure and velocity distribution remained constant during the entire discharge process. Therefore, the pressure values obtained at the end of discharging (t = 1080 s) were used to calculate the pumping power. As shown in Figure 11, the power consumption kept decreasing monotonically with increasing number of channels in the Z-type parallel channel cold plates. This is because increasing number of channel numbers results in a lower mass flow rate in each channel, which reduces the pressure drop through each individual channel. Considering that the total pressure drop across the parallel channel system is equal to that across each single channel, the total energy consumption can therefore be reduced by distributing the cooling medium into a greater number of channels. The minimum power consumption was obtained in design D7 which contains nine branches of parallel channels. Furthermore, the cross-linked channel cold plates, due to their longer, narrower, and more tortuous flow path, require a significantly higher pumping power to achieve the same flow rate compared with cold plates containing Z-shaped parallel channel. However, by adjusting the structure of the cross-linked flow path slightly from design D8 to design D9, the total power consumption can be reduced by more than 20% without significant impact on the cooling performance. Therefore, based on the comprehensive consideration of the thermal-hydraulic performance, cold plate designs D6 and D9 were identified as the optimum channel designs in this study.

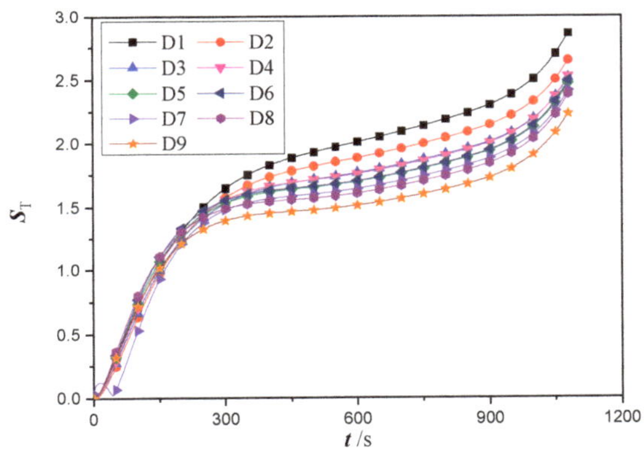

Figure 10. Temporal variation of temperature maldistribution parameter for different designs at Re = 500.

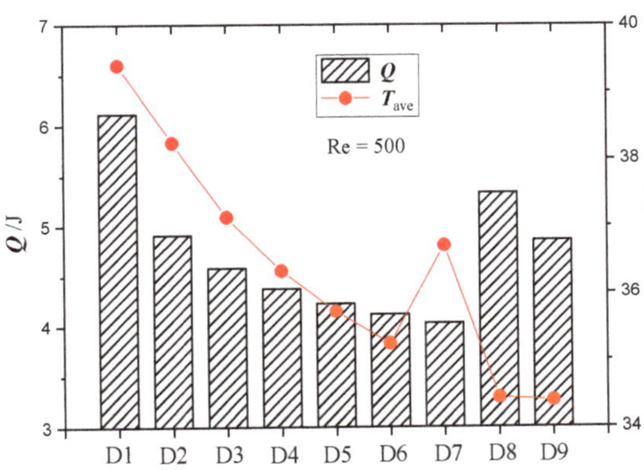

Figure 11. The comparison of the total energy consumption and battery temperature between different cold plate designs at Re = 500.

4.3. Performance Comparison under Different Cooling Strategies

The performances of continuous and delayed cooling strategies were compared using the best performing cold plate designs (i.e., designs D6 and D9). In the delayed cooling scheme, a constant flow of coolant will be provided into the cold plate to cool the battery when the average temperature of battery reaches an upper limit of 40 °C. Once the average temperature of the battery drops below 30 °C, the coolant flow will be stopped (i.e., $v_{in} = 0$). Figure 12a,b shows the temperature change of the battery for designs D6 and D9 at different flow rates when the continuous cooling strategy and delayed cooling strategy were employed. The results indicate that increasing flow rate will cause the battery temperature to rise more slowly for both cooling strategies. When the continuous cooling strategy was employed with the flow rate set at Re = 100, the battery temperature in cold plate designs D6 and D9 was found to exceed the upper limit of 40 °C at t = 591 s and 669 s, respectively. This maximum temperature limit was breached much earlier when adopting the delayed cooling strategy. However, it is interesting to observe that for both designs D6 and D9, the average temperature of the battery measured at the end of discharging process remained almost unchanged for both cooling strategies.

Figure 12c shows the battery temperature in cold plate designs D6 and D9 under both cooling strategy when the discharging cycle completed (t = 1080 s). At the same flow rate, the battery temperature in cold plate design D6 is always higher than that in design D9. In addition, there is very marginal change in the average battery temperature when changing from the continuous to the delayed cooling strategy. The average temperature of the battery at the end of the discharging process is 40.1 °C at Re = 200 in design D6 and 39.8 °C at Re = 180 in design D9. These results suggest that the battery temperature can always be controlled within a safe limit by either operating cold plate D6 at Re = 200 or cold plate D9 at Re = 180 under both cooling strategies.

Figure 13 shows the temperature maldistribution parameter and power consumption for cold plate design D6 at Re = 200 and design D9 at Re = 180 when adopting the two different cooling strategies. Similar to the trend observed for average battery temperature, the temperature maldistribution parameter kept increasing over time under the scheme of continuous cooling. However, when switching to the delayed cooling strategy, the temperature maldistribution parameter remained almost zero during the initial period due to the zero flow rate of the coolant. Subsequently, the battery temperature reached the upper limit and the coolant started being pumped into the cold plate, which resulted in a large temperature difference between the entrance and exit areas. This stage is represented by a sharp rise of the temperature maldistribution parameter shown in Figure 13a. As the coolant got heated up by the battery, the temperature maldistribution parameter experienced a slow drop followed by a gradual rebound towards the end of the discharging process. As shown in the results, the temperature maldistribution parameters of design D6 is always greater than that of design D9 regardless of the cooling strategy being employed. In addition, the power consumption of design D9 is also 13% lower than that of design D6. More importantly, changing from the continuous to the delayed cooling strategy allows the power consumption to be reduced significantly due to the temporal interruption of the cooling flow. In particular, the power consumption was decreased by 23% for both designs D6 and D9 after switching from the continuous to the delayed cooling schemes. All these findings demonstrate that the delayed cooling strategy is a more energy efficient method to control the battery temperature within a safe operating limit compared to the continuous cooling strategy. Furthermore, design D9 was proven to have the best cooling performance than the rest of the designs in terms of a lower battery temperature, a smaller temperature non-uniformity, and a lower power consumption.

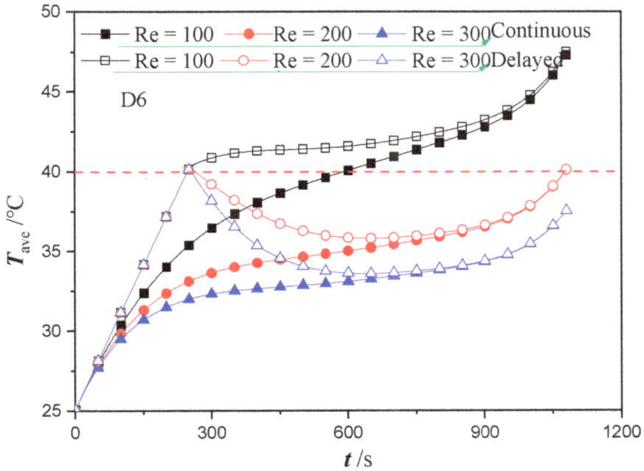

(**a**) Average battery temperature for D6

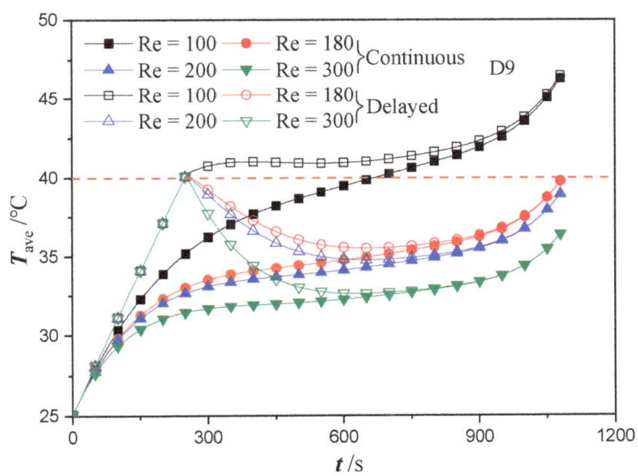

(**b**) Average battery temperature for D9

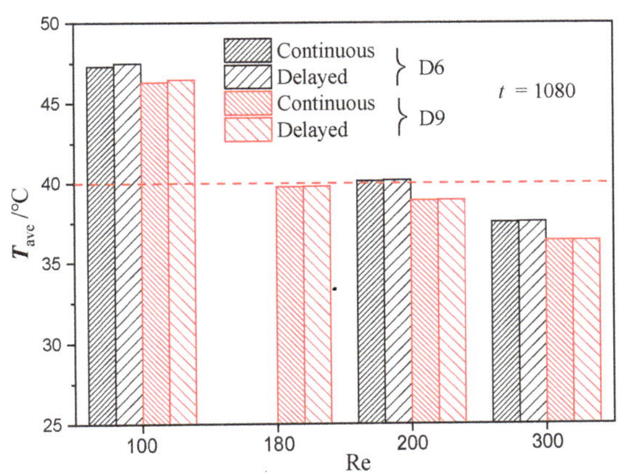

(**c**) Average battery temperature at t = 1080 s

Figure 12. The change in the average temperature of battery for cold plate designs D6 and D9 between the two different cooling strategies: (**a**) D6; (**b**) D9; (**c**) at t = 1080 s.

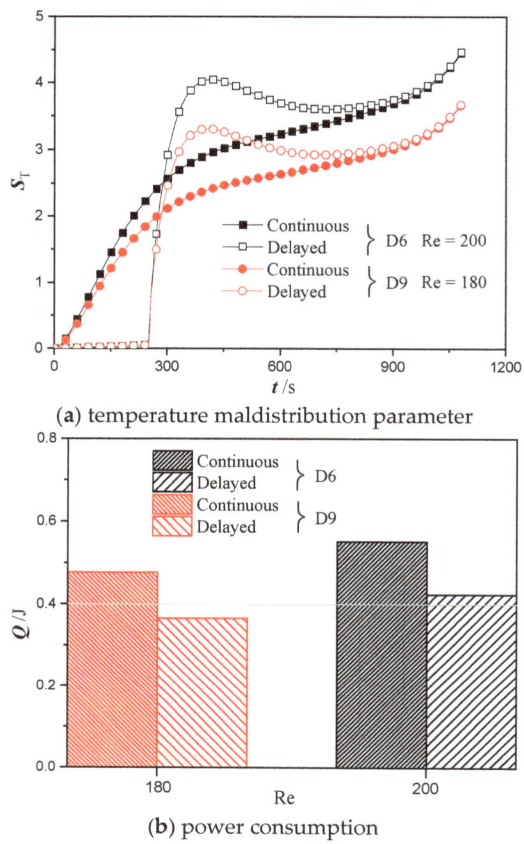

Figure 13. Temporal variation of power consumption for cold plate designs D6 and D9 with the average battery temperature controlled within the allowable range.

The effect of inlet Reynolds number on the power consumption and total duration of coolant flow was further analyzed for designs D6 and D9 under the delayed cooling scheme, as shown in Figure 14. It is evident to notice that the power consumption kept rising with increasing Reynolds number. For cold plate design D9, the total duration of coolant flow was reduced twice when the Reynolds number was increased from 600 to 700 and from 800 to 900. For design D6, however, the total duration of coolant flow was only reduced once when increasing the Reynolds number from 600 to 700. The reduced flow duration provides a direct indication of a better energy efficiency since the battery temperature can be controlled effectively with a lower amount of coolant flow. As shown in Figure 14, the power consumption of design D9 becomes lower than that of D6 at Re = 700, 900, and 1000 due to the shorter duration of coolant flow. However, it should also be noted that a reduced flow duration may also represent an overcooling scenario because the battery temperature has been kept below 30 °C (instead of the upper limit of 40 °C) for a substantial amount of time. Therefore, operating at a higher flow rate for a reduced amount of time does not necessarily provide a better energy efficiency. In fact, a greater pumping power is still required at a higher Reynolds number to ensure that the battery temperature is less than the upper limit of 40 °C. If more stringent operating temperature is required in the practical scenario, then design D9 could potentially provide a much better performance than design D6 by using a high mass flow rate and a shorter cooling period.

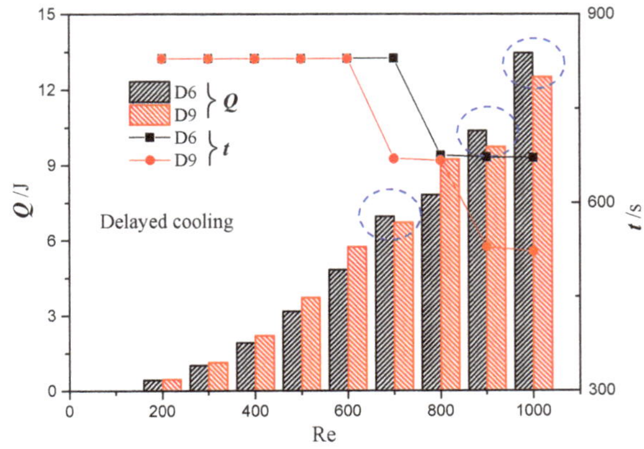

Figure 14. The variation of power consumption and liquid cooling time for designs D6 and D9 under delayed cooling scheme at different Reynolds numbers.

5. Conclusions

In this study, seven Z-type parallel channel cold plates and two cross-linked channel cold plates were proposed for the cooling of high-power lithium-ion batteries under continuous and delayed cooling schemes. The main conclusions made are as follows:

(1) The best thermal performance was obtained at a channel number of eight in the Z-type parallel channel cold plates. Cross-linked channel cold plates show a significantly better cooling performance than Z-type parallel channel cold plates at the price of a higher power consumption.

(2) The operating temperature of a battery can be maintained within a safe range at Re = 200 and 180 for designs D6 and D9, respectively, under both continuous and delayed cooling schemes. At these operating conditions, design D9 requires 13% less pumping power and provides a better temperature uniformity than design D6. Therefore, design D9 has the best thermo-hydraulic performance than the rest of the designs.

(3) Under the premise that the battery temperature is kept within the safe range, the delayed cooling strategy can greatly reduce the power consumption by shortening the duration of liquid cooling. In particular, the power consumption of design D9 can be decreased by 23% after switching from the continuous cooling scheme to the delayed cooling scheme.

Author Contributions: H.Y. wrote the main manuscript text; Z.W., M.L. and F.R. prepared figures; B.M. reviewed the manuscript. All authors have read and agreed to the published version of the manuscript.

Funding: This work was supported by the National Natural Science Foundation of China (No. 52106004), Guangdong Basic and Applied Basic Research Foundation (No. 2020A1515110270), and Science, Technology and Innovation Commission of Shenzhen Municipality (GXWD20220811165249002, GXWD20201230155427003-20200801212235001) for which the authors are thankful.

Data Availability Statement: The datasets generated during and/or analyzed during the current study are available from the corresponding author on reasonable request.

Conflicts of Interest: We declare that we do not have any commercial or associative interest that represents a conflict of interest in connection with the work submitted.

Nomenclature

Latin symbols	
b	inlet and outlet channel width, mm
c, $c1$	fin space of longitudinal channel, mm
cp	heat capacity, $J \cdot K^{-1} \cdot kg^{-1}$
d	fin thickness of longitudinal channel, mm
e	fin thickness of transverse channel, mm
f	fin space of transverse channel, $W \cdot m^{-2} \cdot K^{-1}$
p	pressure, Pa
Q	power consumption, J
Q_{gen}	heat generation rate, $W \cdot m^{-3}$
R	convection thermal resistance, $°C \cdot W^{-1}$
S_T	temperature maldistribution parameter,
t	time, s
T	temperature, °C
v	velocity, $m \cdot s^{-1}$
Greek symbols	
ρ	density, $kg \cdot m^{-3}$
k	thermal conductivity, $W \cdot m^{-1} \cdot K^{-1}$
μ	dynamic viscosity, $kg \cdot m^{-1} \cdot s^{-1}$
Subscripts	
ave	average
b	battery
j	serial number
l	liquid
w	wall

References

1. Feng, X.; Zheng, S.; Ren, D.; He, X.; Wang, L.; Cui, H.; Liu, X.; Jin, C.; Zhang, F.; Xu, C.; et al. Investigating the thermal runaway mechanisms of lithium-ion batteries based on thermal analysis database. *Appl. Energy* **2019**, *246*, 53–64. [CrossRef]
2. Chen, J.; Kang, S.; E, J.; Huang, Z.; Wei, K.; Zhang, B.; Zhu, H.; Deng, Y.; Zhang, F.; Liao, G. Effects of different phase change material thermal management strategies on the cooling performance of the power lithium ion batteries: A review. *J. Power Sources* **2019**, *442*, 227228. [CrossRef]
3. Zheng, C. Examining the Benefits of Using Boron Compounds in Lithium Batteries: A Comprehensive Review of Literature. *Batteries* **2022**, *8*, 187. [CrossRef]
4. Kim, J.; Oh, J.; Lee, H. Review on battery thermal management system for electric vehicles. *J. Appl. Therm. Eng.* **2019**, *149*, 192–212. [CrossRef]
5. Pesaran, A.A. Battery thermal models for hybrid vehicle simulations. *J. Power Sources* **2002**, *110*, 377–382. [CrossRef]
6. Zhang, L.; Zhao, P.; Xu, M.; Wang, X. Computational identification of the safety regime of Li-ion battery thermal runaway. *Appl. Energy* **2020**, *261*, 114440. [CrossRef]
7. Chen, K.; Song, M.; Wei, W.; Wang, S. Design of the structure of battery pack in parallel air-cooled battery thermal management system for cooling efficiency improvement. *Int. J. Heat Mass Transf.* **2019**, *132*, 309–321. [CrossRef]
8. Chen, W.; Hou, S.; Shi, J.; Han, P.; Liu, B.; Wu, B.; Lin, X. Numerical Analysis of Novel Air-Based Li-Ion Battery Thermal Management. *Batteries* **2022**, *8*, 128. [CrossRef]
9. Liu, Y.; Zhang, J. Self-adapting J-type air-based battery thermal management system via model predictive control. *Appl. Energy* **2020**, *263*, 114640. [CrossRef]
10. Lloyd, R.; Akrami, M. A Critical Analysis of Helical and Linear Channel Liquid Cooling Designs for Lithium-Ion Battery Packs. *Batteries* **2022**, *8*, 236. [CrossRef]
11. Park, S.; Jang, D.S.; Lee, D.; Hong, S.H.; Kim, Y. Simulation on cooling performance characteristics of a refrigerant-cooled active thermal management system for lithium ion batteries. *Int. J. Heat Mass Transf.* **2019**, *135*, 131–141. [CrossRef]
12. Lan, C.; Xu, J.; Qiao, Y.; Ma, Y. Thermal management for high power lithium-ion battery by minichannel aluminum tubes. *Appl. Therm. Eng.* **2016**, *101*, 284–292. [CrossRef]
13. Xiong, M.; Wang, N.; Li, W.; Garg, A.; Gao, L. Study on the Heat Dissipation Performance of a Liquid Cooling Battery Pack with Different Pin-Fins. *Batteries* **2023**, *9*, 44. [CrossRef]
14. Ping, P.; Zhang, Y.; Kong, D.; Du, J. Investigation on battery thermal management system combining phase changed material and liquid cooling considering non-uniform heat generation of battery. *J. Energy Storage* **2021**, *36*, 102448. [CrossRef]

15. Yang, H.; Li, Y.; Zhang, L.; Zhu, Y. Thermal performance enhancement of phase change material heat sinks for thermal management of electronic devices under constant and intermittent power loads. *Int. J. Heat Mass Transf.* **2021**, *181*, 121899. [CrossRef]
16. Sun, Z.; Fan, R.; Yan, F.; Zhou, T.; Zheng, N. Thermal management of the lithium-ion battery by the composite PCM-Fin structures. *Int. J. Heat Mass Transf.* **2019**, *145*, 118739. [CrossRef]
17. Yang, H.; Li, Y.; Yang, Y.; Chen, D.; Zhu, Y. Effective thermal conductivity of high porosity open-cell metal foams. *Int. J. Heat Mass Transf.* **2020**, *147*, 118974. [CrossRef]
18. He, L.; Tang, X.; Luo, Q.; Liao, Y.; Luo, X.; Liu, J.; Ma, L.; Dong, D.; Gan, Y.; Li, Y. Structure optimization of a heat pipe-cooling battery thermal management system based on fuzzy grey relational analysis. *Int. J. Heat Mass Transf.* **2022**, *182*, 121924. [CrossRef]
19. Wu, W.; Yang, X.; Zhang, G.; Chen, K.; Wang, S. Experimental investigation on the thermal performance of heat pipe-assisted phase change material based battery thermal management system. *Energy Convers. Manag.* **2017**, *138*, 486–492. [CrossRef]
20. Yang, H.; Li, M.; Wang, Z.; Ma, B. A compact and lightweight hybrid liquid cooling system coupling with Z-type cold plates and PCM composite for battery thermal management. *Energy* **2023**, *263*. [CrossRef]
21. Widyantara, R.D.; Zulaikah, S.; Juangsa, F.B.; Budiman, B.A.; Aziz, M. Review on Battery Packing Design Strategies for Superior Thermal Management in Electric Vehicles. *Batteries* **2022**, *8*, 287. [CrossRef]
22. Akbarzadeh, M.; Jaguemont, J.; Kalogiannis, T.; Karimi, D.; He, J.; Jin, L.; Xie, P.; Van Mierlo, J.; Berecibar, M. A novel liquid cooling plate concept for thermal management of lithium-ion batteries in electric vehicles. *Energy Convers. Manag.* **2021**, *231*, 113862. [CrossRef]
23. Yue, Q.; He, C.; Wu, M.; Zhao, T. Advances in thermal management systems for next-generation power batteries. *Int. J. Heat Mass Transf.* **2021**, *181*, 121853. [CrossRef]
24. Lin, J.; Liu, X.; Li, S.; Zhang, C.; Yang, S. A review on recent progress, challenges and perspective of battery thermal management system. *Int. J. Heat Mass Transf.* **2021**, *167*, 120834. [CrossRef]
25. Mousavi, S.; Siavashi, M.; Zadehkabir, A. A new design for hybrid cooling of Li-ion battery pack utilizing PCM and mini channel cold plates. *Appl. Therm. Eng.* **2021**, *197*, 117398. [CrossRef]
26. E, J.; Han, D.; Qiu, A.; Zhu, H.; Deng, Y.; Chen, J.; Zhao, X.; Zuo, W.; Wang, H.; Chen, J.; et al. Orthogonal experimental design of liquid-cooling structure on the cooling effect of a liquid-cooled battery thermal management system. *Appl. Therm. Eng.* **2018**, *132*, 508–520. [CrossRef]
27. Huo, Y.; Rao, Z.; Liu, X.; Zhao, J. Investigation of power battery thermal management by using mini-channel cold plate. *Energy Convers. Manag.* **2015**, *89*, 387–395. [CrossRef]
28. Qian, Z.; Li, Y.; Rao, Z. Thermal performance of lithium-ion battery thermal management system by using mini-channel cooling. *Energy Convers. Manag.* **2016**, *126*, 622–631. [CrossRef]
29. Li, X.; Zhou, D.; Zhang, G.; Wang, C.; Lin, R.; Zhong, Z. Experimental investigation of the thermal performance of silicon cold plate for battery thermal management system. *Appl. Therm. Eng.* **2019**, *155*, 331–340. [CrossRef]
30. Deng, T.; Zhang, G.; Ran, Y. Study on thermal management of rectangular Li-ion battery with serpentine-channel cold plate. *Int. J. Heat Mass Transf.* **2018**, *125*, 143–152. [CrossRef]
31. Kong, W.; Zhu, K.; Lu, X.; Jin, J.; Ni, M. Enhancement of lithium-ion battery thermal management with the divergent-shaped channel cold plate. *J. Energy Storage* **2021**, *42*, 103027. [CrossRef]
32. Huang, Y.; Mei, P.; Lu, Y.; Huang, R.; Yu, X.; Chen, Z.; Roskilly, A.P. A novel approach for Lithium-ion battery thermal management with streamline shape mini channel cooling plates. *Appl. Therm. Eng.* **2019**, *157*, 113623. [CrossRef]
33. Mo, X.; Zhi, H.; Xiao, Y.; Hua, H.; He, L. Topology optimization of cooling plates for battery thermal management. *Int. J. Heat Mass Transf.* **2021**, *178*, 121612. [CrossRef]
34. Kalkan, O.; Celen, A.; Bakirci, K.; Dalkilic, A.S. Experimental investigation of thermal performance of novel cold plate design used in a Li-ion pouch-type battery. *Appl. Therm. Eng.* **2021**, *191*, 116885. [CrossRef]
35. Amalesh, T.; Narasimhan, N.L. Introducing new designs of minichannel cold plates for the cooling of Lithium-ion batteries. *J. Power Sources* **2020**, *479*, 228775. [CrossRef]
36. Chen, Y.; Chen, K.; Dong, Y.; Wu, X. Bidirectional symmetrical parallel mini-channel cold plate for energy efficient cooling of large battery packs. *Energy* **2021**, *242*, 122553. [CrossRef]
37. Gungor, S.; Cetkin, E.; Lorente, S. Canopy-to-canopy liquid cooling for the thermal management of lithium-ion batteries, a constructal approach. *Int. J. Heat Mass Transf.* **2022**, *182*, 121918. [CrossRef]
38. Sun, H.; Dixon, R. Development of cooling strategy for an air cooled lithium-ion battery pack. *J. Power Sources* **2014**, *272*, 404–414. [CrossRef]
39. Guo, R.; Li, L. Heat dissipation analysis and optimization of lithium-ion batteries with a novel parallel-spiral serpentine channel liquid cooling plate. *Int. J. Heat Mass Transf.* **2022**, *189*, 122706. [CrossRef]
40. Shen, X.; Cai, T.; He, C.; Yang, Y.; Chen, M. Thermal analysis of modified Z-shaped air-cooled battery thermal management system for electric vehicles. *J. Energy Storage* **2023**, *58*, 106356. [CrossRef]
41. Faizan; Pati, S.; Randive, P. Effect of channel configurations on the thermal management of fast discharging Li-ion battery module with hybrid cooling. *Energy* **2023**, *267*, 126358. [CrossRef]
42. Guo, Z.; Xu, Q.; Ni, M. A numerical study on the battery thermal management system with mini-channel cold plate considering battery aging effect. *Appl. Therm. Eng.* **2023**, *219*, 119564. [CrossRef]

43. Chen, K.; Song, M.; Wei, W.; Wang, S. Structure optimization of parallel air-cooled battery thermal management system with U-type flow for cooling efficiency improvement. *Energy* **2018**, *145*, 603–613. [CrossRef]
44. Lu, Z.; Yu, X.; Wei, L.; Qiu, Y.; Zhang, L.; Meng, X.; Jin, L. Parametric study of forced air cooling strategy for lithium-ion battery pack with staggered arrangement. *Appl. Therm. Eng.* **2018**, *136*, 28–40. [CrossRef]
45. Liu, Y.; Zhang, J. Design a J-type air-based battery thermal management system through surrogate-based optimization. *Appl. Energy* **2019**, *252*, 113426. [CrossRef]
46. Cao, J.; Feng, J.; Fang, X.; Ling, Z.; Zhang, Z. A delayed cooling system coupling composite phase change material and nano phase change material emulsion. *Appl. Therm. Eng.* **2021**, *191*, 116888. [CrossRef]
47. Cao, J.; Ling, Z.; Fang, X.; Zhang, Z. Delayed liquid cooling strategy with phase change material to achieve high temperature uniformity of Li-ion battery under high-rate discharge. *J. Power Sources* **2020**, *450*, 227673. [CrossRef]
48. Cao, W.; Zhao, C.; Wang, Y.; Dong, T.; Jiang, F. Thermal modeling of full-size-scale cylindrical battery pack cooled by channeled liquid flow. *Int. J. Heat Mass Transf.* **2019**, *138*, 1178–1187. [CrossRef]
49. Pan, Y.-W.; Hua, Y.; Zhou, S.; He, R.; Zhang, Y.; Yang, S.; Liu, X.; Lian, Y.; Yan, X.; Wu, B. A computational multi-node electro-thermal model for large prismatic lithium-ion batteries. *J. Power Sources* **2020**, *459*, 228070. [CrossRef]
50. Li, T. Study on thermal effects of lithium-ion battery in electric vehicle and battery package dissipation structural optimization. Master's Thesis, Chongqing University, Chongqing, China, 2013.

batteries

MDPI

Article

A Polyacrylonitrile Shutdown Film for Prevention of Thermal Runaway in Lithium-Ion Cells

Jonathan Peter Charles Allen, Marcin Mierzwa, Denis Kramer, Nuria Garcia-Araez and Andrew L. Hector *

School of Chemistry, University of Southampton, Highfield, Southampton SO17 1BJ, UK; jpca1g13@soton.ac.uk (J.P.C.A.); marcin.mierzwa@yahoo.com (M.M.); d.kramer@hsu-hh.de (D.K.); n.garcia-araez@soton.ac.uk (N.G.-A.)

* Correspondence: a.l.hector@soton.ac.uk

Abstract: The electrodeposition of a polymer (polyacrylonitrile, PAN) is used to reduce the risk of thermal runaway in lithium-ion batteries, which is the most important cause of battery accidents and fires. PAN was electrodeposited on a graphite battery electrode, using cyclic voltammetry or chronoamperometry, in a solution with acrylonitrile as the solvent. The electrodeposited PAN film was characterised by Raman spectroscopy, microscopy, energy dispersive X-ray analysis, and thermogravimetric analysis, and it was found that the film thickness could be controlled by the amount of charge passed in the electrochemical experiments. The PAN-coated graphite battery electrode was then tested in lithium half-cells, obtaining capacities close to the uncoated graphite sample (ca. 360 mA h g^{-1}) for thin (<10 µm) polymer coatings at 25 °C. Interestingly, for thicker polymer coatings (>20 µm) it was found that the capacity decreased drastically as the temperature increased beyond 80 °C. Such suppression in capacity has applications for thermal runaway protection since the electrochemical reactions of degradation of the electrolyte in contact with the electrode are the root cause of the thermal runaway process. Further work should look into alternative polymer and liquid electrolyte formulations to achieve the desired suppression of electrochemical capacity at high temperatures while retaining high capacities at the operational temperature range.

Keywords: polyacrylonitrile; electrodeposition; lithium-ion batteries; thermal runaway; thermal shut-down

Citation: Allen, J.P.C.; Mierzwa, M.; Kramer, D.; Garcia-Araez, N.; Hector, A.L. A Polyacrylonitrile Shutdown Film for Prevention of Thermal Runaway in Lithium-Ion Cells. *Batteries* **2023**, *9*, 282. https://doi.org/10.3390/batteries9050282

Academic Editors: Jinsheng Xiao, Hengyun Zhang and Sousso Kelouwani

Received: 6 March 2023
Revised: 12 May 2023
Accepted: 13 May 2023
Published: 21 May 2023

1. Introduction

Lithium-ion batteries are the battery of choice in many devices, such as electric vehicles and portable electronics, so methods to increase their safety are extremely important. Thermal runaway is, by far, the greatest cause of accidents in battery packs [1,2]. The process of thermal runaway occurs when the rate of heat generation, triggered by mechanical/thermal/electrical abuse, is greater than the heat dissipation. Increasing heat generation within a battery leads to higher rates of exothermic reaction, resulting in even greater amounts of heat generation [3]. As a result, battery temperature increases uncontrollably, accelerating battery degradation, which can lead to a fire or explosion. Effective thermal management can be pursued using systems external to a battery pack; however, as heat flux occurs due to battery charge-discharge reaction, thermal management can also be pursued within the materials of a lithium-ion cell [3].

Battery electrode materials are constantly being developed to improve their safety and overall performance [4,5]. One method of safety improvement is via the implementation of functional materials in addition to the battery electrodes, of which a variety have been studied to improve the safety of lithium-ion batteries [6–10]. Particularly effective has been the development of thermal shut-down separators [11–13], whose porosity closes as the temperature of the battery increases, permanently stopping ion transport; ions are unable to reach the electrode-electrolyte interface, halting the battery electrochemistry. This removes one potential source of heat generation. Unfortunately, further heating may still occur which could trigger thermal runaway. This can come internally in the form of gas pressure build-up from the products of thermal degradation or chemical side-reactions with these products. However, external causes can also pose a problem, such as excessive currents or heat sources [3]. Consequently, it is widely agreed that securing battery safety requires the combination of several safety approaches so that if one approach fails the battery would still be protected by other approaches. For instance, venting of flammable gases to reduce pressures alongside a thermal fuse to cut-off the external current to at-risk cells.

Materials with resistance that drastically increases with temperature (positive temperature coefficient of resistivity, or PTCR, materials) have been developed to improve battery safety, including ceramic materials [14] and thermoresponsive polymers [15–23]. An alternative approach using thermoresponsive polymer microspheres that melt as the battery heats, stopping the ion flow and shutting down the battery, has also been reported [24–26]. These materials, in essence, have similar operations to thermal fuses acting to reduce the current to the electrochemical system as temperature increases. However, these materials are present at the source of the electrochemical reaction within the cell allowing for rapid response to increases in temperature.

In this work, we report for the first time the application of electrodeposited polymers to achieve battery shutdown. The working principle is similar to thermal shutdown separators with the exception that the polymer is directly coated onto one of the electrodes. As a result, the thickness and morphology of the polymer coating can be controlled by tuning the electrochemical conditions for the electrodeposition. We also demonstrate that, upon heating the battery, the polymer coating induces the battery shutdown (as seen by a drastic decrease in capacity). Sister batteries run under the same conditions but without polymer coating and still deliver a high capacity; this is undesirable as the high electrochemical activity of the battery at high temperatures could promote a thermal runaway event.

2. Experimental

2.1. Materials

The full list of materials used is given below in Table 1. Information on the usage of each material and the component they formed is also provide along with information on suppliers and their locations.

Table 1. Materials used along with their usage. Supplier information is also provided.

Material	Usage	Supplier
Acrylonitrile (AN, ≥99%, contains the inhibitor monomethyl ether hydroquinone in 35–45 ppm)	Solvent, monomer for electrodeposition of PAN	Sigma-Aldrich, Gillingham, UK
Alumina powder (1 μm, 0.3 μm, and 0.05 μm)	Polishing powder	Buehler, Coventry, UK
Calcium hydride (95%)	Drying agent	Sigma-Aldrich, Gillingham, UK
Carbon powder, (99%, Super C65 carbon black)	Conductive additive	Timcal, Cambridge, UK
Copper metal foil (99.9%, 50 μm thickness)	Current collector	Advent, Oxford, UK
Glass fibre (50 μm thickness, GF/F, Whatman™)	Separator	GE Healthcare Life Sciences, Chicago IL, USA
Graphite powder (99%)	Active material	Hitachi Chemical, Tokyo, Japan
Lithium metal foil (99.9% purity, 120 μm thickness)	Counter-reference electrode	Goodfellow, Cambridge, UK
LP57 (1 mol dm^{-3} $LiPF_6$ in 3:7 ethylene carbonate/ethyl methyl carbonate)	Lithium-ion electrolyte	Soulbrain MI, Northville MI, USA
N-methyl-2-pyrrolidone (NMP, 99.5%)	Solvent	Sigma-Aldrich, Gillingham, UK
PAN, Polyacrylonitrile (PAN, M_w 150,000)	Percolating polymer matrix	Sigma-Aldrich, Gillingham, UK
PVDF, Polyvinylidene difluoride (PVDF, Solef® 5130)	Electrode binder and percolating polymer matrix	Solef, Tavaux, France
Tetrabutylammonium perchlorate (TBAP, ≥99%)	Supporting electrolyte	Sigma-Aldrich, Gillingham, UK

2.2. Graphite Battery Electrode Preparation

The substrates used for the electrodeposition of PAN were graphite battery electrodes and they were prepared using an ink-casting method. For the ink preparation, 10% by weight PVDF was first dissolved in NMP. Graphite, carbon black, and the PVDF solution were then weighed to produce a mixture of mass ratio of 94:3:3 based on the solid components. Additional NMP was added to reduce viscosity until the mixture contained 60% by weight of NMP. A Thinky planetary mixer was used to blend the ink mixture for 15 min at 2000 rpm to produce the composite ink. A coating of the ink was applied to a copper foil, which had been polished using alumina powder of increasing fineness (from 1 to 0.3, and finally 0.05 μm). A TQC Sheen doctor blade was used to perform the coating, the thickness of the coating was set to 0.25 mm and the ink was spread at a speed of 200 mm s^{-1}. The ink was dried at 80 °C for 2 to 3 h in a vacuum oven. Electrodes were sized to 20 × 20 mm and pressed with a manual hydraulic press at 10 tonnes.

2.3. PAN Electrodeposition

Distillation of acrylonitrile was performed at 90 to 100 °C with a nitrogen environment using a drying agent of CaH_2 with a reflux of 8 h before collection. Tetrabutylammonium perchlorate ($[Bu_4N][ClO_4]$) was vacuum dried at ambient temperature for 24 h. A 0.05 mol dm^{-3} $[Bu_4N][ClO_4]$ in acrylonitrile solution was prepared under nitrogen.

A three-electrode arrangement was used for electrodeposition studies. A graphite battery electrode formed the working electrode, platinum gauze the counter, and a silver wire was used as a pseudo reference. This was set up within a Teflon deposition cell with a 14 mm diameter deposition window a Teflon O-ring over the working electrode surface. A volume of 10 mL of electrolyte was added to the Teflon cell which was then saturated with oxygen by bubbling through a sparger for 15 min. The reference and counter electrodes were then placed into the electrolyte and electrodepositions were performed with a Biologics SP-150 potentiostat, employing EC-Lab software. This Teflon cell arrangement was used in a previous electrodeposition study of PAN, which discussed the practicality of the Teflon cell for such depositions and included a preliminary discussion of a cyclic voltammogram and optical microscopy [27]. However, no data were presented on Raman or SEM-EDS to verify PAN deposition nor were TGA studies performed on the electrodeposit. Further studies to tailor film thicknesses or test electrodes in a lithium half-cell were also absent. Work within this article provides all this additional information as proof of concept for PAN electrodepositions onto practical graphite battery electrodes within lithium half-cells.

Two techniques were used for the electrodeposition of PAN: cyclic voltammetry and chronoamperometry. Cyclic voltammetry scans were performed to understand the electrochemistry behind the depositions, using a scan rate of 50 mV s^{-1} in the potential window from 0.0 to −3.0 V vs. Ag. Chronoamperometry studies were utilised to accurately modify the film thickness of the electrodeposits; several different potentials were used to identify the electrochemical behaviour over a potential range, then timescales were varied to achieve the required film thicknesses. Firstly, potential values applied were varied between −0.7 and −3.0 V vs. Ag, with an electrodeposition time of 100 s. Then, different time values were subsequently studied using an applied potential of −3.0 V vs. Ag.

Electrode samples were air-dried following electrodepositions for 24 h. They were then washed in ethanol for 10 min to remove the remaining electrolyte and oligomers, followed by further air drying for 24 h. The mass of electrodes was recorded before and after deposition, and film thicknesses were measured using a digital micrometer.

2.4. Lithium Half-Cell Assembly and Electrochemical Testing

Electrodes of 11 mm diameter were punched from the 14 mm diameter electrodeposited PAN of the graphite electrode substrates with a precision punch (EL-CELL, EL-Cut). The battery graphite electrode (either uncoated or coated with electrodeposited PAN) was used as the working electrode in Swagelok cells. The working electrodes were positioned against counter-reference electrodes of lithium metal foil of 11 mm diameter in a half-cell arrangement. Two 12 mm diameter glass fibre separators were sandwiched between the working and counter-reference electrodes and wetted with 200 μL of LP57 electrolyte. Copper plungers were used for both electrodes as current collectors in the Swagelok cell. Room temperature studies were performed with Swagelok cells with PFA bodies. For elevated temperature studies, Swagelok cells of polymer materials can expand at temperature, and this may affect the seal of the ferrules within the cell and ingress of air. Hence, Swagelok cells of aluminium were used at higher temperatures. Aluminium Swageloks were lined with 125 μm thick FEP film to insulate the inner cell body and prevent a short-circuit between the electrodes.

The Swagelok cells were assembled under the argon environment of a Belle Technology Ltd. glovebox. A Buchi glass oven B-585 was used to dry electrode and separator materials at 120 °C under vacuum for 24 h. These materials were then transported to the glovebox. The same Buchi ovens were used to dry graphite battery composite electrodes with PAN films at ambient temperature under vacuum for an extended time of 72 h; this was done at ambient temperature to avoid any annealing or thermal deformation of the PAN films.

Constant current cycling of the Swagelok cells was performed using either a BCS-805 battery cycler with BT-Lab software or an SP-150 potentiostat (Biologics, Grenoble, France) with EC-Lab software. Current values were kept to a C-rate of C/10. Room temperature studies were repeated for 10 cycles and the potential window was kept within 0.005 and 1.5 V vs. Li/Li^+ in half-cells. The temperature was maintained at 25 °C by making use of an environmental chamber (Memmert, Schwabach, Germany, IPP 55 Plus), and the cells were placed inside the chamber as cycling was performed.

Elevated temperatures were achieved by employing a Genlab Classic oven (Agar Scientific, Stansted, UK, MINO/6), allowing for the cycling of cells at set temperatures. To prevent interference with the electric signal due to the high temperatures and equipment, a coaxial cable insulated with FEP polymer was employed for reliable elevated temperature performance up to 200 °C. Two constant current charge-discharge cycles were performed at each temperature interval at a current C-rate of C/10. The intervals of temperature employed were: 25, 60, 80, 100, and 120 °C. A K-type thermocouple was used to monitor the temperature of the cycling chamber.

2.5. Characterisation

A Philips XL30 ESEM was used to take scanning electron micrographs (SEM images) of the electrodeposited PAN with energy-dispersive X-ray spectroscopy (EDX) from a Thermo Fisher Scientific (Waltham, MA, USA) Ultra Dry 10 mm^2 detector. Samples were mounted onto stubs and secured with double-backed sticky conductive carbon tape and then transferred to the microscope chamber. Secondary electron imaging captured sample topography and EDX measurements were used to identify ratios of carbon and nitrogen within the polymer films.

A 785 nm Renishaw inVia confocal Raman microscope was employed to take the Raman spectra of the polymer films and measurements were taken between 3500 and 150 cm^{-1} where characteristic PAN peaks are identifiable [28,29]. Laser power was set to 1% and data was gathered over a total of three accumulations. Polymer films that had been electrodeposited onto graphite battery electrode substrates were placed under the Raman laser source.

A Netzsch TG 209 F1 Libra was used for thermogravimetric analysis (TGA) studies. Samples were powdered and placed within alumina crucibles that would then be transferred to the TGA chamber. A flow of argon gas at 20 mL min^{-1} maintained an inert atmosphere within the TGA chamber. A temperature gradient of 2 °C min^{-1} was applied from 25 to 250 °C. TGA studies were performed for the materials of the graphite-composite electrode inks: graphite powder active material, carbon black powder conductive additive, and PVDF binder. TGA was performed with a commercially available PAN powder and compared with PAN electrodeposited onto a copper substrate. Samples of the PAN electrodeposit were scratched from the substrate surface for transfer to the TGA alumina crucibles.

3. Results and Discussion

3.1. Formation of Polymer Coating on Graphite Battery Electrode by Electrodeposition

To demonstrate the concept of employing an electrodeposited polymer coating with thermal shutdown properties to improve battery safety, we selected the electrodeposition of polyacrylonitrile (PAN). PAN electrodeposition onto a variety of electrode materials was demonstrated in the 1980s by Lécayon and co-workers [30–32], with electrodeposition being possible on surfaces provided they were electronically conductive. It was, therefore, a reasonable assumption that PAN electrodeposition would be possible onto a range of anode and cathode materials provided the material itself is conductive or part of a conductive composite. More recently, Lacey and co-workers demonstrated the application of the electrodeposition of PAN as a method to form the separator in thin-film batteries [28] and in 3D batteries [29]. It was shown that the electrodeposition enabled the fast formation of uniform polymer thin films and that the amount of polymer deposited could be controlled by the charge involved in the electrodeposition process. It was also shown that the electrodeposition was facilitated by the presence of O_2, the importance of which was attributed to the formation of superoxide species from the reduction of O_2; these superoxide species then initiated the polymerisation of acrylonitrile [29]. Superoxide anions are nucleophiles that react with Michael acceptors that are sufficiently active (in acrylonitrile this is provided by the electron-withdrawing vinyl group). The polymerisation of acrylonitrile via the electronically reduced superoxide species is shown in Figure 1.

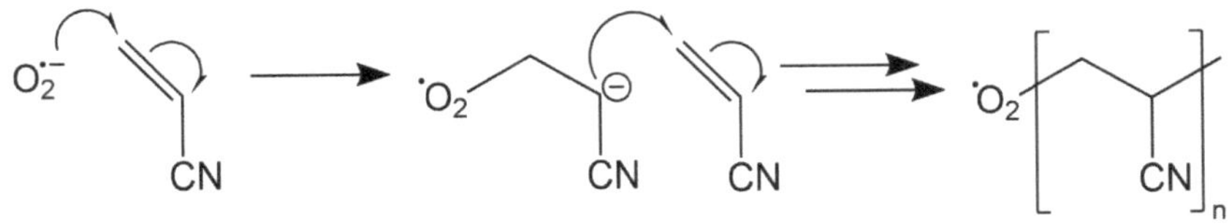

Figure 1. Polymerisation of acrylonitrile via an electro-reduced superoxide anion species. Based on information from [29].

In our work, we employ the conditions reported by Lacey and co-workers for the electrodeposition of PAN in the presence of O_2 [29]. Following their work, we employ acrylonitrile as the solvent as well as the monomer for the formation of PAN, since this facilitates the rapid deposition of the polymer. However, in our work, we employ a graphite battery electrode as the substrate for the electrodeposition rather than glassy carbon or gold disc electrode materials. Moreover, we employ an Ag pseudo-reference electrode instead of an Ag/Ag^+ electrode, to avoid possible Ag^+ contamination.

Figure 2 shows the cyclic voltammogram obtained with the graphite battery electrode in an O_2-saturated acrylonitrile solution. The graphite battery electrode was made by tape casting a viscous ink containing graphite powder, carbon conductive additive, and PVDF binder in a 94:3:3 mass ratio. The voltammogram in Figure 2 shows a peak current at potentials close to -1.5 V vs. Ag, in agreement with the previous work by Lacey and co-workers [29]. The current then decreases markedly and remains low in the following cycles as expected for a passivated electrode surface, again in agreement with previous work [29]. The electrodeposited PAN is an insulator, but when impregnated with electrolyte, as in the present case, PAN becomes a gel-polymer electrolyte [33–37]. The decay in current with time (that is, with each cycle) in the cyclic voltammetry experiments is thus due to the increased ionic resistance that is built upon by the formation of the polymer coating on the electrode. A photo of the electrode after the deposition is included in Figure 2, showing that the polymer coating is transparent. The mass of the PAN electrodeposit in this experiment was 9.3 mg cm^{-2}.

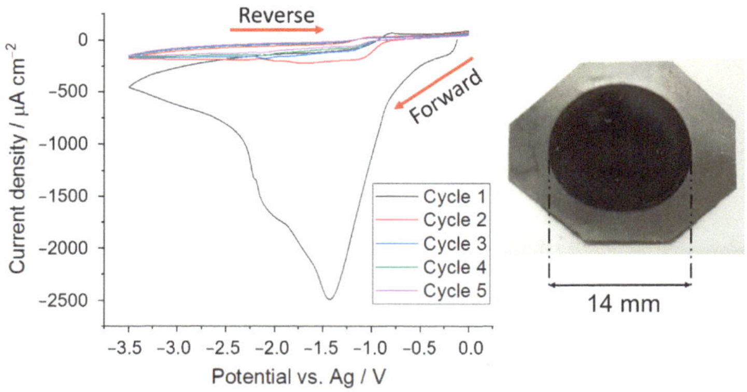

Figure 2. (**Left**) Cyclic voltammogram for the electrodeposition of PAN onto a graphite battery electrode of 14 mm diameter using acrylonitrile electrolyte (0.05 mol dm^{-3} $[Bu_4N][ClO_4]$) saturated with O_2. Deposition was performed at a scan rate of 50 mV s^{-1} from 0 to −3.5 V vs. Ag for 5 cycles. (**Right**) Photo of the graphite electrode after the PAN electrodeposition, showing that PAN is transparent.

The polymer coating on the graphite battery electrode was further characterised by SEM, EDX, and Raman measurements. The SEM image in Figure 3 shows that the polymer film exhibits distinct cracks, which are likely to be formed in the drying process prior to the introduction of the electrode in the SEM (after it has been extracted from the electrochemical cell). The carbon-to-nitrogen ratio measured by EDX (70 wt.% and 30 wt.%, respectively) is close to what is expected for polyacrylonitrile (68 wt.% and 26 wt.%).

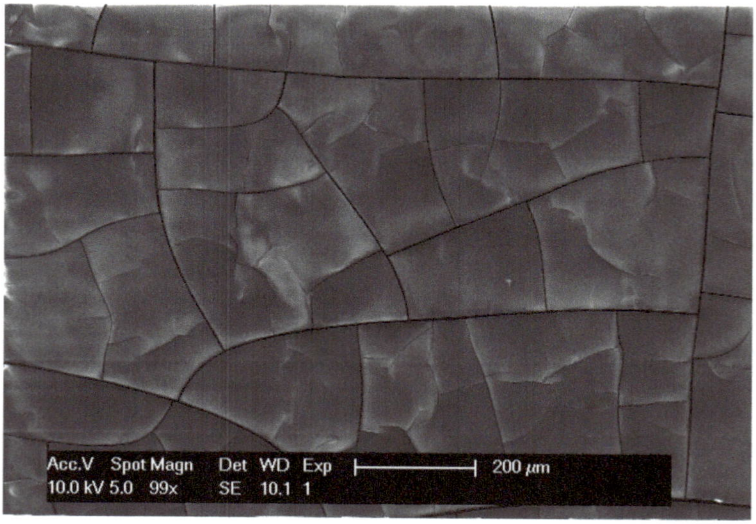

Figure 3. SEM image of electrodeposited PAN on a graphite battery electrode. The electrodeposition conditions are described in Figure 2.

The Raman spectra of the polymer coating on the graphite battery electrode further confirm that the polymer formed is PAN, as shown in Figure 4. The characteristic signals from Raman for the C≡N group at 2250 cm^{-1} and C-H alkyl groups at 3000 cm^{-1} in PAN [38] are clearly observed. In addition, the characteristic Raman signals of graphite at 1600 cm^{-1} and 1350 cm^{-1} [39] are weaker for the PAN-coated graphite electrode, compared to the uncoated one, as expected for the lower penetration of the Raman laser to the graphite substrate in the presence of the polymer coating. The Raman spectra for electrodeposited PAN on graphite also show additional features at lower wavenumber that are likely due to impurities, as previously found [29].

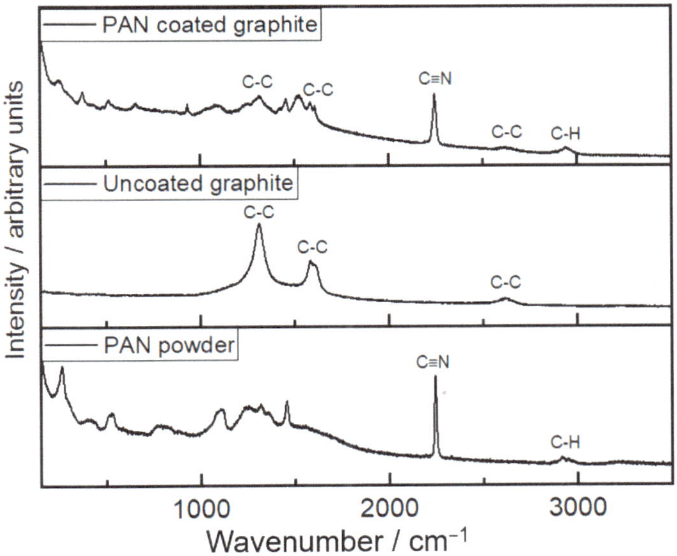

Figure 4. Raman spectra of electrodeposited PAN on a graphite battery electrode (**top**), the uncoated graphite battery electrode (**middle**), and PAN powder purchased from Sigma-Aldrich (**bottom**). The electrodeposition conditions for PAN are described in Figure 2.

Finally, the TGA of the PAN electrodeposit was performed to evaluate its thermal stability. For that purpose, a PAN electrodeposition was performed on a copper substrate to avoid contamination by the graphite battery electrode materials, after which the PAN electrodeposit was scratched off and then characterised by TGA. As shown in Figure 5, the loss of mass from the PAN electrodeposit was larger than that of a commercial PAN powder, and also larger than the variation observed for any of the components that make a graphite battery electrode (graphite, conductive carbon additive, and PVDF binder). No substantial chemical impurities were identified in the PAN electrodeposit with the Raman in Figure 4 and so it was likely that the loss of mass was from materials of similar chemical composition. This is likely to be shorter chain oligomers with boiling points closer to the 77 °C boiling point of the acrylonitrile monomers from which the PAN was electrodeposited; this results in the greater mass loss observed past 120 °C. However, mass loss for the electrodeposited PAN up to 120 °C was only 0.49% of the initial mass, and even at 150 °C the change was only 1.1%. It was thus concluded that the thermal stability of the electrodeposited PAN was sufficient up to 120 °C and in all subsequent experiments the temperature was kept ≤120 °C to avoid complications in the interpretation of the results due to PAN decomposition.

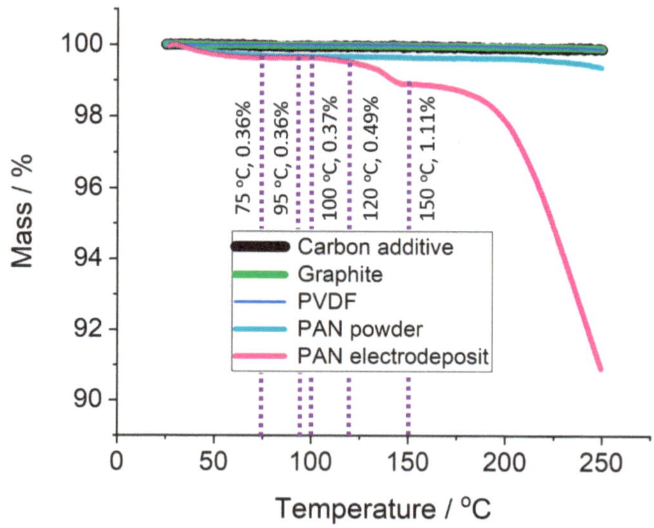

Figure 5. Thermogravimetric analysis of the electrodeposited PAN, commercial PAN power, and the components of the graphite battery electrode (graphite, carbon conductive additive, and PVDF binder). The temperature was ramped at 2 °C min^{-1} from 25 to 250 °C. The purple dotted lines give key temperatures and mass loss of the PAN electrodeposited sample at each temperature. PAN was electrodeposited on copper at −3.0 V vs. Ag for 100 s.

3.2. Controlling the Thickness of the Polymer Coating by Chronoamperometry

After the characterisation of the electrodeposits of PAN on graphite battery electrodes, from which it was confirmed that the electrodeposits were made of PAN, the next step in the development of the method was to optimise electrochemical conditions to control the thickness of the PAN electrodeposits. First, the effects of electrodeposition potential were investigated for a constant deposition time of 100 s; Figure 6 shows the masses of PAN electrodeposits obtained in these experiments (masses were estimated from the mass difference of the dry graphite battery electrode before and after the electrodeposition). It was observed that the electrodeposited mass was very small for potentials higher than −1.0 V vs. Ag. At lower potentials, the mass increases significantly reaching a roughly constant value, within the reproducibility of the measurements, for potentials lower than −1.3 V vs. Ag.

After the study of the effects of the electrodeposition potential, the next experiments were performed by varying the electrodeposition time while keeping the electrodeposition potential constant and equal to −3.0 V vs. Ag. The variation of the mass of the PAN electrodeposits as a function of the charge passed during the experiments is shown in Figure 7. Fitting the experimental data to a straight line with zero intercepts gives a slope of $\Delta m/\Delta Q = 0.049 \pm 0.002$ mg mC^{-1}. From this value of the slope, it is possible to calculate the number of moles of acrylonitrile monomer units deposited per mole of electrons passed using the following equation:

$$n = \frac{\Delta m}{\Delta Q}\frac{F}{M_r}$$

where M_r is the molar mass of the acrylonitrile (CH_2=CH-CN) monomer (M_r = 53.06 g mol^{-1}) and F is Faraday's constant (F = 96,485 C mol^{-1}). In present studies, $n = 86 \pm 6$; this is similar to previous work by Lacey and co-workers that reported values between 70 and 80 [29]. Since the electrochemical polymerisation is initiated by a one-electron process (forming an acrylonitrile radical that polymerises upon reaction with additional acrylonitrile molecules) the value of n calculated above represents an estimation of the degree of polymerization or the average number of acrylonitrile monomer units present in the PAN polymer chain: -(CH_2=CH-CN)$_n$-. The average molecular weight of the electrodeposited PAN is estimated to be close to 5000 g mol^{-1}, which is lower than that of typical PAN polymers [40,41]. This may explain the larger mass loss observed for the electrodeposited PAN by TGA in Figure 5.

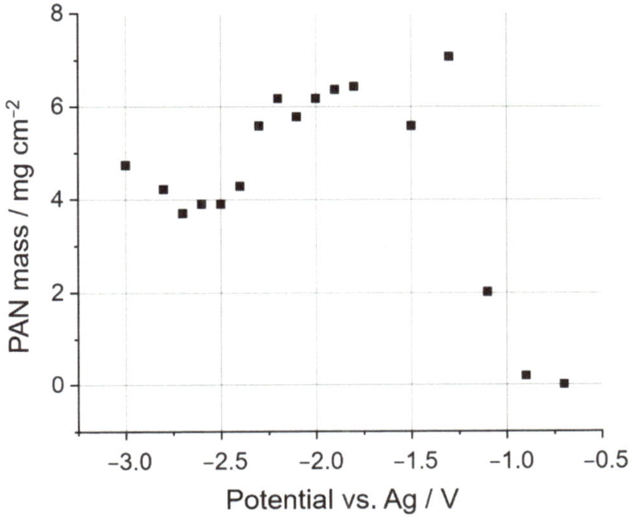

Figure 6. Mass of 14 mm diameter PAN electrodeposits on graphite battery electrodes via the application of a constant electrodeposition potential for 100 s via chronoamperometry experiments, plotted as a function of the electrodeposition potential. All other conditions are as in Figure 2.

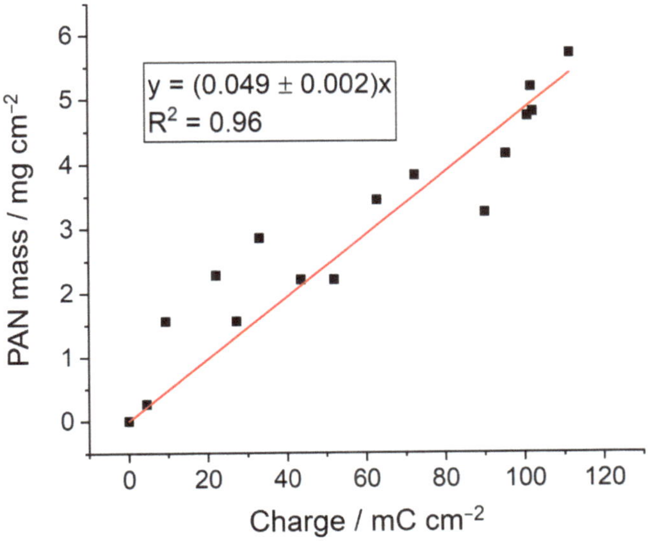

Figure 7. Mass of 14 mm diameter PAN electrodeposits on graphite battery electrodes via the application of a constant electrodeposition potential of −3.0 V vs. Ag for various timescales, between 0.5 and 100 s, and plotted as a function of the charge passed. All other conditions are as in Figure 2. A linear fit with a fixed zero intercept is included in the graph, and the corresponding equation is also included.

The thickness of the PAN electrodeposit on the graphite electrode was estimated by measuring the thickness of the dry electrode before and after the electrodeposition. The results are shown in Figure 8 as a function of the PAN mass deposit. The fit to the experimental data with a straight line with zero intercepts gives a value of the slope of 8.9 ± 0.5 μm mg^{-1} cm^2 or 0.89 ± 0.05 cm^3 g^{-1}. The inverse of the value of the slope is, thus, the density of the dry electrodeposit; in this case: 1.12 ± 0.05 g cm^{-3}. This value is slightly lower than the typical value of 1.2 g cm^{-3} reported for PAN polymers [41], thus suggesting that the electrodeposited PAN has some porosity, as intended to enable good impregnation with the liquid electrolyte and enable fast ion transport.

For the practical purposes of tailoring PAN electrodeposit thickness onto the graphite electrodes gradients were calculated without the dependence on area by multiplying by the value 154 mm^2, the area of the 14 mm diameter depositions. By converting the previously stated relations between thickness, mass, and charge, the gradient of film thickness against charge was found to increase by 0.28 ± 3 μm mC^{-1}. This uncertainty is attributed in part to the surface roughness of the films but is also impacted by the instrumental errors of ± 1 μm for the micrometer employed. The film thickness was, therefore, more accurate for thicker films.

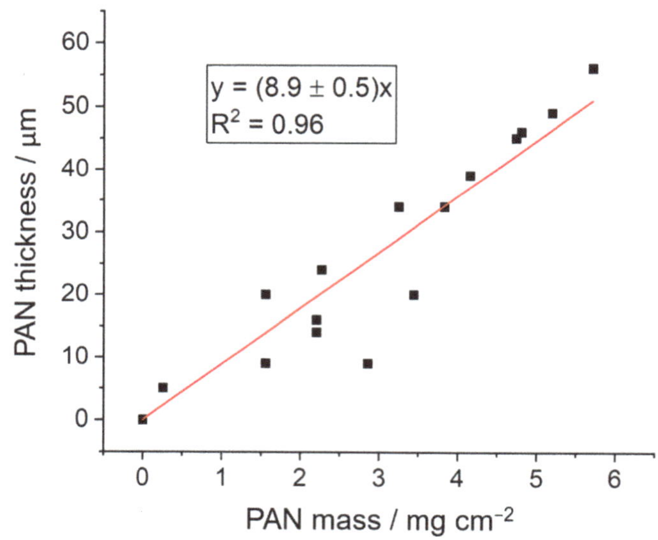

Figure 8. Thickness of 14 mm diameter PAN electrodeposits on graphite battery electrodes, plotted as a function of the mass of PAN electrodeposits. All other conditions are as in Figure 2. A linear fit with a fixed zero intercept is included in the graph, and the corresponding equation is also included.

3.3. Application of PAN-Coated Graphite Electrodes for Thermal Runaway Prevention

First, the electrochemical performance of the PAN-coated graphite battery electrodes was tested in lithium half-cells at 25 °C and the results were compared to the uncoated graphite battery electrode in Figure 9. It is seen that the capacity of the graphite electrode decreases substantially in the presence of the electrodeposited PAN coating. The uncoated electrode reached delithiation capacities of 366–361 mA h g^{-1} in the first two cycles, whereas in the presence of the electrodeposited PAN coating, the delithiation capacity was 193–223 mA h g^{-1} in the first two cycles. The drop in capacity from the uncoated to the PAN-coated samples is likely due to an increase in internal ionic resistance introduced by the PAN coating. The experiments in Figure 9 show the results with a PAN coating of 32 μm, but it will be shown below (Figure 10) that the decrease in capacity, with respect to the uncoated electrode, is lower with thinner coatings. Figure 9 also shows that, for the PAN-coated battery electrode, the capacity increases in cycle 2, with respect to cycle 1. This could be due to the time required for the impregnation/swelling of the electrolyte in the PAN deposit, which could be complete in cycle 2 but only partial in cycle 1.

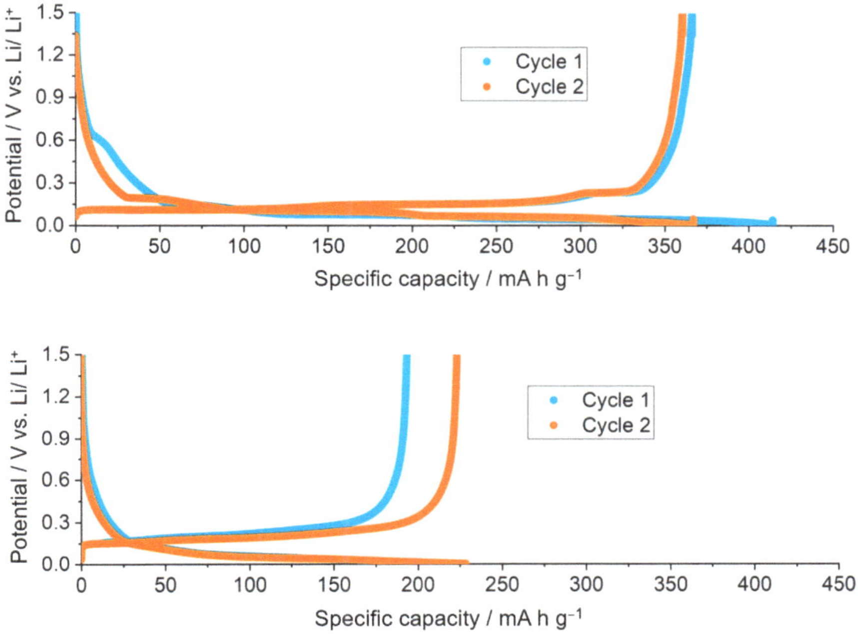

Figure 9. Results of the galvanostatic cycling at C/10 of uncoated (**top**) and PAN-coated (**bottom**) graphite electrodes in lithium half-cells containing LP57 electrolyte (1 M $LiPF_6$ in EC:EMC 3:7) at 25 °C. Conditions of PAN electrodeposition were using chronoamperometry at −3.0 V vs. Ag as in Figures 7 and 8. The thickness of the PAN electrodeposit was 32 μm.

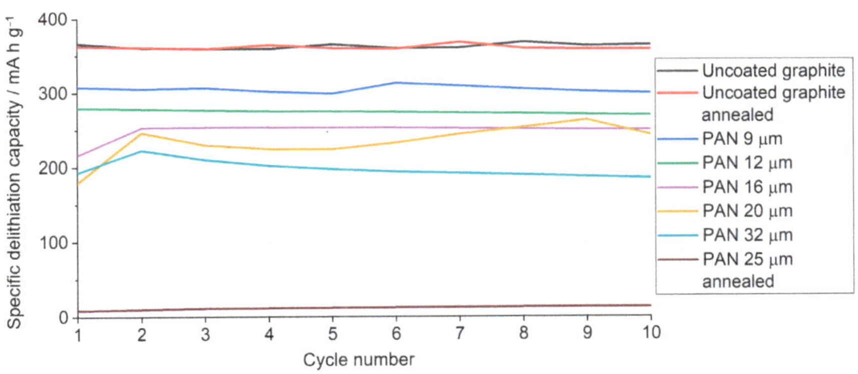

Figure 10. Delithiation capacity of uncoated and PAN-coated battery electrodes as a function of the cycle number, as measured in lithium half-cells. All conditions as in Figure 9.

A systematic investigation of the effect of the thickness of the PAN electrodeposit on graphite was then undertaken. Figure 10 shows the variation of the delithiation capacity measured in lithium half-cells for uncoated and PAN-coated graphite battery electrodes, with PAN coatings of different thicknesses, as indicated. Consistent with the results in Figure 9, the PAN-coated electrodes experience an increase in capacity when the cycle number increases from 1 to 2, and then the capacity remains roughly constant in subsequent cycles. It is also seen that, as the thickness of the PAN deposit decreases, the capacity approaches that of the uncoated graphite electrode.

Figure 10 also includes the results obtained with uncoated and PAN-coated electrodes that were annealed, prior to the introduction of the electrodes in the cell, at 120 °C for 144 h under an argon environment. It is observed that the annealing does not affect the capacity of the uncoated graphite electrode, whereas it decreases dramatically the capacity of the PAN-coated graphite electrode. The former results indicate that an elevated temperature environment has little effect on the uncoated graphite itself and so any change in capacity would be due to the presence of the PAN coating. The latter results indicate that the PAN coating can effectively suppress the electrochemical activity of the graphite electrode at high temperatures, as is required to suppress the electrochemical degradation reactions that sustain or exacerbate the thermal runaway process. The TGA data in Figure 5 show that the electrodeposited PAN was thermally stable until 120 °C and changes in capacity are therefore unlikely to be a result of PAN thermal degradation. The capacity was below 5% of the theoretical capacity for the PAN annealed sample and this could not have been caused by the thermal degradation of other cell materials since cycling was performed at room temperature and annealing of the PAN coating occurred prior to cell assembly. As such, it was concluded that the capacity loss was due to the rearrangement of polymer chains during annealing and not due to thermal degradation. PAN has a glass-transition temperature of around 95 °C, allowing the rearrangement of polymer chains at the annealing temperatures of 120 °C. The annealing time used in these experiments is too long to be relevant to prevent a thermal runaway event, but below we show that the suppression of the electrochemical activity by the PAN electrodeposit, as the cell heats, is also observed with short heating times and more moderate temperatures.

Figure 11 shows the evolution of the delithiation capacities of uncoated and PAN-coated graphite battery electrodes as cell temperature is systematically increased from 25 to 60, 80, 100, and, finally, 120 °C. The electrochemical measurements were carried out, as in Figures 9 and 10, at a C-rate of C/10 and the measurements at each temperature started as soon as the temperature inside the oven and the cells reached the target temperature (which took around 5 min). It is seen that for the uncoated graphite electrode, the delithiation capacity remains high at all studied temperatures, this means that the graphite electrode would remain active in a thermal runaway event, which is undesirable. On the other hand, for the PAN-coated samples, it is seen that the capacities increase when the temperature is increased from 25 to 60 °C, likely a result of faster lithium-ion conduction through the PAN film with increased temperature. PAN-coated sample capacities then markedly decrease when the temperature is increased further. The latter suppression of the graphite electrochemical activity is tentatively ascribed to the thermal expansion and rearrangement of the PAN polymer chains in an annealing process as the temperature increases; this would then decrease the porosity of the PAN film and, thus, also decrease the lithium-ion conductivity, similarly to the operation mechanism of thermal shut-down separators. This suppression of electrochemical activity is more marked for the thicker PAN coatings, where a coating of 24 μm is seen to completely suppress the capacity at 100 °C. Unfortunately, such thick PAN coating compromises the capacity at room temperature, which is only around 49% of the capacity that is seen with the uncoated graphite electrodes.

Clearly, further optimisation of the ion-conduction properties of the PAN electrodeposit on graphite battery electrodes is necessary in order to make this thermal protection approach advantageous for commercial applications, which require high capacities. For example, increasing the lithium-ion concentration and tuning the solvent of the electrolyte to produce a good plasticiser effect to form a high conductivity PAN-based gel electrolyte appears a promising avenue that could be explored in further work; other types of electrodeposited polymers could be studied as well. The present study demonstrates proof of the principle of using an electrodeposited polymer to effectively suppress the electrochemical activity of a battery electrode, which could be employed to reinforce other protection measures against thermal runaway.

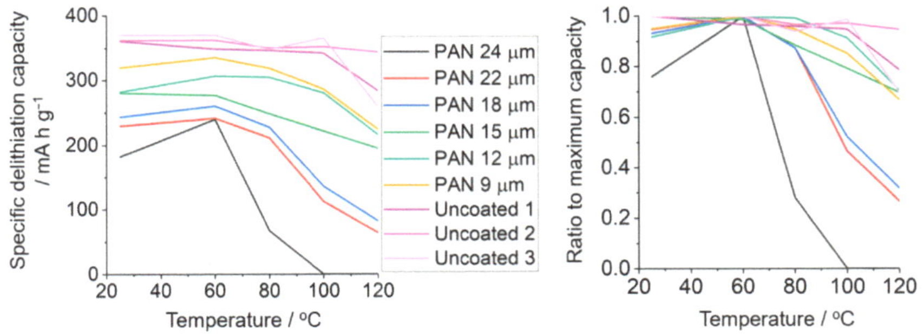

Figure 11. Delithiation capacity of uncoated and PAN-coated battery electrodes as a function of the cell temperature. Two galvanostatic cycles were performed at each temperature, and the cell temperature was controlled by hosting the cells inside an oven. All conditions as in Figure 9.

Competitive technologies presented in other literature share similar difficulties. For instance, a 4 μm coating of polyethylene microspheres onto graphite reported by Baginska and co-workers [24] resulted in capacity reduction of more than 98% at 110 °C. Ji and co-workers [18] investigated the application of poly(3-cotylthiophene) as a coating for lithium cobalt oxide cathodes. These materials were promising in their shutdown response, dropping the capacity at 90 °C. Comparatively, commercial Celgard shutdown separators see shutdown from 110 to 130 °C. All these materials share the same working principle of experiencing a phase transition that results in irreversible structural changes to the polymer; ionically conductive pores then seal resulting in the shutdown.

The temperature of the shutdown for the electrodeposited PAN films was promising and competitive with other technologies, occurring in the range from 80 to 100 °C. The simple one-step electrodeposition that requires only a dry room to operate also allows for simplified manufacturing routes, especially compared to other stated technologies such as the poly(3-cotylthiophene) coatings and polyethylene microspheres that require several stages of mixing and extraction or Celgard that requires forming and calendaring several laminated layers. Future work is required to improve the ionic conductivity of the thicker PAN coatings to alleviate the significant capacity loss experienced. Shutdown to negligible capacities at 100 °C was experienced for 24 μm samples, but at ambient conditions, capacities of only around 182 mA h g^{-1} were found. An alternative avenue to improve shutdown for thinner films is therefore also advisable. To this end, it is suggested that future work pursue a deeper understanding of the PAN pore microstructure and the interplay with liquid electrolytes. For example, Yaya Wang and co-workers [42] described a surface treatment strategy for current collectors to etch them and provide a porous electrode with improved current density whilst maintaining the lithium-ion concentration over the surface of the electrode. Tao Wang and co-workers [43] showed that 3D-holey PAN-graphene composite improved lithium-ion conduction by shortening the ion transport path. This was highlighted to be particularly useful when improving ionic conduction through thicker films.

4. Conclusions

PAN was electrodeposited onto a graphite battery electrode, producing a transparent polymer coating that could be identified as PAN based on the characteristic Raman signatures and on the chemical composition estimated by EDS. TGA measurements showed that the polymer coating expressed good thermal stability up to a temperature of 120 °C. Using chronoamperometry, it was discovered that the polymer film thickness could be controlled by the amount of charge passed and was quantifiable at 0.28 ± 3 μm mC^{-1}. The selected method of electrodeposition was carried out in the presence of O_2, and thus, it could be easily translated to a large-scale process with open-air cells to be carried out in a dry room. The simple one-step manufacturing route to produce a uniform electrodeposited layer is highly promising, especially compared to other multi-step commercial alternatives. Differing two-dimensional deposition areas are also possible by altering the channel of the deposition cell.

The PAN-coated graphite electrode was characterised by lithium half-cells. It was observed that increasing the thickness of the PAN coating decreased the lithium insertion/extraction capacities. To overcome this problem, further formulations of polymer and liquid electrolytes could be studied. On the other hand, increasing thickness produced an increasing suppression of the capacity at high temperatures resulting in a shutdown effect. This effectively prevents the transport of ions between electrodes removing a source of heat generation from exothermic charge/discharge reactions, similar in performance to commercially available shutdown separators. This suppression of the electrochemical activity would also stop the interfacial degradation reactions that trigger the process of thermal runaway, and thus, the use of an electrodeposited polymer coating is identified as a promising approach to enhance battery safety.

Suppression of capacity was found to be linked to the annealing of the PAN coating around its glass-transition of 95 °C, with shutdown occurring between 80 and 100 °C, likely due to rearrangements of the polymer microstructure preventing infiltration of lithium-ions. The competitiveness of the PAN coatings with current technology is significant, with shutdown responses occurring below 100 °C and a demonstrable complete shutdown. As such, enhancement of the pore structure is an area of improvement that could yield significant results producing greater shutdown responses for thinner polymer films that have inherent greater capacity. Improvements to capacity whilst maintaining the high shutdown response for thicker films would also allow for possible application in the industry.

An alternative application of the same PAN coatings could be to improve the energy density of the cells by substituting the need for a separator within the system, in essence, the electrodeposited PAN would act as the separator between the two electrodes. The tailorable mass and thickness of the PAN would also be of benefit, potentially providing a protective coating of lower mass and thickness than a conventional separator, which would further improve the energy density. Automated shutdown may also be achievable for thinner films with lesser shutdown effects via external battery management systems to measure internal resistance during cell operation. Fluctuations in resistance could be detected allowing the cells to be isolated.

Author Contributions: Conceptualization, J.P.C.A., N.G.-A., A.L.H. and D.K.; methodology, J.P.C.A. and M.M.; manuscript writing, J.P.C.A.; review and editing, J.P.C.A., N.G.-A. and A.L.H. All authors have read and agreed to the published version of the manuscript.

Funding: The work was supported by EPSRC through an Industrial Strategy Challenge Fund grant (EP/R021295/1) and an early career Fellowship to NGA (EP/N024303/1).

Institutional Review Board Statement: Not applicable.

Informed Consent Statement: Not applicable.

Data Availability Statement: Raw data used to produce the figures in this manuscript are freely available at https://doi.org/10.5258/SOTON/D2617.

Conflicts of Interest: The authors declare no conflict of interest.

References

1. Feng, X.; Ouyang, M.; Liu, X.; Lu, L.; Xia, Y.; He, X. Thermal runaway mechanism of lithium ion battery for electric vehicles: A review. *Energy Storage Mater.* **2018**, *10*, 246–247. [CrossRef]
2. Sun, P.; Bisschop, R.; Niu, H.; Huang, X. A Review of Battery Fires in Electric Vehicles. *Fire Technol.* **2020**, *56*, 1361–1410. [CrossRef]
3. Börger, A.; Mertens, J.; Wenzl, H. Thermal runaway and thermal runaway propagation in batteries: What do we talk about? *J. Energy Storag.* **2019**, *24*, 100649. [CrossRef]
4. Liu, Q.; Hu, Y.; Yu, X.; Qin, Y.; Meng, T.; Hu, X. The pursuit of commercial silicon-based microparticle anodes for advanced lithium-ion batteries: A review. *Nano Res. Energy* **2022**, *1*, e9120037. [CrossRef]

5. Sui, Y.; Guan, J.; Li, K.; Feng, Y.; Peng, S.; Maximov, M.; Liu, Q.; Yang, J.; Geng, H. Synergy of oxygen defects and structural modulation on titanium niobium oxide with a constructed conductive network for high-rate lithium-ion half/full batteries. *Inorg. Chem. Front.* **2023**, *10*, 2304–2313. [CrossRef]
6. Liu, K.; Liu, Y.; Lin, D.; Pei, A.; Cui, Y. Materials for lithium-ion battery safety. *Sci. Adv.* **2018**, *4*, eaas9820. [CrossRef]
7. Kong, L.; Li, C.; Jiang, J.; Pecht, M. Li-Ion Battery Fire Hazards and Safety Strategies. *Energies* **2018**, *11*, 2191. [CrossRef]
8. Chombo, P.; Laoonual, Y. A review of safety strategies of a Li-ion battery. *J. Power Sources* **2020**, *478*, 228649. [CrossRef]
9. Zhang, L.; Li, X.; Yang, M.; Chen, W. High-safety separators for lithium-ion batteries and sodium-ion batteries: Advances and perspective. *Energy Storag. Mater.* **2021**, *41*, 522–545. [CrossRef]
10. Feng, X.; Ren, D.; He, X.; Ouyang, M. Mitigating Thermal Runaway of Lithium-Ion Batteries. *Joule* **2020**, *4*, 743–770. [CrossRef]
11. Arora, P.; Zhang, Z. Battery separators. *Chem. Rev.* **2004**, *104*, 4419–4462. [CrossRef] [PubMed]
12. Huang, X. Separator technologies for lithium-ion batteries. *J. Solid State Electrochem.* **2011**, *15*, 649–662. [CrossRef]
13. Orendorff, C. The Role of Separators in Lithium-Ion Cell Safety. *Electrochem. Soc. Interface* **2012**, *21*, 61. [CrossRef]
14. Zhang, M.; Fop, S.; Kramer, D.; Garcia-Araez, N.; Hector, A. A La and Nb co-doped $BaTiO_3$ film with positive-temperature-coefficient of resistance for thermal protection of batteries. *J. Mater. Chem. A* **2022**, *10*, 11587–11599. [CrossRef]
15. Xia, L.; Li, S.; Ai, X.; Yang, H.; Cao, Y. Temperature-sensitive cathode materials for safer lithium-ion batteries. *Energy Environ. Sci.* **2011**, *4*, 2845–2848. [CrossRef]
16. Zhong, H.; Kong, C.; Zhan, H.; Zhan, C.; Zhou, Y. Safe positive temperature coefficient composite cathode for lithium ion battery. *J. Power Sources* **2012**, *216*, 273–280. [CrossRef]
17. Xia, L.; Zhu, L.; Zhang, H.; Ai, X. A positive-temperature-coefficient electrode with thermal protection mechanism for rechargeable lithium batteries. *Chin. Sci. Bull.* **2012**, *57*, 4205–4209. [CrossRef]
18. Ji, W.; Wang, F.; Liu, D.; Qian, J.; Cao, Y.; Chen, Z.; Yang, H.; Ai, X. Building thermally stable Li-ion batteries using a temperature-responsive cathode. *J. Mater. Chem. A* **2016**, *4*, 11239–11246. [CrossRef]
19. Li, H.; Wang, F.; Zhang, C.; Ji, W.; Qian, J.; Cao, Y.; Yang, H.; Ai, X. A temperature-sensitive poly(3-octylpyrrole)/carbon composite as a conductive matrix of cathodes for building safer Li-ion batteries. *Energy Storag. Mater.* **2019**, *17*, 275–283. [CrossRef]
20. Zhang, H.; Pang, J.; Ai, X.; Cao, Y.; Yang, H.; Lu, S. Poly(3-butylthiophene)-based positive-temperature-coefficient electrodes for safer lithium-ion batteries. *Electrochim. Acta* **2016**, *187*, 173–178. [CrossRef]
21. Chen, Z.; Hsu, P.; Lopez, J.; Li, Y.; To, J.; Liu, N.; Wang, C.; Andrews, S.; Liu, J.; Cui, Y.; et al. Fast and reversible thermoresponsive polymer switching materials for safer batteries. *Nat. Energy* **2016**, *1*, 15009. [CrossRef]
22. Li, H.; Zhang, X.; Zhang, C.; Cao, Y.; Yang, H.; Ai, X.; Zhong, F. Building a Thermal Shutdown Cathode for Li-Ion Batteries Using Temperature-Responsive Poly(3-Dodecylthiophene. *Energy Technol.* **2020**, *8*, 7. [CrossRef]
23. Li, M.; Shi, Y.; Gao, H.; Chen, Z. Bio-Inspired Nanospiky Metal Particles Enable Thin, Flexible, and Thermo-Responsive Polymer Nanocomposites for Thermal Regulation. *Adv. Funct. Mater.* **2020**, *30*, 23. [CrossRef]
24. Baginska, M.; Blaiszik, B.; Merriman, R.; Sottos, N.; Moore, J.; White, S. Autonomic shutdown of lithium-ion batteries using thermoresponsive microspheres. *Adv. Energy Mater.* **2012**, *2*, 583–590. [CrossRef]
25. Baginska, M.; Blaiszik, B.; Rajh, T.; Sottos, N.; White, S. Enhanced autonomic shutdown of Li-ion batteries by polydopamine coated polyethylene microspheres. *J. Power Sources* **2014**, *269*, 735–739. [CrossRef]
26. Zhang, C.; Li, H.; Wang, S.; Cao, Y.; Yang, H.; Ai, X.; Zhong, F. A polyethylene microsphere-coated separator with rapid thermal shutdown function for lithium-ion batteries. *J. Energy Chem.* **2020**, *44*, 33–40. [CrossRef]
27. Allen, J.; Hector, A.; Garcia-Araez, N. Cell design for the electrodeposition of polyacrylonitrile onto graphite composite electrodes for use in lithium-ion cells. *Energy Rep.* **2021**, *7*, 15–19. [CrossRef]
28. El-Enany, G.; Lacey, M.; Johns, P.; Owen, J. In situ growth of polymer electrolytes on lithium ion electrode surfaces. *Electrochem. Commun.* **2009**, *11*, 2320–2323. [CrossRef]
29. Lacey, M.; Sosna, M.; Owen, J. An electrochemical quartz crystal microbalance study of poly(acrylonitrile) deposition initiated by electrogenerated superoxide. *Electrochim. Acta* **2013**, *29*, 23–26. [CrossRef]
30. Lécayon, G.; Bouizem, Y.; Le Gressus, C.; Reynaud, C.; Boiziau, C.; Juret, C. Grafting and growing mechanisms of polymerised organic films onto metallic surfaces. *Chem. Phys. Lett.* **1982**, *91*, 506–510. [CrossRef]
31. Boiziau, C.; Lécayon, G. Adhesion of polymers to metals: A review of the results obtained studying a model system. *Surf. Interface Anal.* **1988**, *12*, 475–485. [CrossRef]
32. Boiziau, C.; Leroy, S.; Reynaud, C.; Lécayon, G.; Le Gressus, C.; Viel, P. Elementary Mechanisms in the Interaction of Organic Molecules with Mineral Surfaces. *J. Adhes.* **1987**, *23*, 21–44. [CrossRef]
33. Abraham, K.; Alamgir, M. Li+-Conductive Solid Polymer Electrolytes with Liquid-Like Conductivity. *J. Electrochem. Soc.* **1990**, *137*, 1657. [CrossRef]
34. Huang, B.; Wnag, Z.; Li, G.; Huang, H.; Xue, R.; Chen, L.; Wang, F. Lithium ion conduction in polymer electrolytes based on PAN. *Solid State Ion.* **1996**, *85*, 79–84. [CrossRef]
35. Abraham, K.; Choe, H.; Pasquariello, D. Polyacrylonitrile electrolyte-based Li ion batteries. *Electrochim. Acta* **1998**, *43*, 2399–2412. [CrossRef]
36. Watanabe, M.; Kanba, M.; Matsuda, H.; Tsunemi, K.; Mizoguchi, K.; Tsuchida, E.; Shinohara, I. High lithium ionic conductivity of polymeric solid electrolytes. *Die Makromol. Chem. Rapid Commun.* **1981**, *2*, 741–744. [CrossRef]

37. Watanabe, M.; Kanba, M.; Nagaoka, K.; Shinohara, I. Ionic conductivity of hybrid films composed of polyacrylonitrile, ethylene carbonate, and LiClO4. *J. Polym. Sci. Polym. Phys. Ed.* **1983**, *21*, 939–948. [CrossRef]
38. Huang, Y.; Koenig, J. Raman Spectra of Polyacrylonitrile. *Appl. Spectrosc.* **1971**, *25*, 620–622. [CrossRef]
39. Tuinstra, F.; Koenig, J. Raman Spectrum of Graphite. *J. Chem. Phys.* **1970**, *53*, 1126. [CrossRef]
40. Krigbaum, W.; Kotliar, A. The molecular weight of polyacrylonitrile. *J. Polym. Sci.* **1958**, *32*, 323–341. [CrossRef]
41. Beevers, R. The physical properties of polyacrylonitrile and its copolymers. *J. Polym. Sci. Macromol. Rev.* **1968**, *3*, 113–254. [CrossRef]
42. Wang, Y.; Zhao, Z.; Zhong, J.; Wang, T.; Wang, L.; Xu, H.; Cao, J.; Li, J.; Zhang, G.; Fei, H.; et al. Hierarchically Micro/Nanostructured Current Collectors Induced by Ultrafast Femtosecond Laser Strategy for High-Performance Lithium-ion Batteries. *Energy Environ. Mater.* **2022**, *5*, 969–976. [CrossRef]
43. Wang, T.; Zhang, Q.; Zhong, J.; Chen, M.; Deng, H.; Cao, J.; Wang, L.; Peng, L.; Zhu, J.; Lu, B. 3D Holey Graphene/Polyacrylonitrile Sulfur Composite Architecture for High Loading Lithium Sulfur Batteries. *Adv. Energy Mater.* **2021**, *11*, 16. [CrossRef]

batteries
MDPI

Article

Thermal Management of Lithium-Ion Batteries Based on Honeycomb-Structured Liquid Cooling and Phase Change Materials

Tianqi Yang [1], Shenglin Su [1], Qianqian Xin [1], Juan Zeng [1,2,*], Hengyun Zhang [3,*], Xianyou Zeng [4] and Jinsheng Xiao [1,2]

1 Hubei Research Center for New Energy & Intelligent Connected Vehicle, School of Automotive Engineering, Wuhan University of Technology, Wuhan 430070, China; tqyang@whut.edu.cn (T.Y.); shenglinsu@whut.edu.cn (S.S.); xin11280429@whut.edu.cn (Q.X.); jinsheng.xiao@whut.edu.cn (J.X.)
2 Chongqing Research Institute, Wuhan University of Technology, Chongqing 401135, China
3 School of Mechanical and Automotive Engineering, Shanghai University of Engineering Science, Shanghai 201620, China
4 Shanghai Marine Diesel Engine Research Institute, Shanghai 201108, China; zengxianyou@csic711.com
* Correspondence: zengjuan1973@whut.edu.cn (J.Z.); zhanghengyun@sues.edu.cn (H.Z.)

Abstract: Batteries with high energy density are packed into compact groups to solve the range anxiety of new-energy vehicles, which brings greater workload and insecurity, risking thermal runaway in harsh conditions. To improve the battery thermal performance under high ambient temperature and discharge rate, a battery thermal management system (BTMS) based on honeycomb-structured liquid cooling and phase change materials (PCM) is innovatively proposed. In this paper, the thermal characteristics of INR18650/25P battery are studied theoretically and experimentally. Moreover, the influence of structure, material and operating parameters are studied based on verifying the simplified BTMS model. The results show that the counterflow, honeycomb structure of six cooling tubes and fins, 12% expanded graphite mass fraction and 25 mm battery spacing give a better battery thermal performance with high group efficiency. The maximum temperature and temperature difference in the battery in the optimal BTMS are 45.71 °C and 4.4 °C at the 40 °C environment/coolant, as against 30.4 °C and 4.97 °C at the 23.6 °C environment/coolant, respectively. Precooling the coolant can further reduce the maximum battery temperature in high temperature environments, and the precooling temperature difference within 5 °C could meet the uniformity requirements. Furthermore, this study can provide guidance for the design and optimization of BTMS under harsh conditions.

Keywords: lithium-ion battery; honeycomb structure; thermal management; phase change material; liquid cooling

Citation: Yang, T.; Su, S.; Xin, Q.; Zeng, J.; Zhang, H.; Zeng, X.; Xiao, J. Thermal Management of Lithium-Ion Batteries Based on Honeycomb-Structured Liquid Cooling and Phase Change Materials. *Batteries* **2023**, *9*, 287. https://doi.org/10.3390/batteries9060287

Academic Editor: Carlos Ziebert

Received: 17 April 2023
Revised: 15 May 2023
Accepted: 22 May 2023
Published: 24 May 2023

1. Introduction

Aside from material compositions, temperature is one of the most important parameters to significantly affect the performance of lithium-ion batteries. Too low a temperature will reduce the diffusion capacity of lithium ions, resulting in the attenuation of performance indicators [1]. Charging at low temperatures even leads to Li plating, which increases the degradation/aging rate [2]. Too high a temperature will cause severe side reactions inside the battery, resulting in the decline of cycle life [3]. High energy density power batteries could be adopted with more compact grouping efficiency, leading to instability and even safety issues in the battery systems, risking thermal runaway at high temperatures [4]. Therefore, it is vital to control the battery temperature within a reasonable range for its safety and high performance.

Generally speaking, the operating temperature range of lithium-ion batteries is between −20 °C and 60 °C [5]. To ensure the optimal performance and safety of batteries,

50 °C is often set as the highest temperature acceptable for battery thermal management systems (BTMSs) [6]. In addition, to prevent the occurrence of internal short circuits, the temperature distribution of BTMSs should be relatively uniform; i.e., the maximum temperature difference inside the battery should not exceed 5 °C [7]. Furthermore, 25 °C to 35 °C has been proved to be the optimal operating temperature range for lithium-ion batteries, resulting in the highest electrochemical efficiency and the lowest aging rate [8]. Other studies also show that when the battery pack works at 30 °C to 40 °C, battery life will be reduced by about 60 days for each degree of increase [9].

To control the maximum temperature and temperature difference in the battery pack, a variety of cooling methods are widely used, such as air cooling, liquid cooling, heat pipe and phase change materials (PCM), etc., whose application in practical engineering could be found in the literature [10]. The design objectives of BTMSs are different for diverse working conditions, and so are the cooling methods. Among them, air cooling is the most traditional and direct heat dissipation method, utilizing airflow to take away the generated heat of batteries. Air cooling is characterized by a simple structure and low cost, but its cooling ability is weak, so it is often used for vehicles with low maximum temperature restriction in practical engineering [11]. Recently, heat pipes have been introduced as a cooling method for power batteries because of their high heat transfer efficiency and long service life. Its liquid working medium absorbs heat at the evaporation end, flows to the condensing section under the drive of pressure difference, and releases the heat. In addition, heat pipes can also realize the circulating cooling process through the action of gravity and capillary force. However, this cooling method has not been widely developed in practical engineering applications [12] due to its high cost. However, BTMSs must have a much stronger active heat dissipation capacity in high ambient temperatures, high discharge rates and cycle conditions to prevent heat accumulation. So, other cooling methods are investigated in detail.

Liquid cooling, one of the most commonly used cooling methods in practical engineering, adopts liquid as the cooling medium to take away the heat of batteries through direct or indirect contact. Compared with air cooling, the cooling medium of liquid cooling has higher density, specific heat capacity and thermal conductivity, which can take away the heat from the surface more quickly and make the battery pack layout more compact. Due to the different shapes of batteries, the structure and arrangement of liquid cooling devices are also diverse. For square batteries, Panchal et al. [13] adopted a water-cooling BTMS and obtained the temperature distribution on the battery surface and the cooling effect of the water-cooling system by comparing and analyzing different boundary conditions at 1C and 3C discharge rates. To improve the cooling effect, Gungor et al. [14] proposed a liquid cooling plate with canopy-shaped micro-flow channels. According to the simulation results, this structure could use the lower mass flow rate coolant to reduce the maximum battery temperature and maintain better temperature uniformity. Wei et al. [15] established a numerical model of square cells with liquid cooling and analyzed factors such as the number of cells between cold plates, the thickness of cold plates, the flow rate of coolant and so on. The results showed that the cooling effect was worse when the number of cells between cold plates increased. Oppositely, the cooling effect was enhanced when the thickness of the liquid cooling plates and the flow rate of the coolant increased within a certain range. For the cylindrical battery pack, a new BTMS based on the classical serpentine liquid cooling channel was proposed in the literature [16], with a good cooling effect and compact structure. Compared with the original structure, the cooling performance was obviously improved under vibration conditions. In addition, liquid cooling tubes are commonly used because they can be well arranged in the gaps between cylindrical cells. In the literature [17], a cooling structure of longitude-arranged liquid cooling pipes surrounded by three curved contact surfaces was proposed. The influences of mass flow rate, inner diameter, height and angle of the contact surface were studied through numerical simulation. The results showed that the optimal structure could control the maximum temperature and temperature difference within a reasonable range at a high discharge rate of 5C. In the literature [18], a cooling structure in which the liquid cooling pipes were placed horizontally between two adjacent rows of batteries was presented. The influences of different structures, different types of coolant and pump power consumption were discussed. It can be concluded that in a certain volume flow range, the structure with fewer copper pipes but a higher flow rate had a better cooling performance.

As a passive cooling method, PCM can absorb a large amount of heat through physical state changes as the battery temperature rises. It not only has a simple structure and no extra energy consumption but also can significantly improve the temperature uniformity of batteries due to the existence of an isothermal heat absorption during the phase change. Wang et al. [19] proposed a BTMS based on pure PCM and conducted detailed studies on thermal conductivity, viscosity, latent heat, the thickness of PCM and the thermal conductivity of cell shells. The results showed that the PCM unit could effectively reduce the maximum temperature and temperature difference in the battery at the end of discharge and can also realize temperature preservation after heat dissipation. Lazrak et al. [20] conducted relevant studies on the cooling effect of a PCM-based BTMS and found that the higher the thermal conductivity of the PCM used, the better the temperature uniformity of batteries, but the influence on the maximum temperature was little. However, when the battery is subjected to harsh conditions such as high temperatures, a single kind of PCM may not be able to adequately cool the battery due to its low thermal conductivity. To solve this issue, carbon-based or metal-based materials are often added to PCM to form composite PCM (CPCM) with high thermal conductivity [21]. He et al. [22] successfully prepared a new type of CPCM by constructing a thermal conduction skeleton of expanded graphite (EG) and copper foam (CF). This structure has a better cooling effect and temperature uniformity than PCM with only EG or CF. Although PCM with metal foam has excellent thermal conductivity, it has a high cost, large weight and complex processing, so the EG with lighter weight, good adsorption and thermal conductivity is more inclined to be used in practical projects. Jiang et al. [23] prepared paraffin (PA) and EG to form CPCM and performed experiments to measure the thermal conductivity and liquid phase PCM leakage of CPCM. The results showed that the addition of EG significantly increases the thermal conductivity of PCM and reduces the battery temperature rise. To improve the heat transfer characteristics of PCM after the addition of high thermal conductivity materials, Mauro et al. [24] utilized mono- and multi-objective optimization methods to obtain the optimal heat transfer effect and pressure drop of the graded metal foam-filled channels. In addition, fins with high thermal conductivity can be used in conjunction with PCM to enhance heat dissipation. In refs. [25,26], heat sinks based on PCM and fins are presented, and the influence of structural and operational parameters, such as PCM thickness, length and thickness of fins, etc., is studied in detail. Furthermore, Bianco et al. [27] proposed finned heat sinks combined with PCM and metal foam to improve the overall thermal conductivity and utilized multi-objective Pareto optimization with genetic algorithms to achieve an optimal trade-off between cost and operation time.

Although the BTMS based on PCM can absorb a significant amount of heat in the first cycle, it may not meet the performance requirement at an ambient temperature higher than the PCM melting temperature or under cyclic conditions. Therefore, hybrid cooling methods that combine PCM with other active cooling methods are often used to improve the overall thermal performance of BTMS. Lebrouhi et al. [28] proposed a BTMS combining horizontal liquid cooling pipes with PCM and analyzed the effects of different cooling methods, cooling pipes' number, coolant temperature and environmental conditions on the thermal performance of battery packs. The results showed that compared with a BTMS of natural convection and passive PCM, the hybrid cooling method has a better heat dissipation effect. Similarly, Fan et al. [29] combined longitudinal liquid cooling plates and PCM to form a hybrid cooling BTMS and constructed a nonlinear mathematical model among cell spacing, coolant velocity and system thermal performance. The simulation and experimental results showed that the optimal BTMS processed a better thermal performance in different application scenarios, including liquid cooling and heating. Compared with the horizontal cooling pipes and longitudinal cooling plates, the longitudinal cooling pipes can better solve the problem of heat concentration generated at the positive and negative electrodes of batteries. Xin et al. [30] adopted the combination of longitudinal liquid cooling tubes and PCM for the conditions of high temperature and 5C high discharge rate and discussed the influence of PCM thickness, EG mass fraction and coolant operating parameters on the thermal performance of batteries. The results showed that when the EG mass fraction was 12%, the maximum temperature and maximum temperature difference under the optimal BTMS were 45.25 °C and 3.49 °C, respectively. Moreover, Weng et al. [31] proposed a BTMS which embedded liquid cooling tubes in spiral copper tubes to strengthen the heat dissipation of PCM. The experimental results showed that this BTMS had an excellent cooling effect at high temperatures, and the dynamic liquid cooling mode was able to reduce energy consumption. In addition to liquid cooling, air cooling and heat pipe cooling are often combined with PCM, which can also meet the heat dissipation requirements of the battery pack under conventional working conditions [32,33].

According to the review of previous studies, it is found that the active BTMS combined with liquid cooling and PCM has a good cooling effect under harsh conditions of high temperature and high discharge rate. To further optimize the performance of the hybrid cooling method, a new honeycomb-shaped frame is proposed in this paper, formed by the liquid cooling tubes arranged in the battery gaps and the connected fins. Compared with previous studies [34,35], this structure not only improves the thermal performance of the single battery but also enables the extended battery pack to have a good temperature consistency. In addition, the influence of structural parameters, including the number of fins, number of liquid cooling tubes and cell spacing, material parameters such as the mass fraction of EG and operating parameters such as coolant flow direction and temperature on the thermal performance of BTMS based on honeycomb-structured liquid cooling and PCM will be analyzed in detail in this study.

2. Thermal Characteristic Experiments of Lithium-Ion Battery

2.1. Experimental Setup

In this experiment, the research object was selected for the INR18650/25P power battery, which can withstand high discharge currents and ambient temperatures. It has a nominal capacity of 2500 mAh in standard charging/discharging mode and can be charged at ambient temperatures ranging from 0 °C to 45 °C and discharged at ambient temperatures ranging from −20 °C to 60 °C. Detailed parameters are shown in Table 1.

Table 1. Relevant parameters of INR18650/25P cylindrical battery.

Key Indicators	Data
Nominal capacity	2500 mAh
Nominal voltage	3.6 V
Standard charging mode	1.25 A constant current charge to 4.20 V, constant voltage charge to 100 mA
Charge/Discharge cut-off voltage	4.2 V/2.5 V
Maximum charge/discharge current	4 A/30 A
Battery height	65.00 ± 0.15 mm
Battery diameter	18.35 ± 0.10 mm
Battery weight	48 g

As shown in Figure 1, the experimental platform includes a thermostat chamber, a DP100 data acquisition recorder, several T-type thermocouples, a high-performance battery test system (voltage measurement range is 0~5 V and the current measurement range is 0.1~30 A) and a host computer. This test platform has the functions of controlling the ambient temperature, controlling the battery charging and discharging process, obtaining real-time voltage, current and temperature readings, and acquiring data, which can meet the requirements of experiments in this study. Moreover, the accuracies of each equipment in the experimental platform are shown in Table 2.

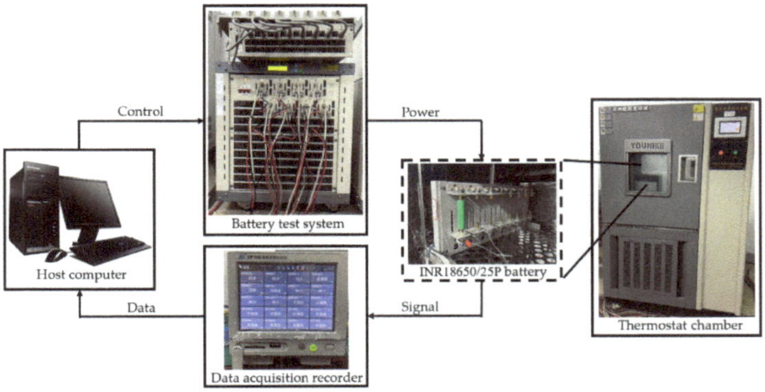

Figure 1. Experimental platform for internal resistance and temperature rise tests.

Table 2. Test accuracy of experimental instruments.

Experimental Instruments	Accuracy
Thermostat chamber	0.5%
Data acquisition recorder	0.05%
T-type thermocouples	1%
Battery test system	0.2%

2.2. Internal Resistance Characteristic Test

Internal resistance is one of the most important indexes to measure the power, life and thermal characteristics of batteries. Due to the complexity of internal and external conditions during the battery utilization, the internal resistance is an unfixed value and changes nonlinearly. So, it is necessary to conduct experimental research for the establishment of a more accurate simulation model. In this paper, hybrid pulse power characterization (HPPC) was used to test the internal resistance of the experimental battery. The test principle is that when the battery is discharged with a pulse current, its voltage will have a step decrease mainly caused by the ohmic resistance. Then, the voltage has a slightly slower drop due to the internal resistance of polarization [36]. The specific experimental steps to measure the internal resistance of lithium-ion batteries at different discharge rates are as follows:

1. The fully discharged battery is placed in a thermostatic chamber with the temperature set at 23.6 °C and then charged completely in a standard charging mode of constant-current and constant-voltage (CCCV). After charging, the battery stands for one hour to keep its temperature consistent with the ambient temperature.
2. Discharge the battery with a pulse current of 1C for 10 s, stop the current and stand for 40 s, and then charge the battery with a pulse current of 0.75C for 10 s; stop the loading current and stand for 40 s again. The change in the open-circuit voltage of the battery is collected and recorded during this process.
3. Discharge the battery at a constant current of 1C until its state of charge (SOC) = 0.9, and stand for one hour to restore the battery temperature to the ambient temperature;
4. Repeat Steps 2 and 3, so that when the SOC of the battery is at 0.1, 0.2, 0.3, 0.4, 0.5, 0.6, 0.7, 0.8, 0.9 and 1, the corresponding voltage abrupt changes can be recorded.
5. Change the discharge current in Step 2 to other discharge rates of 2C, 3C and 5C, and repeat steps 2~4 to measure the total discharge internal resistance under different discharge rates and SOC.

The total internal resistance of the battery under different discharge rates and SOC can be calculated by the voltage changes caused by accepting pulse current in the experiment. The experimental results are shown in Figure 2. On the one hand, it can be concluded that, under the condition of the same ambient temperature and different discharge rates, the corresponding internal resistance of the battery decreases as the discharge rate increases. Because the ionic activity and electrochemical reaction rate of the battery increase with the increase in discharge rate, the ohmic and polarization internal resistance of the battery will decrease. In addition, the relationship between the total discharge internal resistance and SOC at the same ambient temperature and discharge rate can be obtained. When SOC is between 0.1 and 0.3, the internal resistance of the battery evidently decreases with the increase in SOC. When SOC is from 0.3 to 1.0, the internal resistance of the battery is stable. The reason for this phenomenon is that when the battery possesses medium or high capacity, the ohmic resistance takes up a large percent of the battery's internal resistance and does not noticeably change with the variation in SOC. Furthermore, when the battery is at a low capacity, the ratio of battery polarization resistance increases, which obviously varies with the change in SOC. This HPPC test is carried out under a controlled environmental temperature condition. To obtain more accurate results, the establishment of isothermal conditions (uniform cell surface temperature) in the literature [8,37] can be referred to.

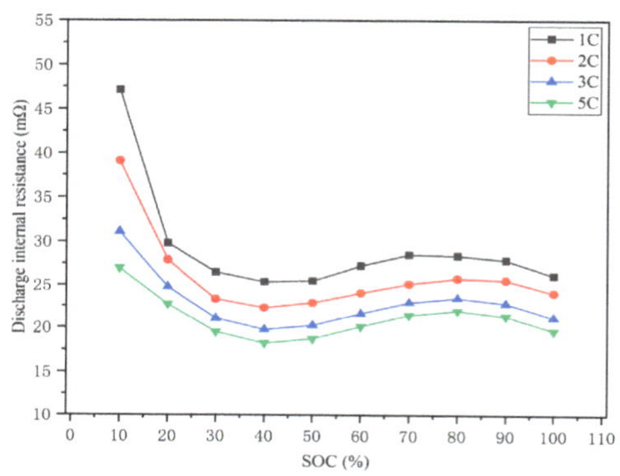

Figure 2. Relationship between discharge internal resistance and SOC of battery at different rates.

2.3. Temperature Rise Characteristic Test

The temperature rise characteristic test of the battery was carried out to obtain the temperature rise in the experimental battery at different discharge rates. Three thermocouples were placed at the positive, negative and intermediate positions of the battery to prevent accidental error in the process of temperature measurement. In this way, the difference in temperature change at different positions of the battery and the average temperature rise in the battery surface could be obtained and estimated. Using the experimental platform built above, the temperature increment and current and voltage changes in the battery discharging at different rates under the 23.6 °C ambient temperature were measured. The detailed steps are as follows:

1. The fully discharged battery is placed in a thermostatic chamber at 23.6 °C for one hour to make its temperature stable and consistent with the ambient temperature.
2. Charge the battery with a constant current of 1.6C rate until the voltage reaches the charging cut-off voltage of 4.2 V, and then charge with a constant voltage of 4.2 V until the current achieves the charging cut-off current of 0.1 A.
3. Place the fully charged battery in the thermostat for one hour again so that the battery temperature keeps in line with the ambient temperature of 23.6 °C.
4. Discharge the battery at 1C constant current until the voltage reaches the discharge cut-off voltage of 2.5 V. Meanwhile, the temperature change in the battery at each thermocouple is monitored and recorded in real time.
5. Repeat steps 1~4 and adjust the intensity of constant current in step 4 to 1C, 2C, 3C and 5C, respectively, until all experiments are completed.

According to the experimental results, although there are some differences in the temperature values measured by the thermocouples at the three positions, represented in the scatter diagram in Figure 3, the range of variation is small. Therefore, the mean value of the temperature measured by the three thermocouples is used as the experimental result in this paper. The average temperature rise in the battery when discharging at different current rates can be obtained as the line graph shown in Figure 3. It can be found that, under the same ambient temperature, the higher the discharge rate of the battery is, the higher the temperature rise will be because the heat generation of the battery is the square term of the current. When the discharge current rate increases, the heat generation of the battery increases sharply. In addition, it shows that the maximum battery temperature rises rapidly, then gently, and finally rapidly in the discharge process. The cause of this change is as follows. In the early stage of discharge, the heat generation rate of the battery is large, and its temperature is low, so the temperature rise trend is obvious in the early stage. In the middle stage of discharge, the battery temperature has a significant difference from the ambient temperature, so the convective heat transfer between the battery and the environment is strong, which makes the temperature rise trend more gentle. When the discharge enters the late stage, the battery's internal polarization resistance increases rapidly, increasing the total internal resistance, so the battery temperature rises rapidly again.

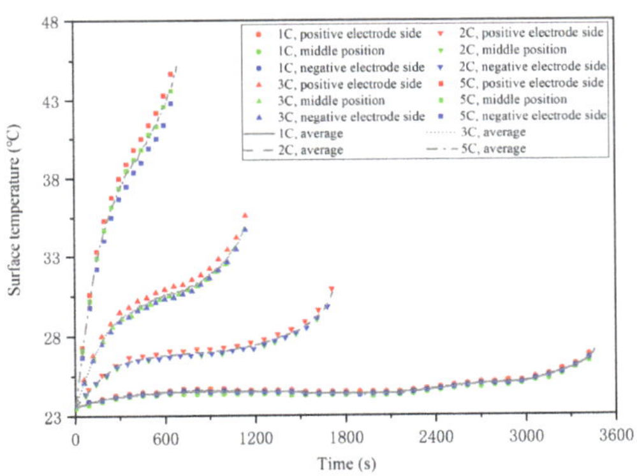

Figure 3. Surface temperature at different thermocouple locations and average temperature of battery during the discharge process at 1C, 2C, 3C and 5C.

2.4. Heat Generation Rate of Single Battery

The volume heat generation rate of lithium-ion batteries is necessary for the simulation of the thermal behavior of the battery pack. Therefore, the parameters at different discharge rates were obtained through the above experiments in this section. However, the heat generation rate of the battery is affected by many indicators, such as the battery SOC, charging and discharging current rates, etc., and it has a high degree of nonlinearity. Generally, the heat generation model proposed by Bernardi is used to quantitatively describe the heat generation of lithium-ion batteries [38]:

$$Q_v = \frac{I}{V_b}(U_{\text{ocv}} - U) + \frac{I}{V_b} T \frac{dU_{\text{ocv}}}{dT} \quad (1)$$

where Q_v is the volume heat generation rate of the battery, W/m^3; I is the current flowing through the battery, A; V_b is the battery volume, m^3; U_{ocv} is the battery open circuit voltage, V; U is the battery terminal voltage, V; and dU_{ocv}/dT is the temperature entropy coefficient of the battery, which can be approximately regarded as a constant in the working process of the battery.

The relationship between the total internal resistance of the battery and SOC at different discharge rates could be obtained by fitting the previous results of internal resistance characteristic tests. Moreover, the corresponding data of current, terminal voltage, temperature and time at different discharge rates could be obtained from the temperature rise experiment. The battery volume could be calculated from Table 1, and the open-circuit voltage could be obtained by adding the terminal voltage and the voltage caused by the internal resistance. Therefore, the volume heat generation rates of the battery under different discharge rates can be obtained in Figure 4.

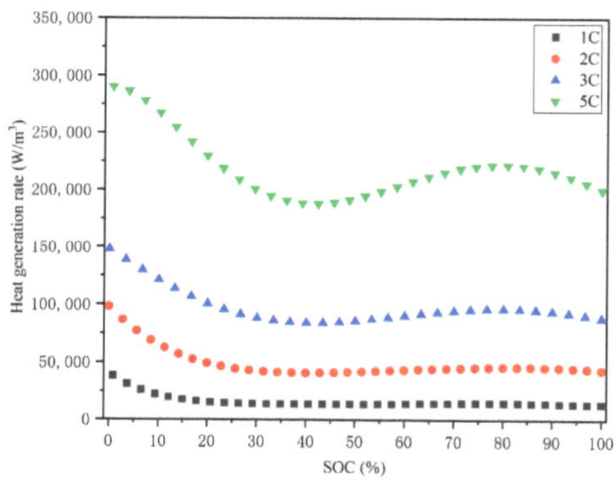

Figure 4. Volume heat generation rate of the battery with different SOC at 1C, 2C, 3C and 5C.

The relationship between the battery SOC and discharging time is shown as follows:

$$\mathrm{SOC} = \left(1 - \frac{\eta \int_0^t Idt}{Q_c}\right) \times 100\% \tag{2}$$

where t is the full charge/discharge time of the battery, s; η is the battery charge/discharge efficiency; Q_c is the battery's nominal capacity, mAh; SOC can be defined as the ratio of the current remaining capacity of a battery to its nominal capacity; and 0% and 100% indicate the fully discharged and charged state of the battery.

Combined with Equation (2), the expression and coefficients of the heat generation rate equation can be conversed and obtained by polynomial fitting, as shown in Table 3. Then, the corresponding heat transfer simulation will be completed by COMSOL software, which can carry out multi-physical field coupling simulation.

Table 3. Expression and coefficients of volume heat generation rate of battery at different discharge rates.

Discharge Rate	Q_v=k_0+k_1t+k_2t^2+k_3t^3+k_4t^4+k_5t^5+k_6t^6						
	k_0	k_1	k_2	k_3	k_4	k_5	k_6
1C	13,752.01	−0.491	0.0117	-2.047×10^{-5}	1.403×10^{-8}	-4.368×10^{-12}	5.202×10^{-16}
2C	43,722.65	20.752	−0.0496	3.731×10^{-5}	-1.376×10^{-8}	-2.123×10^{-12}	5.01×10^{-15}
3C	88,659.77	60.361	−0.0496	-3.859×10^{-4}	5.395×10^{-7}	-5.914×10^{-11}	-7.776×10^{-14}
5C	200,078.4	280.647	−0.5359	-3.14×10^{-3}	3.359×10^{-6}	1.409×10^{-8}	-1.651×10^{-11}

3. Numerical Model

3.1. Geometric Model

To better study the parameters' influence of liquid cooling and PCM on batteries, the shell of the battery pack is removed, and the geometric model of the honeycomb-structured BTMS is presented in Figure 5a. Due to its strong regularity and repeatability, a simplified heat transfer model representing the thermal performance of the central region of the battery pack can be obtained in Figure 5b. Furthermore, the detailed structure and composition of the simplified model are shown in Figure 5c.

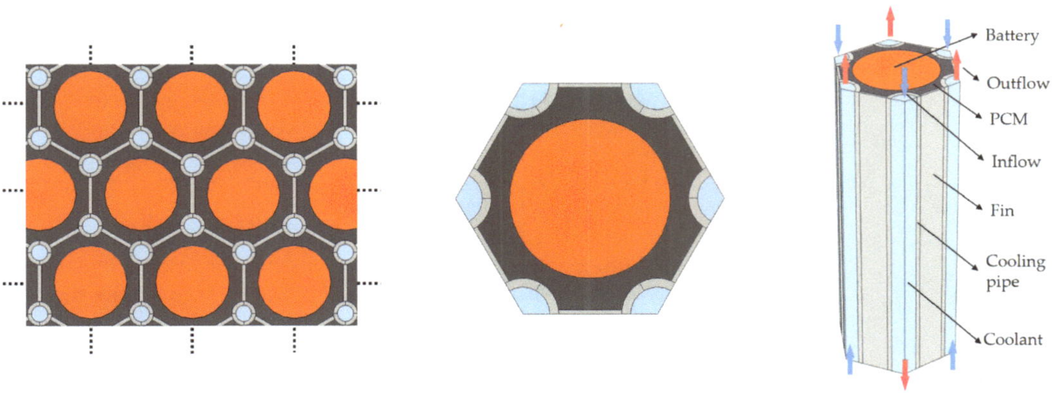

(a) (b) (c)

Figure 5. BTMS model based on the honeycomb structure: (**a**) battery pack, (**b**) simplified model and (**c**) schematic components.

The proposed BTMS model has a compact structure and uniform heat dissipation and can prevent leakage of liquid PCM [39]. To some extent, it can also reduce the harm of battery thermal runaway [40]. The battery is surrounded by PCM, whose outer layer is the proposed honeycomb-shaped structure which consists of liquid cooling tubes (with an inner diameter of 6 mm and an outer diameter of 8 mm) and fins (with a thickness of 1 mm). Moreover, the honeycomb structure is made of aluminum, the coolant used in liquid cooling is water, and the external environment is air. Meanwhile, it is assumed that the material distribution is uniform, the specific heat capacity and thermal conductivity are constant, and the heat generation rate inside the battery and at the pole ear is consistent. The calculation of the equivalent thermal conductivity of the battery can refer to the calculation method of equivalent resistance. The relevant thermal physical property parameters of each part of the BTMS simulation model can be obtained, as shown in Table 4.

Table 4. Physical property parameters of each part of the proposed BTMS.

Materials	**Density** kg/m^3	**Specific Heat Capacity** J/(kg·K)	**Thermal Conductivity** W/(m·K)	**Dynamic Viscosity** Pa·s
Battery	2755.9	1129.95	Radial 1.6 Axial 27	-
Aluminum	2719	871	238	-
Water	998	4180	0.599	1.01×10^{-3}

The current BTMS module can be fabricated by the following steps. First, the aluminum honeycomb structure can be casted by a dedicated mold. Then, the batteries are installed in each honeycomb unit, and housed in the module casing with the baseplate, side panels and top cover. After that, the composite PCM will be filled in the void space between the batteries and the honeycomb structure.

3.2. Mathematical Model

3.2.1. Lithium-Ion Battery Model

Mathematical models of lithium-ion batteries include heat generation and heat transfer models. The total heat generation of batteries comprises Joule heat, reaction heat, polarization heat and side reaction heat. Among them, Joule heat is generated by the current passing through the inherent and contact resistances of battery parts. Reaction heat refers to the energy change during the occurrence of electrochemical reactions. Polarization heat means the heat generated by the internal polarization resistance of the battery when passing through current. When use conditions are relatively routine, the proportion of side reactions is small, so the corresponding side reaction heat can be ignored [41]. In practical applications, the Bernardi formula is usually used to establish the heat generation model of the battery, as shown in Equation (1).

A certain proportion of the heat generated by lithium-ion batteries will be absorbed by itself, while the other part will be transferred to the environment through three ways of heat transfer, including convection, conduction and radiation. The radiation heat transfer is relatively small and can be ignored, and the convective heat transfer can also be neglected due to the poor fluidity of electrolytes. Therefore, heat conduction is mainly considered. Combining Fourier's law with the law of energy conservation, the thermal conductivity differential equation can be obtained as follows:

$$\rho_b c_{p,b}\frac{\partial T_b}{\partial t} = \nabla\cdot\left(\begin{bmatrix}\lambda_{bx} & 0 & 0\\ 0 & \lambda_{by} & 0\\ 0 & 0 & \lambda_{bz}\end{bmatrix}\nabla T_b\right) + Q_v \tag{3}$$

where ρ_b is the battery equivalent density, kg/m^3; $c_{p,b}$ is the battery specific heat capacity at constant pressure, J/(kg·K); λ_{bx}, λ_{by} and λ_{bz} are the battery thermal conductivity in spatial coordinate axis, W/(m·K); and Q_v is the heat generation rate of the battery, W/m^3.

3.2.2. Composite Phase Change Material Model

PCM absorbs and stores the heat released by the battery through two-phase conversion. The phase transition process is divided into three stages: (1) when the temperature of PCM is lower than the melting temperature, the absorbed heat will be stored in the form of sensible heat; (2) when the temperature of PCM is higher than the melting temperature, PCM will undergo the phase transition and consume its latent heat; (3) when the latent heat of PCM is exhausted, the phase transition process ends. Due to the high viscosity and the small temperature difference in PCM under the working conditions involved, the natural convection heat transfer term is neglected. The energy balance equation involved is as follows:

$$\rho_{\mathrm{PCM}} c_{\mathrm{PCM}}\frac{\partial T_{\mathrm{PCM}}}{\partial t} = \nabla\cdot(\lambda_{\mathrm{PCM}}\nabla T_{\mathrm{PCM}}) \tag{4}$$

where T_{PCM} is PCM temperature, K; ρ_{PCM} is PCM density, kg/m^3; c_{PCM} is the specific heat capacity of PCM, J/(kg·K); and λ_{PCM} is the thermal conductivity of PCM, W/(m·K).

Moreover, because the polyphase transformation of PCM is difficult to describe accurately by experiments, the mathematical model of CPCM should be established. The following assumptions are made: (1) The physical properties of each part of CPCM are constant. (2) CPCM is incompressible in the liquid phase and has Newtonian properties. (3) The material inside the CPCM is uniform, and the thermal conductivity in all directions is the same. For the thermal conductivity of CPCM, the Ling model [42] and the Maxwell-Eucke model [23] can be used for calculation. This paper adopted the Ling model, which considered the CPCM density and the EG mass fraction, and its formula is as follows:

$$\lambda_{\mathrm{eff,CPCM}} = \frac{\lambda_{\mathrm{EG}}\rho_{\mathrm{CPCM}}\omega_{\mathrm{EG}}}{\rho_{\mathrm{EG}}} \tag{5}$$

where $\lambda_{eff,CPCM}$ and λ_{EG} are the thermal conductivity of composite PCM and EG, respectively, W/(m·K); ρ_{CPCM} and ρ_{EG} are the densities of composite PCM and EG, respectively, kg/m^3; and ω_{EG} represents the mass fraction of EG.

In addition, the latent heat of CPCM can be obtained by the mass weighting method, and its calculation formula is as follows:

$$L_{CPCM} = L_{PA}(1 - \omega_{EG}) \tag{6}$$

where L_{CPCM} and L_{PA} are the latent heat of CPCM and paraffin, respectively, kJ/kg.

Moreover, the specific heat capacity of CPCM can be calculated by the formula:

$$c_{p,CPCM} = c_{p,PA}(1 - \omega_{EG}) + c_{p,EG}\omega_{EG} \tag{7}$$

where $c_{p,CPCM}$,$c_{p,PA}$ and $c_{p,EG}$ are the specific heat capacity of CPCM, paraffin and EG, respectively, J/(kg·K).

Based on the experimental studies on the preparation of CPCM and the Ling prediction model in the literature, the thermophysical parameters of a PA/EG-type CPCM with different EG mass fractions used in this study can be obtained in Table 5.

Table 5. Thermophysical PA/EG-type CPCM with different EG mass fractions [23].

CPCM	Specific Heat Capacity J/(kg·K)	Thermal Conductivity W/(m·K)	Latent Heat kJ/kg	Melting Temperature K	Density kg/m^3
PA/0 wt%EG	2000	0.2	275	314–317	800
PA/3 wt%EG	1963	0.58	266.8	314–317	825
PA/6 wt%EG	1926	1.23	258.5	314–317	832
PA/9 wt%EG	1889	3.15	250.3	314–317	845
PA/12 wt%EG	1852	5.74	242	314–317	897
PA/20 wt%EG	1754	10.6	220	314–317	913

3.2.3. Liquid Cooling Model

In this paper, cooling pipes and water are selected as the structure and coolant of liquid cooling, respectively. According to the calculation, the Reynolds number of each simulation condition is less than 2300, so the laminar flow model is adopted in the simulation. The governing equations related to liquid cooling, including the conservation equations of mass, momentum and energy, are shown as follows:

$$\frac{\partial \rho_c}{\partial t} + \nabla \cdot \left(\rho_c \vec{u}\right) = 0 \tag{8}$$

$$\rho_c \frac{\partial \vec{u}}{\partial t} + \rho_c\left(\vec{u} \cdot \nabla\right)\vec{u} = -\nabla p + \nabla \cdot \left[\mu\left(\nabla \vec{u} + \left(\nabla \vec{u}\right)^T\right)\right] \tag{9}$$

$$\rho_c c_{p,c} \frac{\partial T}{\partial t} + \nabla \cdot \left(\rho_c c_{p,c} \vec{u} T\right) = \nabla \cdot (\lambda_c \nabla T) \tag{10}$$

where ρ_c, $c_{p,c}$ and λ_c represents the density, specific heat capacity and thermal conductivity of coolant, whose units are kg/m^3, J/(kg·K) and W/(m·K), respectively; $\vec{u}$, p and μ represent the vector velocity, static pressure and dynamic viscosity of coolant, whose units are m/s, Pa and N·s/m^2, respectively. The cooling medium of liquid cooling is water, which has incompressibility.

3.3. Boundary Conditions

As for the heat transfer part, the boundary conditions of each side surface of the simplified BTMS model are symmetric because the battery unit is surrounded by the same cooling unit before simplification. The heat transfer modes among batteries, PCM, honeycomb-shaped structure and coolant are heat conduction and convection, and the interface and boundary conditions are as follows:

$$-\lambda_b \frac{\partial T_b}{\partial n_b} = -\lambda_{\text{PCM}} \frac{\partial T_{\text{PCM}}}{\partial n_{\text{PCM}}} \tag{11}$$

$$-\lambda_{\text{PCM}} \frac{\partial T_{\text{PCM}}}{\partial n_{\text{PCM}}} = -\lambda_h \frac{\partial T_h}{\partial n_h} \tag{12}$$

$$-\lambda_h \frac{\partial T_h}{\partial n_h} = h_c(T_h - T_c) \tag{13}$$

where λ_b, λ_{PCM} and λ_h are the thermal conductivity of battery, PCM and honeycomb structure, respectively, W/(m·K); h_c is the convective heat transfer coefficient between the coolant and the wall of liquid cooling tubes, W/(m^2·K); T_b, T_{PCM}, T_h and T_c are the temperatures of battery, PCM, honeycomb structure and coolant, respectively, K; and $\partial T/\partial n$ is the temperature gradient of each material along the normal direction of the outer surface.

As for the boundary conditions of laminar flow, since the coolants in the simplified BTMS model are divided into three parts, symmetrical boundary conditions should be set for the two sides of each fluid region to simulate the flow effect in the whole tube. The result shows that the flow velocity of coolant presents high velocity in the center area and low velocity on the walls along the radial direction, which conforms to the flow rate distribution law caused by the viscous action of liquid flowing in the pipe.

3.4. Model Verification

The simplified BTMS model based on honeycomb-structured liquid cooling and PCM involves complex structures and simulation modules such as solid and fluid heat transfer, laminar flow and fluid-solid coupling. Hence, each region needs to finely mesh to ensure calculation accuracy. The fluid region meshes first, and then the solid regions of the end face are divided, and finally, the whole end face is used as the source plane to carry out the sweep operation. A computational grid model with 99,200 cells is obtained in Figure 6, which is evenly distributed in vertical and horizontal directions.

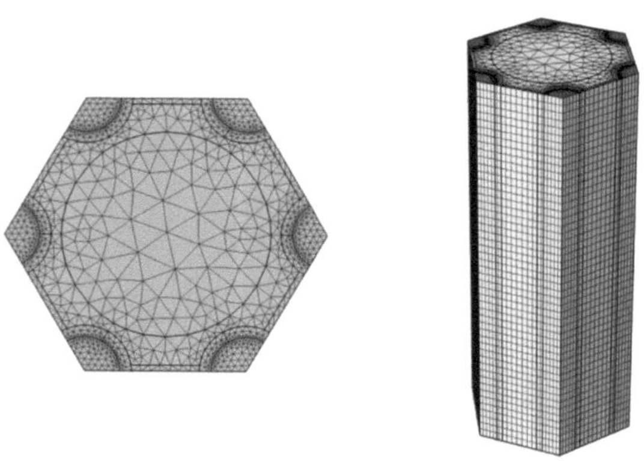

Figure 6. Grid model of the simplified BTMS based on the honeycomb structure.

The simulation model of the single battery is established by COMSOL to predict the heat generation and heat transfer characteristics. To verify this model, the simulations are conducted at 23.6 °C ambient temperature and 1C, 2C, 3C and 5C discharge rates. Moreover, the simulation results are compared with the consequences of temperature rise characteristic tests in Section 3.2, as shown in Figure 7a. It indicates that the maximum relative errors of the simulation model for the battery temperature prediction are around 4% during the universe discharging process, which meets the accuracy requirements. When the discharge rate is 1C, 2C, 3C and 5C, the values of root mean square error (RMSE) between the experimental and simulation results is 0.45, 0.60, 0.75 and 0.38, respectively. In addition, to verify the grid independence of the simplified BTMS model, different grid numbers are also simulated and compared under an ambient/coolant temperature of 23.6 °C, discharge rate of 5C, mass fraction of EG of 12% and coolant flow velocity of 0.1 m/s. The relationship between the number of grids (6060, 22,116, 57,478, 99,200, 123,700, 172,700), and the maximum temperature of the battery is obtained in Figure 7b. When the number of grids increases from 6060 to 99,200, the maximum battery temperature changes noticeably, nearly 0.27 °C. However, when the number of grids increases from 99,200 to 123,700 or 172,700, the maximum battery temperature does not change and stabilizes at 33.5 °C. Therefore, a simulation model with a grid number of 99,200 is chosen to balance the calculation accuracy and time consumption.

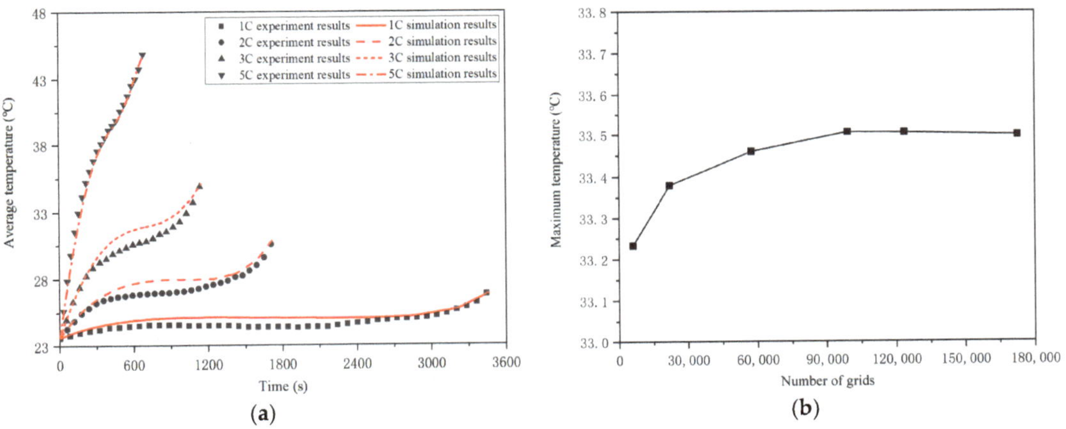

Figure 7. Verification of single battery simulation model for the average surface temperature (**a**) and grid independence of simplified BTMS based on the honeycomb structure (**b**).

4. Results and Discussion

4.1. *Effects of the Honeycomb Structure*

The simulation and comparative analysis of battery packs with different honeycomb cooling structures (three/six liquid cooling tubes, with or without fins) under diverse environment/coolant temperatures (23.6 °C, 30 °C, 35 °C and 40 °C) are conducted to study the influence of the proposed honeycomb structure on the maximum temperature, maximum temperature difference and temperature distribution of batteries. In addition, the basic simulation conditions were set as 25 mm battery spacing, counterflow coolant, 0.1 m/s flow rate of coolant and a 5C high discharge rate. Battery spacing is the distance between two adjacent battery centers in the honeycomb-structured BTMS, mm.

The maximum temperature and temperature difference in the battery cooled by each honeycomb structure at different ambient temperatures are shown in Figure 8a,b. When the ambient temperature is the same, the maximum temperature and maximum temperature difference in the BTMS with honeycomb fins are significantly better than those without fins, regardless of whether three or six liquid cooling tubes are used. The fin structure can lead to a decrease in the maximum temperature and temperature difference in the battery by as much as 3.3 °C and 2.8 °C at 23.6 °C ambient temperature, respectively, and a decrease of 0.15 °C and 0.2 °C at 40 °C ambient temperature, respectively. Figure 9a,c show the phenomenon of high temperature and temperature differences caused by local heat concentration. To solve this issue, the high thermal conductivity honeycomb fins are directly connected to the liquid cooling pipes, which allows the surrounding area of the battery to be cooled evenly, as shown in Figure 9b,d. In addition, the maximum temperature and temperature difference in the battery corresponding to the six liquid cooling tubes is lower than those of the three tubes, which is caused by increased cooling sources around the battery. Additionally, it can be seen that with the increase in the external temperature, the corresponding maximum temperature of the battery under different structures will increase. In contrast, the maximum temperature difference will decrease. Because PCM plays a leading cooling role at high temperatures and possesses a large melting temperature range, the BTMS has a better temperature uniformity.

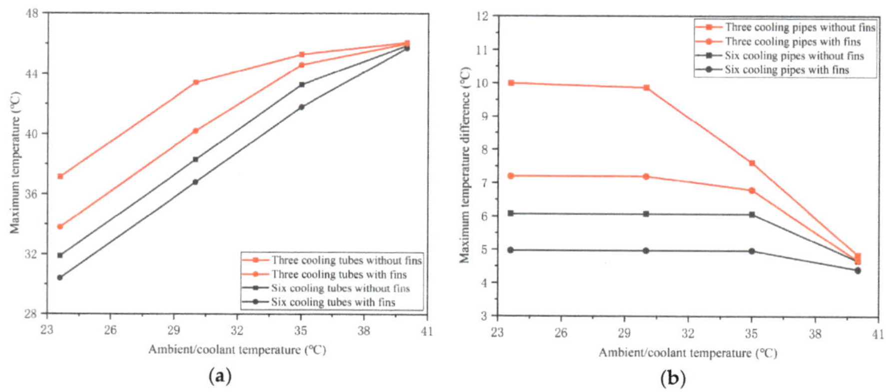

Figure 8. Maximum temperature (**a**) and temperature difference (**b**) of battery at different ambient/coolant temperatures and honeycomb cooling structures.

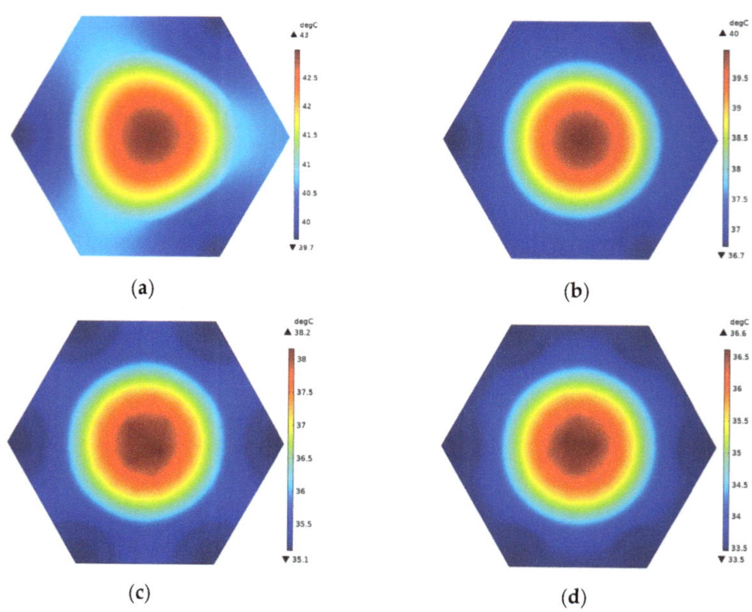

Figure 9. Temperature distribution at 1/2 height of battery packs with different honeycomb structures at the ambient temperature of 30 °C: (**a**) three cooling tubes without fins, (**b**) three cooling tubes with fins, (**c**) six cooling tubes without fins and (**d**) six liquid cooling tubes with fins.

According to the above analysis, the honeycomb structure proposed in this paper can significantly improve the maximum temperature and maximum temperature difference in the battery. Moreover, with the decrease in ambient temperature and the enhancement of the liquid cooling effect, the comprehensive cooling effect of the BTMS will be stronger. In the following research, the honeycomb-structured BTMS with good thermal performance will be used as the basis to analyze the influence of other structures, materials and operating parameters on the thermal performance of batteries.

4.2. Effects of the Coolant Flow Directions

The BTMS based on honeycomb-structured liquid cooling and PCM involves the coupling of two cooling modes, and many parameters will affect the cooling effect of this system. In this section, the basic operating parameters, i.e., the flow direction of coolant, will be studied. The simulation conditions were set as the coolant flow rate of 0.1 m/s, the EG mass fraction of 12% and the battery spacing of 25 mm. Then, the cases of different flow directions (six/three cooling tubes with parallel-/counter-flow coolant) at diverse environment/coolant temperatures (23.6 °C, 30 °C, 35 °C and 40 °C) were carried out for the simulation and result analysis. Figure 10 shows the cases of different liquid cooling pipe numbers and flow directions.

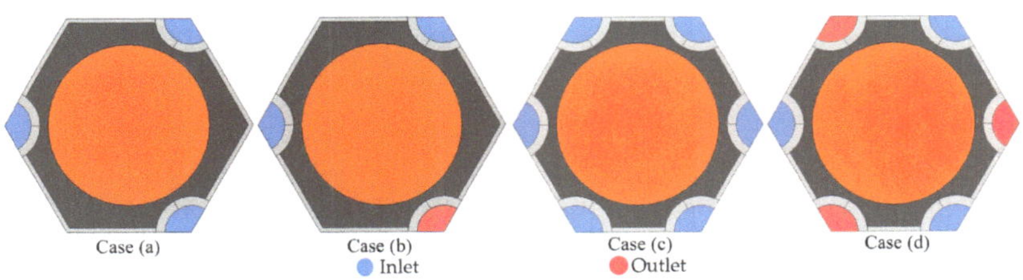

Figure 10. Different coolant flow directions at the end face of the BTMS model: Case (**a**): three tubes with parallel flow coolant; Case (**b**): three tubes with counter-flow coolant; Case (**c**): six tubes with parallel flow coolant; and Case (**d**): six tubes with counter-flow coolant.

The maximum battery temperature and temperature difference in three and six liquid cooling pipes with different coolant flow directions at 23.6 °C, 30 °C, 35 °C and 40 °C are shown in Figure 11. At a constant ambient/liquid cooling temperature, the counter-flow coolant can better limit the maximum temperature and temperature difference in batteries than the parallel flow. The counter flow can result in a decline in maximum temperature and temperature difference in the battery by as much as 3 °C and 3 °C at 23.6 °C ambient temperature, and a decline of 0.45 °C and 0.7 °C at 40 °C ambient temperature, respectively. When the coolant flows in a parallel direction, the coolant flowing through the outlet has absorbed part of the heat, and its cooling effect decreases, resulting in a higher battery maximum temperature at this position. Oppositely, when the coolant flow directions of the adjacent cooling pipes are different, i.e., the liquid cooling outlet and inlet are distributed alternately, there will be less uneven heat accumulation at a particular position. Similarly, the corresponding heat dissipation effect of six cooling tubes is better than that of three cooling tubes at a certain flow direction because the six cooling tubes as cooling sources are more densely and uniformly distributed around the battery. However, in individual cases, at high ambient/coolant temperatures, the highest temperature and temperature difference in the battery unit using six cooling tubes are higher than those using three cooling tubes because six cooling tubes occupy the larger space and decrease the amount of PCM, which absorbs most of the heat generated by the battery at high temperatures.

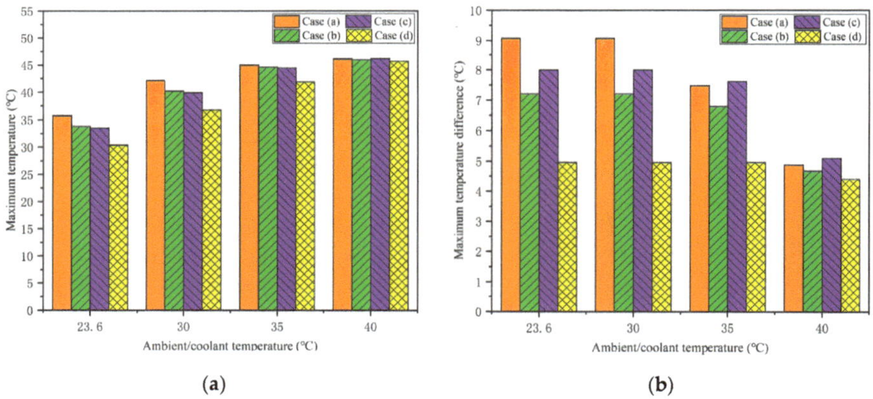

Figure 11. Maximum temperature (**a**) and maximum temperature difference (**b**) of battery under different ambient/coolant temperatures, cooling pipe numbers and flow directions.

In summary, it can be concluded that the cooling effect of counterflow coolant is better than that of parallel flow, no matter the performance of the maximum temperature or the temperature difference. This phenomenon is more obvious when the coolant temperature is lower. In addition, when the flow direction is the same, the lower the environment/coolant temperature is, the lower the maximum battery temperature is, and the higher the maximum temperature difference is. Six liquid cooling tubes with counter-flow direction is the best scheme with the highest temperature and temperature difference of 45.71 °C and 4.4 °C, respectively, in the 40 °C environment. Meanwhile, the thermal performance of the battery in this scheme will be furtherly improved and can meet the requirements of uniformity with the decrease in ambient temperature. Therefore, in the subsequent research optimization, the counterflow coolant is taken as a determined condition to continue the discussion about the influence of other parameters.

4.3. Effects of the Mass Fraction of EG

Except for the operating and structural parameters, the change in material parameters also greatly impacts the heat transfer of hybrid cooling BTMS. The proposed structure is mainly filled with CPCM, whose thermal conductivity and latent heat directly affect the thermal performance of the battery pack. Therefore, this section will simulate the cooling effect of BTMS with different EG mass fractions (0, 3%, 6%, 9%, 12%, 20%) at ambient/coolant temperatures of 40 °C and 23.6 °C. In addition, battery spacing (23 mm, 25 mm, 27 mm, 29 mm, 31 mm) is also considered in this section for further analysis because of the strong relationship between battery spacing and EG mass fraction on the thermal performance of battery packs. Other simulation conditions were set: six liquid cooling pipes, counter-flow and a coolant flow rate of 0.1 m/s.

The maximum temperature of the battery in BTMS with different EG mass fractions at environment/coolant temperatures of 40 °C and 23.6 °C are presented in Figure 12. The lower ambient/coolant temperature can reduce the maximum battery temperature at a constant EG mass fraction. In addition, at a constant temperature and battery spacing, the maximum battery temperature will decrease rapidly first and then flatten out with the increase in EG mass fraction. When the EG mass fraction is small, ranging from 0 to 6%, its thermal conductivity is insufficient so that only part of PCM around the battery takes effect. At this time, increasing the EG mass fraction can significantly enhance the role of PCM or liquid cooling, leading to a rapid decline in the maximum battery temperature. Then, when the EG mass fraction increases to 12%, its thermal conductivity is enough to bring most of the heat from the battery, and the PCM at the far end can play a full role. However, when the EG mass fraction rises to 20%, the thermal conductivity of PCM is sufficient, but its latent heat is low, resulting in the decline of the heat dissipation effect under high temperatures. The temperature distribution of BTMS under different EG mass fractions due to the different thermal conductivity of PCM can be obtained in Figure 13.

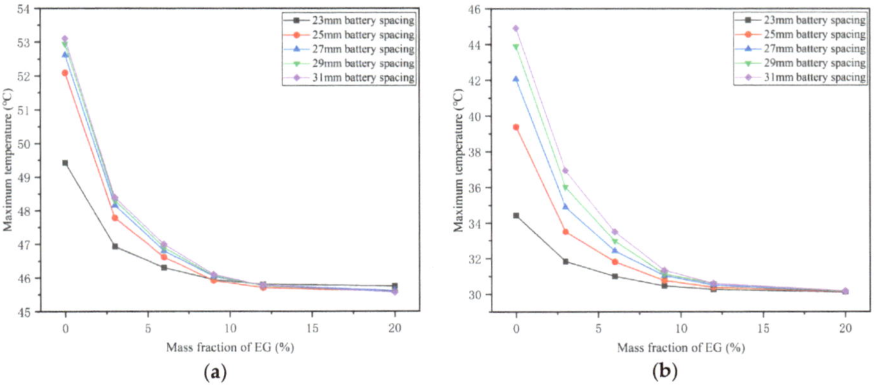

Figure 12. Maximum temperature of the battery in BTMS with different EG mass fractions and cell spacing at (**a**) 40 °C and (**b**) 23.6 °C ambient/coolant temperatures.

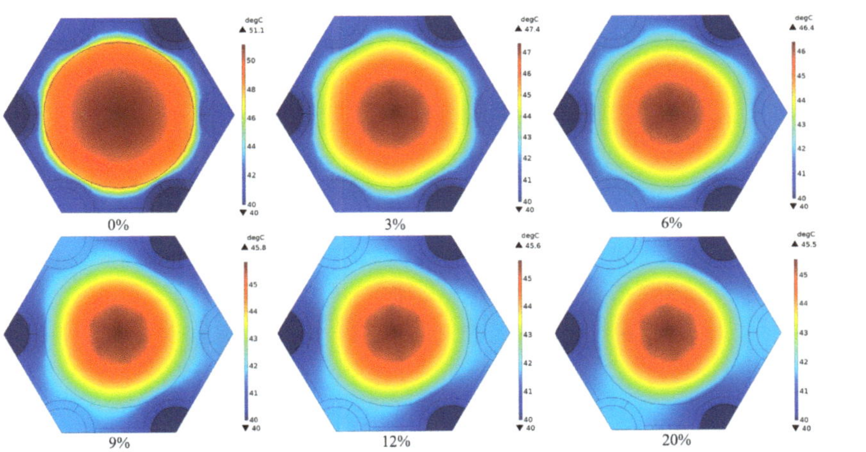

Figure 13. Temperature distribution at the upper-end surface of BTMS at 40 °C ambient/coolant temperature, 24 mm battery spacing and different EG mass fractions.

The results of the maximum temperature difference in batteries in BTMS with different EG mass fractions at ambient/coolant temperatures of 40 °C and 23.6 °C are shown in Figure 14. The maximum temperature difference in the battery at a high temperature of 40 °C is smaller when the EG mass fraction is constant, which is caused by the fact that PCM can melt and play a major role in the dissipation effect at high temperatures. Figure 15 is a supplementary illustration, i.e., as the ambient/coolant temperature decreases, the maximum temperature difference in the battery will rise, which is caused by the increased liquid cooling effect. Moreover, the corresponding maximum temperature difference is stable at a high value until the coolant temperature is low enough. In addition, when the EG mass fraction increases, the maximum temperature difference in the battery rises rapidly at first and then gradually flattens out. The reason for this trend can be explained according to Figure 13. When the EG mass fraction is low, the heat inside the battery cannot be dissipated in time, so its temperature is high, but the temperature gradient is small. With the rise in the EG mass fraction, the thermal conductivity is enhanced so that the phase change is increased, or the liquid cooling effect is obvious, which can reduce the minimum temperature of the battery and lead to the increase in temperature gradient. Until the thermal conductivity is sufficient, the battery heat generation and external heat dissipation are balanced, and the maximum temperature difference in the battery tends to flatten out.

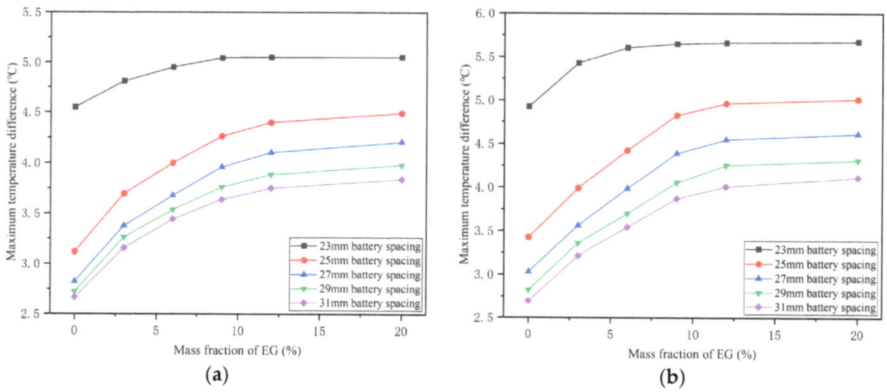

Figure 14. Maximum temperature difference in battery in BTMS with different EG mass fractions and cell spacing at (**a**) 40 °C and (**b**) 23.6 °C ambient/coolant temperatures.

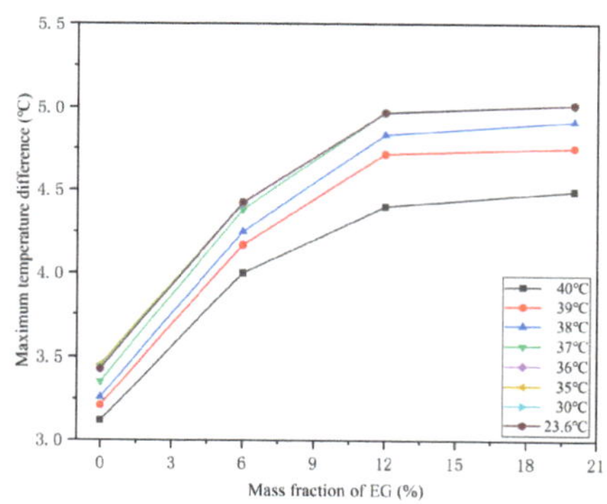

Figure 15. Maximum temperature difference in battery in 25 mm battery spacing BTMS with different EG mass fractions at various ambient/coolant temperatures.

According to the above results, at a constant EG mass fraction, the lower the ambient/coolant temperature is, the smaller the maximum temperature of the battery is, and the larger the maximum temperature difference is. In addition, increasing the mass fraction of EG can decrease the maximum temperature of the battery and increase the maximum temperature difference. Specifically, this phenomenon is quite apparent when the value of EG mass fraction is 0~9% and less obvious between 9 and 12%. However, furtherly increasing the mass fraction of EG to 20% will have little effect on the thermal performance and may cause the increase in maximum temperature. Therefore, on the premise of meeting the maximum temperature difference, a larger EG mass fraction of 12% can be selected to obtain a lower maximum battery temperature.

4.4. Effects of the Battery Spacing

The distance between battery cells greatly influences the two cooling methods used in this study. With the increase in the distance, the amount of PCM will increase and it can absorb more heat, while the cooling tubes will be farther away from the battery, resulting in a weak cooling effect. Moreover, the battery spacing also affects the grouping efficiency of battery packs. Therefore, this section provides heat transfer simulation and comparative analysis for BTMS with different battery spacing (23 mm, 25 mm, 27 mm, 29 mm, and 31 mm) at different ambient temperatures. At the same time, the mass fraction of EG was also used as a reference variable to draw a more comprehensive conclusion. The basic conditions were set as follows: six cooling pipes, counter flow and 0.1 m/s flow rate.

The simulation results of the maximum battery temperature in the BTMS with different battery spacing at 40 °C and 23.6 °C ambient/coolant temperatures are shown in Figure 16. It can be seen from the comparison that the maximum battery temperature varies significantly under different ambient/coolant temperatures, i.e., the lower the ambient/coolant temperature is, the lower the maximum battery temperature is. At high temperatures, with the change in cell spacing, the corresponding maximum temperature of the battery varies in a small range. This is because the PCM melts and controls the battery temperature in the small temperature range of the phase change process, and the convective heat transfer between PCM and the coolant is weak in the high temperature environment. However, with the decrease in ambient/coolant temperature, the proportion of phase transition of the PCM decreases, while the cooling effect of low temperature liquid cooling gradually increases. In this situation, the maximum temperature of the battery is greatly affected by the distance between the cooling tubes and the battery. In addition, at a constant ambient/coolant temperature, the influence of different cell spacing on the maximum battery temperature is mainly determined by EG mass fractions. When the EG mass fraction is low, the maximum temperature of the battery increases with the increase in spacing. While the EG mass fraction is high, the corresponding maximum temperature is not affected by the spacing. The reason for this phenomenon is that the thermal conductivity of PCM with various EG mass fractions is different. When the heat conduction is insufficient, the closer the cooling source is, the stronger the heat dissipation is. However, when the heat conduction is enough, the cooling effect is not obviously affected by battery spacing.

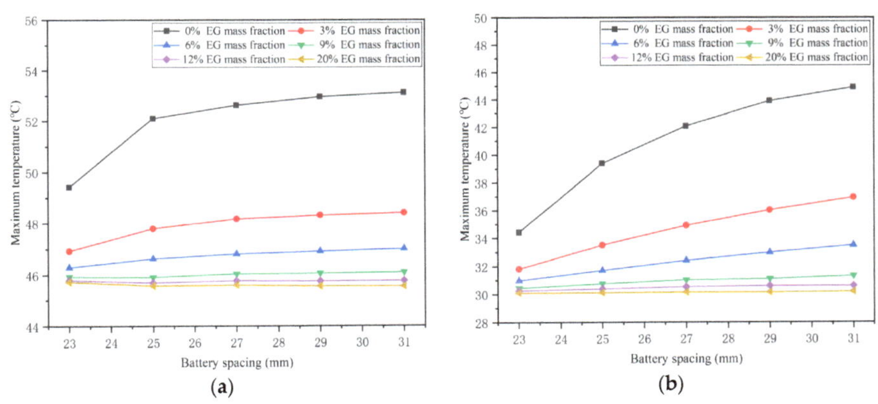

Figure 16. Maximum temperature of the battery in BTMS with different battery spacing and EG mass fractions at (**a**) 40 °C and (**b**) 23.6 °C ambient/coolant temperatures.

The simulation results of the maximum temperature difference in the battery in the BTMS with different battery spacing at 40 °C and 23.6 °C ambient/coolant temperatures are presented in Figure 17. The comparison shows that, at the same battery spacing, the maximum temperature difference in the battery at 40 °C is smaller than that at 23.6 °C. This is because PCM plays a leading role at high temperatures, wherein the cooling effect is more uniform, and liquid cooling plays a leading role at low temperatures, wherein the corresponding temperature uniformity is poor. This phenomenon can be further supplemented by Figure 18a. With the decrease in temperature, the liquid cooling effect is enhanced, and the melting part of PCM decreases, which leads to the gradual increase in the maximum temperature difference in the battery. While the ambient/liquid cooling temperature continues to reduce to a low level, the maximum temperature difference in the battery no longer increases. Moreover, when the battery spacing rises to 25 mm, the maximum temperature difference in the battery decreases to 5 °C below at each external temperature, as shown in Figure 18b. Further increasing the battery spacing will reduce the compactness of the battery pack, so 25 mm spacing is considered the optimal configuration of the BTMS unit with a total volume of 3.5×10^{-5} m^3 and mass of 0.064 kg.

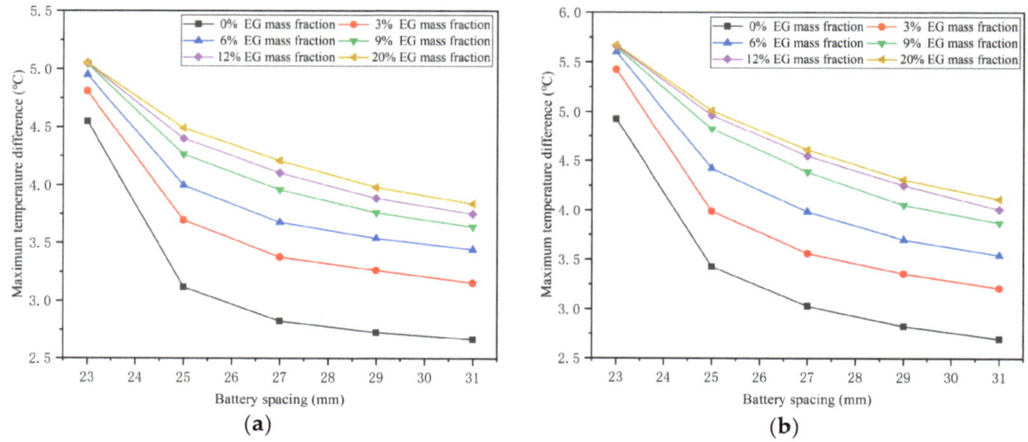

Figure 17. Maximum temperature difference in battery in BTMS with different cell spacing and EG mass fractions at (**a**) 40 °C and (**b**) 23.6 °C ambient/coolant temperatures.

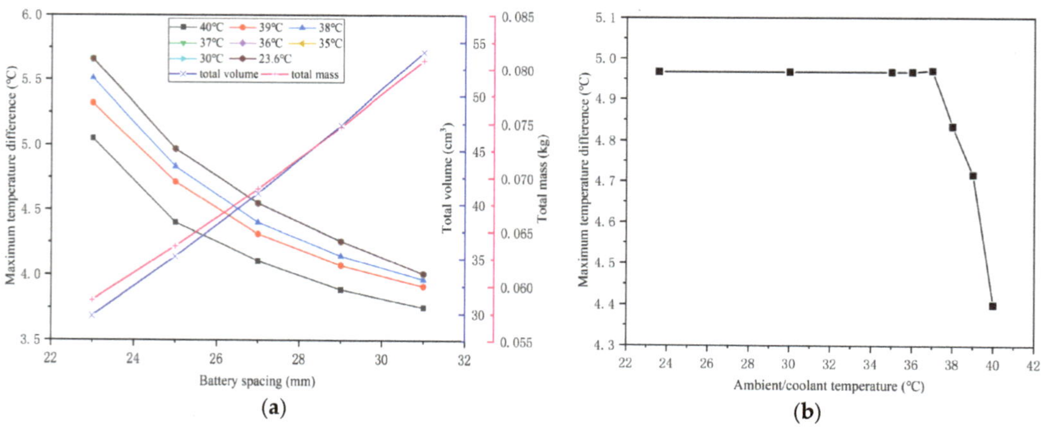

Figure 18. Maximum temperature difference in battery and total volume/mass of BTMS unit with 12% EG mass fraction at various ambient/coolant temperatures, (**a**) different spacing and (**b**) optimal battery spacing of 25 mm.

In addition, at a constant ambient/coolant temperature, the maximum temperature difference decreases with the increase in battery spacing. This is because liquid cooling has the characteristics of uneven temperature distribution at specific locations such as coolant inlets. The closer the distance between the cooling tubes and the battery, the more obvious the local low temperature of the battery, resulting in a larger maximum temperature difference. As shown in Figure 19, at the ambient/coolant temperature of 23.6 °C, when the battery spacing is small, some of the low temperature zone caused by liquid cooling is inside the battery, resulting in a lower minimum temperature. Until the spacing increases to 29 mm, the low temperature zone gradually moves away from the battery.

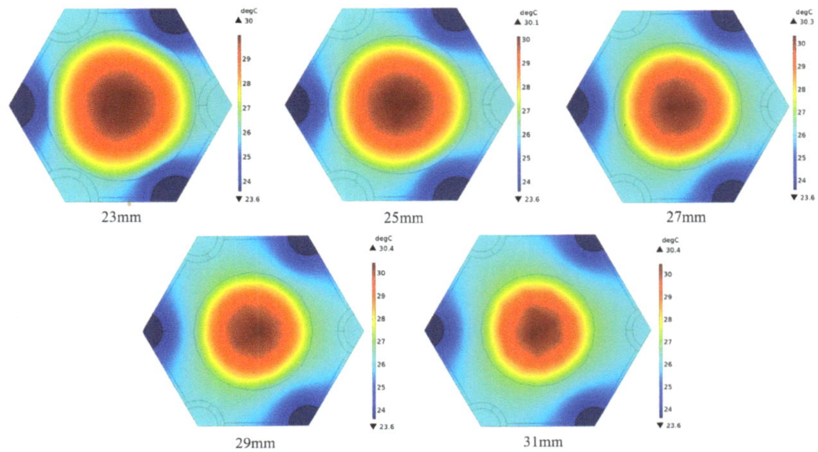

Figure 19. Temperature distribution at the upper-end surface of BTMS at 23.6 °C ambient/coolant temperature, 12% EG mass fraction and different battery spacing.

Overall, at the same ambient/liquid cooling temperature, the distance between batteries has a great impact on the maximum temperature difference, i.e., the corresponding maximum temperature difference decreases with the spacing increases. However, the influence of battery spacing on the maximum temperature is determined by the EG mass fraction. At the high EG mass fraction, the maximum temperature is not affected by battery spacing. At the low EG mass fraction, the distance between batteries has a significant effect; i.e., the maximum temperature increases with the increase in the distance. In addition, PCM plays a major role in cooling, with good temperature uniformity at high ambient/coolant temperatures. With the temperature decrease, the effect of liquid cooling gradually becomes stronger, and the maximum temperature difference becomes larger. Furthermore, a higher battery spacing of 25 mm is selected to optimize the maximum temperature difference in batteries to meet the uniformity requirements.

4.5. Effects of the Coolant Temperature

Based on the above discussion, six cooling pipes and counterflow coolant can be used to improve the maximum temperature and temperature difference in the battery. A large EG mass fraction of 12% is adopted to ensure a lower maximum temperature, and a larger battery spacing of 25 mm can reduce the maximum temperature difference to meet the requirements. Furtherly, to improve the cooling capacity of hybrid cooling BTMS in a high temperature environment, the method of precooling the coolant temperature below the ambient temperature is proposed. Therefore, this section will simulate the BTMS using different coolant temperatures (40 °C, 39 °C, 38 °C, 37 °C, 36 °C, 35 °C, 30 °C) under the environment temperature of 40 °C and discuss the maximum temperature and temperature difference in the battery under the conditions of different temperatures and cell spacing (23 mm, 24 mm, 25 mm, 26 mm).

Figure 20 shows the maximum battery temperature and temperature difference with different temperatures of precooled coolant and battery spacing at the 40 °C ambient temperature. It can be concluded that, with the decrease in coolant temperature, the maximum temperature of the battery decreases considerably, and the maximum temperature difference increases first and then remains unchanged at a constant external temperature and the battery spacing. Specifically, when the coolant temperature begins to drop from 40 °C, the corresponding maximum temperature difference in the battery will gradually increase. When the coolant precools to 37 °C, the maximum temperature difference in the battery will keep stable at a certain high value. The reason for the significant drop in maximum battery temperature at a low coolant temperature is that the heat transfer effect is strong due to the large temperature difference between the coolant and battery. In addition, the reasons for the changes in maximum temperature difference are as follows. When the temperature of the coolant is 37~40 °C, PCM and liquid cooling work together, and the control effect of liquid cooling on the battery temperature gradually becomes stronger as the coolant temperature decreases. Until the coolant temperature is low in the range of 30~37 °C, or even lower, PCM does not reach the melting temperature and only conducts heat, and the liquid cooling plays a leading cooling role, so the corresponding maximum temperature difference in the battery maintains no obvious change within a certain range.

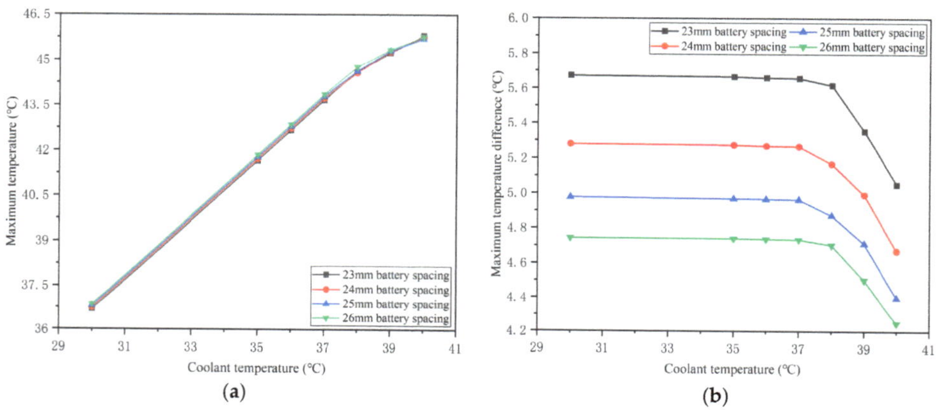

Figure 20. Maximum temperature (**a**) and maximum temperature difference (**b**) of battery under the ambient temperature of 40 °C, different coolant temperatures and battery spacing.

At the same time, at a constant coolant temperature, the battery spacing has little effect on the maximum temperature of the battery but has a significant impact on the maximum temperature difference. When the battery spacing is 23 mm, the maximum temperature difference between the battery and coolant temperature exceeds 5 °C, which does not meet the uniformity requirements. Then, increasing the battery spacing to 24 mm, the maximum temperature difference in the battery is slightly improved, but only the 40 °C coolant without precooling can satisfy the requirements. However, until the spacing is increased to 25 mm, the maximum temperature difference in the battery at each coolant temperature is less than 5 °C. Further expanding the spacing will not be conducive to the compactness of the battery pack. Therefore, 25 mm is also used as the optimal battery spacing for the proposed honeycomb-structured BTMS under the condition of precooling.

The maximum temperature and temperature difference curves of the battery discharge process with different coolant temperatures under the selected optimized battery spacing are shown in Figure 21. As for the maximum temperature difference curve of the battery, it increases rapidly at first, and then it remains flat and finally rises. The lower the coolant temperature is, the higher the temperature difference is at the turning point in the early discharge process. When the battery starts to generate heat, a temperature gradient will form from the inside out in the early discharging stage. The lower the coolant temperature, the lower the minimum temperature of the battery surface, resulting in a rapid increase in the maximum temperature difference. Then, in the middle period of discharging, liquid cooling and PCM act the cooling effect, respectively, according to the different coolant temperatures, which not only makes the maximum battery temperature decrease or stable but also results in the temperature difference decreasing and reaching a balance. In the final stage of discharging, the temperature difference expands again due to the increase in the heat generation rate of the battery. It is worth noting that when the temperature difference between the coolant and the environment is more than 5 °C, the maximum temperature difference in the battery exceeds 5 °C in the discharge process. Therefore, when the liquid cooling can control the maximum battery temperature well, precooling the coolant to a too-low temperature is not recommended to avoid the maximum temperature difference in the discharge process not meeting the conditions. In addition, the maximum battery temperature increases rapidly first, and then it changes gently and finally rises sharply, because the PCM does not reach melting temperature and cannot absorb heat in the early discharge stage, and the heat generation rate of the battery increases rapidly in the last period of discharging. However, whether the maximum temperature rises or falls gently in the middle discharge process depends on the difference between the instantaneous heat generation of the battery and the heat dissipation of cooling measures. With the decrease in coolant temperature, the effect of liquid cooling becomes stronger, making the maximum temperature trend change from flat to falling.

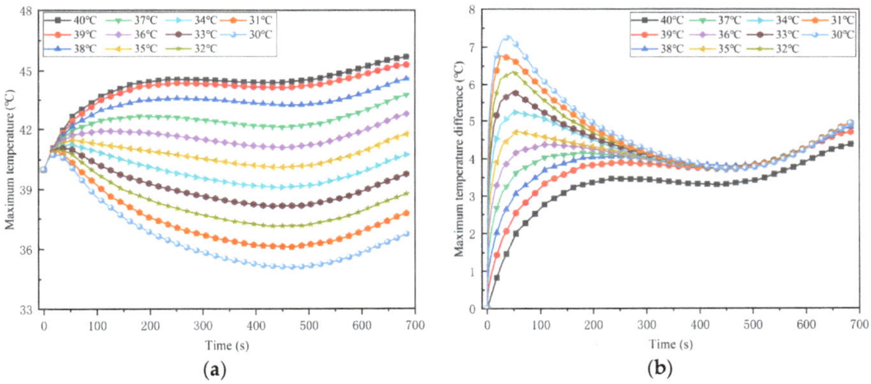

Figure 21. Maximum temperature (**a**) and maximum temperature difference (**b**) of battery during discharging process under 40 °C ambient temperature, 25 mm battery spacing and different coolant temperatures.

Based on the above results, it can be concluded that precooling coolant has an obvious effect on reducing the maximum temperature of the battery. At the ambient temperature of 40 °C, the coolant temperature in the range of 37~40 °C has a significant effect on the maximum temperature difference, while a coolant temperature lower than 37 °C has little effect. In addition, too low a coolant temperature will cause the maximum temperature difference in the battery discharge process to be too high to meet the requirements. Considering the above rules and the structure compactness, the 25 mm battery spacing is selected as the optimal choice of the honeycomb-structured BTMS. The precooling coolant of 40 °C, 39 °C, 38 °C, 37 °C, 36 °C and 35 °C can be selected according to the battery requirements, compared with the corresponding pure PCM-based BTMS, the maximum temperature can be decreased by 0.48 °C, 0.88 °C, 1.55 °C, 2.38 °C, 3.39 °C and 4.39 °C, respectively. Meanwhile, this BTMS can satisfy the requirements of the maximum temperature difference in the battery under all the operating conditions mentioned in this study.

5. Conclusions

To improve the heat dissipation effect of the battery pack at high temperatures and high discharge rates, a battery thermal management system (BTMS) based on honeycomb-structured liquid cooling and phase change materials (PCM) is innovatively proposed in this paper. Moreover, the heat transfer simulations of the simplified model under different structural, material and operation parameters are carried out and analyzed. Through experimental and simulation results, the important findings can be described as follows:

1. The heat source expression of the battery can be obtained by combining the internal resistance and temperature rise experiment results with the Bernardi heat generation equation. According to the experiment results, the higher the discharge rate, the lower the total internal resistance and the higher the temperature rise at the same ambient temperature. In addition, with the decrease in the remaining energy, the total internal resistance of the battery is unchanged at first, and then it increases at the same ambient temperature and discharge rate.
2. The proposed honeycomb structure composed of six cooling tubes and fins using the counter-flow coolant is the optimal basic scheme, which can significantly improve the thermal performance of the battery. In the environment of 40 °C, the maximum temperature and temperature difference in the battery can reach 45.71 °C and 4.4 °C, respectively. Moreover, the maximum battery temperature in this scheme will be further improved and can meet the uniformity requirements by decreasing the ambient temperature.
3. The battery spacing and the mass fraction of expanded graphite (EG) jointly affect the thermal performance of the battery at different temperatures. As the spacing between batteries increases, the corresponding maximum temperature difference in the battery will decrease. However, the effect of battery spacing on the maximum temperature is determined by the mass fraction of EG. At a high EG mass fraction, battery spacing has little impact on the maximum temperature. While the EG mass fraction is low, the maximum temperature increases as the battery spacing rises. In addition, increasing the mass fraction of EG can lower the maximum temperature of the battery and increase the maximum temperature difference.
4. The BTMS based on the honeycomb structure is relatively compact, and PCM plays the leading role of heat absorption at high temperatures, while liquid cooling has a significant effect at low temperatures. Meanwhile, precooling the coolant at high ambient temperatures can not only significantly reduce the battery temperature but also restore the heat storage capacity of PCM. Under the conditions of 40 °C ambient temperature and 5C discharge rate, the maximum battery temperature in the optimal BTMS cooled by 35 °C coolant is 41.79 °C, which is 4.39 °C lower than that of the corresponding pure PCM-based BTMS. Furthermore, when the temperature difference between the coolant and the environment is within 5 °C, the maximum temperature difference during the whole discharge process meets the uniformity requirements.

The proposed BTMS model has a compact structure, uniform heat dissipation, and can prevent the leakage of liquid PCM. To some extent, it can also reduce the harm of thermal runaway of the battery. However, due to the complexity of the structure and cooling measures, high manufacturing costs may be incurred. In addition, the simulation studies of the BTMS under conventional and high temperature and discharge rate conditions are carried out in this paper. To comprehensively consider the thermal performance of the proposed BTMS, more situations such as the actual cycle conditions, thermal runaway and other extreme conditions can be considered. Moreover, generalized optimization and parameter study are presented in this paper. For subsequent work, the response surface model, genetic algorithm and other mathematical models can be used to obtain more accurate optimal parameters.

Author Contributions: Conceptualization, J.X. and T.Y.; methodology, J.Z. and H.Z.; software, T.Y. and Q.X.; validation, S.S. and Q.X.; formal analysis, T.Y. and S.S.; investigation, S.S.; resources, J.X. and J.Z.; data curation, T.Y. and X.Z.; writing—original draft preparation, T.Y. and S.S.; writing—review and editing, J.X., J.Z. and H.Z.; visualization, Q.X.; supervision, X.Z.; project administration, J.Z. and J.X.; funding acquisition, J.X. and H.Z. All authors have read and agreed to the published version of the manuscript.

Funding: This research was funded by the Research Project of Wuhan University of Technology Chongqing Research Institute (YF 2021-08), the Science and Technology Development Foundation of CMVR from China Merchants Testing Certification Vehicle Technology Research Institute Co., Ltd. (20AKC3), the National Natural Science Foundation of China (51876113), the 111 Project of China (B17034), and the Innovative Research Team Development Program of Ministry of Education of China (IRT_17R83).

Data Availability Statement: The data are contained within the article.

Acknowledgments: Thanks should be given to Kodjo Agbossou for his helpful discussions.

Conflicts of Interest: The authors declare no conflict of interest.

References

1. Landini, S.; Leworthy, J.; O'Donovan, T.S. A Review of Phase Change Materials for the Thermal Management and Isothermalisation of Lithium-Ion Cells. *J. Storage Mater.* **2019**, *25*, 100887. [CrossRef]
2. Min, H.; Zhang, Z.; Sun, W.; Min, Z.; Yu, Y.; Wang, B. A thermal management system control strategy for electric vehicles under low-temperature driving conditions considering battery lifetime. *Appl. Therm. Eng.* **2020**, *181*, 115944. [CrossRef]
3. Yang, X.; Deng, G.; Cai, Z.; Li, H.; Zeng, J.; Yang, H. Experimental study on novel composite phase change materials with room-temperature flexibility and high-temperature shape stability in a battery thermal management system. *Int. J. Heat Mass Transfer* **2023**, *206*, 123953. [CrossRef]
4. Shelkea, A.V.; Buston, J.E.H.; Gill, J.; Howard, D.; Williams, R.C.E.; Read, E.; Abaza, A.; Cooper, B.; Richards, P.; Wen, J.X. Combined numerical and experimental studies of 21,700 lithium-ion battery thermal runaway induced by different thermal abuse. *Int. J. Heat Mass Transfer* **2022**, *194*, 123099. [CrossRef]
5. Kang, D.; Lee, P.-Y.; Yoo, K.; Kim, J. Internal thermal network model-based inner temperature distribution of high-power lithium-ion battery packs with different shapes for thermal management. *J. Storage Mater.* **2020**, *27*, 101017. [CrossRef]
6. Liu, H.; Wei, Z.; He, W.; Zhao, J. Thermal issues about Li-ion batteries and recent progress in battery thermal management systems: A review. *Energy Convers. Manag.* **2017**, *150*, 304–330. [CrossRef]
7. Rao, Z.; Qian, Z.; Kuang, Y.; Li, Y. Thermal performance of liquid cooling based thermal management system for cylindrical lithium-ion battery module with variable contact surface. *Appl. Therm. Eng.* **2017**, *123*, 1514–1522. [CrossRef]
8. Landini, S.; O'Donovan, T.S. Experimental investigation of lithium-ion cells ageing under isothermal conditions for optimal lifetime performance. *J. Storage Mater.* **2022**, *48*, 103680. [CrossRef]
9. Lisbona, D.; Snee, T. A review of hazards associated with primary lithium and lithium-ion batteries. *Process Saf. Environ. Prot.* **2011**, *89*, 434–442. [CrossRef]
10. Wang, Q.; Jiang, B.; Li, B.; Yan, Y. A critical review of thermal management models and solutions of lithium-ion batteries for the development of pure electric vehicles. *Renew. Sustain. Energy Rev.* **2016**, *64*, 106–128. [CrossRef]
11. Akinlabi, A.A.H.; Solyali, D. Configuration, design, and optimization of air-cooled battery thermal management system for electric vehicles: A review. *Renew. Sustain. Energy Rev.* **2020**, *125*, 109815. [CrossRef]
12. Weragoda, D.M.; Tian, G.; Burkitbayev, A.; Lo, K.-H.; Zhang, T. A comprehensive review on heat pipe based battery thermal management systems. *Appl. Therm. Eng.* **2023**, *224*, 120070. [CrossRef]
13. Panchal, S.; Dincer, I.; Agelin-Chaab, M.; Fraser, R.; Fowler, M. Experimental and theoretical investigation of temperature distributions in a prismatic lithium-ion battery. *Int. J. Therm. Sci.* **2016**, *99*, 204–212. [CrossRef]
14. Gungor, S.; Cetkin, E.; Lorente, S. Canopy-to-canopy liquid cooling for the thermal management of lithium-ion batteries, a constructal approach. *Int. J. Heat Mass Transfer* **2022**, *182*, 121918. [CrossRef]
15. Tong, W.; Somasundaram, K.; Birgersson, E.; Mujumdar, A.S.; Yap, C. Numerical investigation of water cooling for a lithium-ion bipolar battery pack. *Int. J. Therm. Sci.* **2015**, *94*, 259–269. [CrossRef]
16. Yin, B.; Zuo, S.; Xu, Y.; Chen, S. Performance of liquid cooling battery thermal management system in vibration environment. *J. Storage Mater.* **2022**, *53*, 105232. [CrossRef]
17. Lai, Y.; Wu, W.; Chen, K.; Wang, S.; Xin, C. A compact and lightweight liquid-cooled thermal management solution for cylindrical lithium-ion power battery pack. *Int. J. Heat Mass Transfer* **2019**, *144*, 118581. [CrossRef]
18. Liu, Z.; Wang, H.; Yang, C.; Zhao, J. Simulation study of lithium-ion battery thermal management system based on a variable flow velocity method with liquid metal. *Appl. Therm. Eng.* **2020**, *179*, 115578. [CrossRef]
19. Wang, Y.; Wang, Z.; Min, H.; Li, H.; Li, Q. Performance investigation of a passive battery thermal management system applied with phase change material. *J. Storage Mater.* **2021**, *35*, 102279. [CrossRef]
20. Lazrak, A.; Fourmigué, J.-F.; Robin, J.-F. An innovative practical battery thermal management system based on phase change materials: Numerical and experimental investigations. *Appl. Therm. Eng.* **2018**, *128*, 20–32. [CrossRef]
21. Luo, J.; Zou, D.; Wang, Y.; Wang, S.; Huang, L. Battery thermal management systems (BTMs) based on phase change material (PCM): A comprehensive review. *Chem. Eng. J* **2022**, *430*, 132741. [CrossRef]

22. He, J.; Yang, X.; Zhang, G. A phase change material with enhanced thermal conductivity and secondary heat dissipation capability by introducing a binary thermal conductive skeleton for battery thermal management. *Appl. Therm. Eng.* **2019**, *148*, 984–991. [CrossRef]
23. Jiang, G.; Huang, J.; Fu, Y.; Cao, M.; Liu, M. Thermal optimization of composite phase change material/expanded graphite for Li-ion battery thermal management. *Appl. Therm. Eng.* **2016**, *108*, 1119–1125. [CrossRef]
24. Mauro, G.M.; Iasiello, M.; Bianco, N.; Chiu, W.K.S.; Naso, V. Mono- and Multi-Objective CFD Optimization of Graded Foam-Filled Channels. *Materials* **2022**, *15*, 968. [CrossRef] [PubMed]
25. Verma, A.; Rakshit, D. Performance analysis of PCM-fin combination for heat abatement of Li-ion battery pack in electric vehicles at high ambient temperature. *Therm. Sci. Eng. Prog* **2022**, *32*, 101314. [CrossRef]
26. Arshad, A.; Jabbal, M.; Sardari, P.T.; Bashir, M.A.; Faraji, H.; Yan, Y. Transient simulation of finned heat sinks embedded with PCM for electronics cooling. *Therm. Sci. Eng. Prog* **2020**, *18*, 100520. [CrossRef]
27. Bianco, N.; Busiello, S.; Iasiello, M.; Mauro, G.M. Finned heat sinks with phase change materials and metal foams: Pareto optimization to address cost and operation time. *Appl. Therm. Eng.* **2021**, *197*, 117436. [CrossRef]
28. Lebrouhi, B.E.; Lamrani, B.; Ouassaid, M.; Abd-Lefdil, M.; Maaroufi, M.; Kousksou, T. Low-cost numerical lumped modelling of lithium-ion battery pack with phase change material and liquid cooling thermal management system. *J. Storage Mater.* **2022**, *54*, 105293. [CrossRef]
29. Fan, Y.; Wang, Z.; Xiong, X.; Zhu, J.; Gao, Q.; Wang, H.; Wu, H. Novel concept design of low energy hybrid battery thermal management system using PCM and multistage Tesla valve liquid cooling. *Appl. Therm. Eng.* **2023**, *220*, 119680. [CrossRef]
30. Xin, Q.; Xiao, J.; Yang, T.; Zhang, H.; Long, X. Thermal management of lithium-ion batteries under high ambient temperature and rapid discharging using composite PCM and liquid cooling. *Appl. Therm. Eng.* **2022**, *210*, 118230. [CrossRef]
31. Weng, J.; Xiao, C.; Yang, X.; Ouyang, D.; Chen, M.; Zhang, G.; Lee Waiming, E.; Kit Yuen, R.K.; Wang, J. An energy-saving battery thermal management strategy coupling tubular phase-change-material with dynamic liquid cooling under different ambient temperatures. *Renew. Energ.* **2022**, *195*, 918–930. [CrossRef]
32. Peng, P.; Wang, Y.; Jiang, F. Numerical study of PCM thermal behavior of a novel PCM-heat pipe combined system for Li-ion battery thermal management. *Appl. Therm. Eng.* **2022**, *209*, 118293. [CrossRef]
33. Qi, X.; Sajadi, S.M.; Mahmoud, M.Z.; Li, Z.; Shamseldin, M.A.; Aybar, H.Ş. Study of circular, horizontal and vertical elliptical enclosures filled with phase change material in thermal management of lithium-ion batteries in an air-cooled system. *J. Storage Mater.* **2022**, *53*, 105041. [CrossRef]
34. Yang, W.; Zhou, F.; Liu, Y.; Xu, S.; Chen, X. Thermal performance of honeycomb-like battery thermal management system with bionic liquid mini-channel and phase change materials for cylindrical lithium-ion battery. *Appl. Therm. Eng.* **2021**, *188*, 116649. [CrossRef]
35. Zhao, D.; Lei, Z.; An, C. Research on battery thermal management system based on liquid cooling plate with honeycomb-like flow channel. *Appl. Therm. Eng.* **2023**, *218*, 119324. [CrossRef]
36. Wang, S.; Fernandez, C.; Yu, C.; Fan, Y.; Cao, W.; Stroe, D.-I. A novel charged state prediction method of the lithium ion battery packs based on the composite equivalent modeling and improved splice Kalman filtering algorithm. *J. Power Sources* **2020**, *471*, 228450. [CrossRef]
37. Landini, S.; O'Donovan, T.S. Novel experimental approach for the characterisation of Lithium-Ion cells performance in isothermal conditions. *Energy* **2021**, *214*, 118965. [CrossRef]
38. Newman, J.; Bernardi, D.; Pawlikowski, E. A General Energy-Balance for Battery Systems. *J. Electrochem. Soc.* **1985**, *132*, 5.
39. Liu, F.; Wang, J.; Liu, Y.; Wang, F.; Yang, N.; Liu, X.; Liu, H.; Li, W.; Liu, H.; Huang, B. Performance analysis of phase change material in battery thermal management with biomimetic honeycomb fin. *Appl. Therm. Eng.* **2021**, *196*, 117296. [CrossRef]
40. Weng, J.; He, Y.; Ouyang, D.; Yang, X.; Chen, M.; Cui, S.; Zhang, G.; Yuen, R.K.K.; Wang, J. Honeycomb-inspired design of a thermal management module and its mitigation effect on thermal runaway propagation. *Appl. Therm. Eng.* **2021**, *195*, 117147. [CrossRef]
41. Lu, J. The Research and Simulation Analysis of Heat Dissipation with Liquid Cooling for the Power Battery Pack of Pure Electric Vehicles. Master's Thesis, University of Electronic Science and Technology of China, Chengdu, China, 2018.
42. Ling, Z.; Chen, J.; Xu, T.; Fang, X.; Gao, X.; Zhang, Z. Thermal conductivity of an organic phase change material/expanded graphite composite across the phase change temperature range and a novel thermal conductivity model. *Energy Convers. Manag.* **2015**, *102*, 202–208. [CrossRef]

batteries

MDPI

Article

Simulation and Optimization of Lithium-Ion Battery Thermal Management System Integrating Composite Phase Change Material, Flat Heat Pipe and Liquid Cooling

Qianqian Xin [1,2], Tianqi Yang [1,*], Hengyun Zhang [3,*], Juan Zeng [1,2] and Jinsheng Xiao [1,2]

1 Hubei Research Center for New Energy & Intelligent Connected Vehicle, School of Automotive Engineering, Wuhan University of Technology, Wuhan 430070, China; xin11280429@whut.edu.cn (Q.X.); zengjuan1973@whut.edu.cn (J.Z.); jinsheng.xiao@whut.edu.cn (J.X.)
2 Chongqing Research Institute, Wuhan University of Technology, Chongqing 401135, China
3 School of Mechanical and Automotive Engineering, Shanghai University of Engineering Science, Shanghai 201620, China
* Correspondence: tqyang@whut.edu.cn (T.Y.); zhanghengyun@sues.edu.cn (H.Z.)

Abstract: A large-capacity prismatic lithium-ion battery thermal management system (BTMS) combining composite phase change material (CPCM), a flat heat pipe (FHP), and liquid cooling is proposed. The three conventional configurations analyzed in this study are the BTMSs using only CPCM, CPCM with aluminum thermal diffusion plates, and CPCM with FHPs. In addition, a CPCM–FHP assisted with liquid cooling at the lateral sides is established to enhance the thermal performance of large-capacity batteries. Moreover, the influences of coolant temperature, the number of FHPs and cooling pipes, and the coolant direction on the temperature field of a BTMS are discussed. Finally, the orthogonal design method is used for the multi-level analysis of multiple factors to improve the light weight of the system. The optimal parameter combination is obtained to achieve the best thermal performance of the BTMS, with the maximum temperature and the temperature difference at 43.17 °C and 3.36 °C, respectively, under a maximum discharge rate of 2C and a high-temperature environment of 37 °C. The optimal scheme is further analyzed and affirmed through the comprehensive balance method.

Keywords: lithium-ion battery; thermal management; phase change material; flat heat pipe; liquid cooling; orthogonal design method

Citation: Xin, Q.; Yang, T.; Zhang, H.; Zeng, J.; Xiao, J. Simulation and Optimization of Lithium-Ion Battery Thermal Management System Integrating Composite Phase Change Material, Flat Heat Pipe and Liquid Cooling. *Batteries* **2023**, *9*, 334. https://doi.org/10.3390/batteries9060334

Academic Editor: Carlos Ziebert

Received: 23 May 2023
Revised: 13 June 2023
Accepted: 15 June 2023
Published: 20 June 2023

1. Introduction

Electric vehicles can significantly reduce dependence on fossil energy and have the advantages of energy conservation and emission reduction, compared with traditional fuel vehicles, which are the critical link to achieving the double carbon goal and green development. As the power source of most electric vehicles, the lithium-ion battery (LIB) is easily affected by operation temperature during work. Therefore, the battery thermal management system (BTMS), which is energy-saving, efficient, and safe, could effectively enhance the thermal safety and stable operation of electric vehicles.

The cooling mediums of BTMSs mainly include air, liquid, heat pipe, and phase change material (PCM). Li et al. [1] established an electric thermal model based on fluid dynamics and studied the model under natural ventilation conditions. Then, the output training and optimization design of the battery module were carried out through a model based on an artificial neural network (ANN). The results showed better optimization of the battery layout and cooling efficiency. Yamanaka et al. [2] proposed an equivalent circuit model and applied it to a BTMS based on liquid cooling. The simulation results were fitted well with the experiment results. Furthermore, the accuracy of the relationship between the coolant flow rate and the pressure drop or heat transfer coefficient was high. Chen et al. [3]

put forward three structures of BTMSs: Structure 1, which is based on hollow spoiler prisms with batteries; Structure 2, which adds a PCM to Structure 1; and Structure 3, which adds aluminum fins based on Structure 2. The first two structures have better thermal performance than conventional air cooling systems, and Structure 3 possesses the best cooling performance.

The requirements for the BTMS are more urgent with the continuous development of electric vehicles and the increase in energy density of battery modules or the driving of cars under high-temperature and rapid discharging rate conditions. Currently, a BTMS that combines active and passive cooling is a very effective cooling method. The most common ways are to combine composite phase change material (CPCM) with air cooling [4–7], liquid cooling [8–11], or heat pipes [12–14].

Khan et al. [15] used air with different flow rates to cool cylindrical batteries inside PCM and studied the effects of cross-sectional shapes on PCMs at room temperature. The results indicated that the increase in air flow rate could decrease the battery's temperature and enhance the heat transfer performance of the PCM. Faizan et al. [16] designed two structures of BTMSs with PCM and liquid cooling under high discharge rates and room temperature. The first is the PCM sandwiched between the batteries, and the second is the PCM wrapped around the batteries. The effects of the coolant velocity, the volume fraction of the nanofluid, and the flow direction were analyzed. The results showed that the latter had better thermal performance than the BTMS. Abbas et al. [12] proposed two types of BTMS based on the PCM and HPs under room temperature and a high heat generation rate, including the indirect/direct contact between the heat pipe evaporation section and the battery. It was verified through experiments that the latter had a better temperature control effect on the BTMS. For the HP-based BTMS, the range of the temperature difference was small, and the range of the maximum temperature was significantly lower than the heat transfer characteristics of the HP during the actual discharging. In addition, the heat transfer coefficient of the HP is generally stable within a specific field and has no significant volatility. Therefore, the simplified heat transfer model in this work is adopted [17–20], simplifying the HP into a flat plate 2 mm in width with the same length and height as the battery and high thermal conductivity as the actual HP. The heat transfer mode is adopted during the heat transfer, ignoring the complex internal heat transfer, mass transfer, vapor-liquid flow, and momentum transfer, and the copper HP is selected [21].

Xu et al. [22] compared the heat generation of a single large-capacity prismatic battery and a small-capacity cylindrical battery with the influence of the depth of discharge (DOD) at different discharge rates and environmental temperatures. The results showed that the heat production of the prismatic battery was higher than that of the cylindrical battery when the DOD was below 0.7 and lower in the later stage of discharge due to the large polarization internal resistance of the cylindrical battery. The thermal performance of the BTMS based on 106 large-sized batteries [23,24] was studied under room temperature and high discharge rates using the hybrid cooling of a PCM and a liquid. However, a high-temperature environment was not considered. In addition, thermoelectric cooling using thermoelectric coolers is a novel method for battery cooling applications. Sun et al. [25] pointed out that thermoelectric coolers have a wide range of potential applications in miniaturized electronics and have a better temperature control effect on battery cooling than liquid cooling and natural convection methods at room temperature.

Most studies focus on the thermal characteristics of small-capacity batteries under low environmental temperatures and high discharge rates or high environmental temperatures and low discharge rates. Compared with small-capacity batteries, the thermal management requirements are higher with grouped large-capacity batteries. There are few studies on large-capacity battery modules under high-temperature environments and high discharge rate conditions. In addition, many factors affect the heat dissipation of batteries in BTMSs. Quantitative analysis based solely on parameter research cannot comprehensively evaluate system performance. Appropriate optimization methods should be selected to conduct a qualitative analysis of the influencing factors to design the optimal structure and improve the light weight, economy, and safety of a BTMS.

In this work, a large-capacity prismatic BTMS integrating CPCM, FHPs, and liquid cooling is established. The thermal characteristics of the BTMS based on only CPCM, CPCMs with aluminum thermal diffusion plates (ATDPs), and CPCMs with FHPs were compared first. Then, the liquid cooling was assisted in the CPCM–FHP model to further improve thermal performance at the maximum discharge rate of 2C and the high temperature of 37 °C. In addition, the effects of the number of FHPs, the coolant temperature, the number of cooling pipes, and the coolant flow direction were optimized through the orthogonal design method. Finally, the parameter combination that enabled the BTMS to have optimal thermal performance was obtained.

2. Physical Model and Mathematical Formulation

2.1. Physical Model of Battery Modules

In this work, the thermal performance of the combined CPCM, FHPs, and liquid cooling are examined through numerical simulation compared with the schemes without combined liquid cooling techniques. Four schemes of BTMSs are proposed: Scheme 1 with CPCM cooling only; Scheme 2 with combined CPCMs and ATDPs; Scheme 3 with combined CPCMs and FHPs; and Scheme 4 with combined CPCMs, FHPs, and liquid cooling, as shown in Figure 1a–d, respectively.

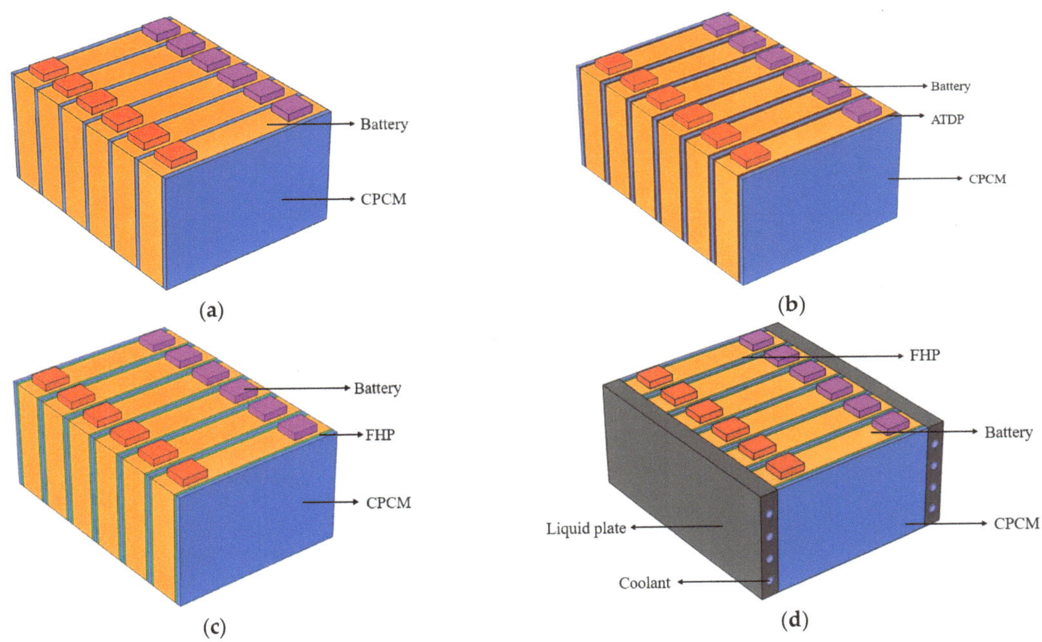

Figure 1. Four schemes of battery module structures: Scheme 1 with CPCM only (a), Scheme 2 with CPCMs and ATDPs (b), Scheme 3 with CPCMs and FHPs (c), and Scheme 4 combined CPCMs, FHPs, and liquid cooling (**d**).

Six nickel–cobalt–manganese LIBs with a 50 Ah capacity were connected electrically in series to form the battery module. The width, height, and depth of the battery were 148 mm, 92 mm, and 27 mm, respectively. The CPCMs (the blue part) 4 mm in thickness, made from a mixture of paraffin wax and EG by Shenzhen HM New Material Technology Co., Ltd., were sandwiched between batteries in Scheme 1. A total of 12 ATDPs (the brown part) 4 mm thickness were sandwiched between batteries in Scheme 2. FHPs (the green part) 2 mm thickness were set up to study the impact of the heat pipe on the thermal performance of the BTMS in Scheme 3. Furthermore, Scheme 4 added liquid cooling on both sides of the battery module based on Scheme 3, with four cooling channels with a diameter of 6 mm inside each Al plate. Furthermore, the comparison of the cooling methods in the four schemes is shown in Table 1. The parallel flow on the same side of the battery module was set inlet while the other side was set outlet, and water was chosen as the coolant. In addition, the thermophysical properties of the components in the battery module are shown in Table 2. The melting temperature and the latent heat of the CPCM were 37 °C~40 °C and 150 kJ/kg, respectively. Considering that 37 °C is the critical temperature of phase change of PCM, this work studied the thermal performance of large-capacity BTMSs under a high-temperature environment of 37 °C and a maximum discharge rate of 2C to prevent the occurrence of thermal runaway.

Table 1. Comparison of cooling methods of the four schemes.

Design Schemes	Cooling Methods
Scheme 1	Only 7 CPCMs
Scheme 2	7 CPCMs and 12 ATDPs
Scheme 3	7 CPCMs and 12 FHPs
Scheme 4	7 CPCMs, 12 FHPs, and liquid cooling

Table 2. Thermophysical properties of components in the battery module.

Materials	Density (kg/m^3)	Specific Heat Capacity (J/(kg·K))	Thermal Conductivity (W/(m·K))
Battery	2300	1072	18.5, 18.5, 1.5
Aluminum	2719	871	202.4
Water	998.2	4128	0.6
Heat pipe	8978	381	6000
CPCM	1000	2000	1.2

2.2. Mathematical Formulation

Solid heat transfer of the BTMS models involving batteries, CPCMs, ATDPs, FHPs and the energy conservation equations involved is as follows:

$$\rho_b c_{p,b} \frac{\partial T_b}{\partial t} = \nabla \cdot (k_b \nabla T_b) + Q_{\mathrm{VOL}} \tag{1}$$

$$\rho_{\mathrm{PCM}} c_{p,\,\mathrm{eff}} \frac{\partial T_{\mathrm{PCM}}}{\partial t} = \nabla \cdot (k_{\mathrm{PCM}} \nabla T_{\mathrm{PCM}}) \tag{2}$$

$$\rho_{\mathrm{ATDP}} c_{p,\,\mathrm{ATDP}} \frac{\partial T_{\mathrm{ATDP}}}{\partial t} = \nabla \cdot (k_{\mathrm{ATDP}} \nabla T_{\mathrm{ATDP}}) \tag{3}$$

$$\rho_{\mathrm{FHP}} c_{p,\,\mathrm{FHP}} \frac{\partial T_{\mathrm{FHP}}}{\partial t} = \nabla \cdot (k_{\mathrm{FHP}} \nabla T_{\mathrm{FHP}}) \tag{4}$$

where ρ, c_p, and k represent the density, specific heat capacity, and thermal conductivity of the different materials, respectively, and Q_{VOL} represents the heat generation rate of the battery.

Moreover, the velocity of cooling water is 0.1 m/s, the corresponding Reynolds number is 598.92, and the laminar model is employed for cooling flow. The mass, momentum, and energy equations of the liquid cooling are shown as follows, respectively:

$$\frac{\partial \rho_w}{\partial t} + \nabla \cdot \left(\rho_w \vec{v} \right) = 0 \tag{5}$$

$$\frac{\partial}{\partial t}\left(\rho_w \vec{v}\right) + \nabla \cdot \left(\rho_w \vec{v}\vec{v}\right) = -\nabla P + \nabla \cdot \left(\mu \nabla \vec{v}\right) \tag{6}$$

$$\frac{\partial}{\partial t}\left(\rho_w C_{p,w} T_w\right) + \nabla \cdot \left(\rho_w C_{p,w} \vec{v} T_w\right) = \nabla \cdot \left(k_w \nabla T_w\right) \tag{7}$$

where ρ_w, $\vec{v}$, $C_{p,w}$, k_w, and P are the density, velocity vector, specific heat capacity, thermal conductivity, and static pressure, respectively.

2.3. Boundary and Initial Conditions

We employed COMSOL Multiphysics [25] to set the interfacial boundary conditions. The interfacial boundary conditions between the battery and the CPCM, the battery and the ATDP, the battery and the FHP, the FHP and the CPCM, the Al plate and the cooling pipe were set as follows:

$$-k_b \frac{\partial T_b}{\partial n_b} = -k_{\mathrm{PCM}} \frac{\partial T_{\mathrm{PCM}}}{\partial n_{\mathrm{PCM}}} \tag{8}$$

$$-k_b \frac{\partial T_b}{\partial n_b} = -k_{\mathrm{ATDP}} \frac{\partial T_{\mathrm{ATDP}}}{\partial n_{\mathrm{ATDP}}} \tag{9}$$

$$-k_b \frac{\partial T_b}{\partial n_b} = -k_{\mathrm{FHP}} \frac{\partial T_{\mathrm{PCM}}}{\partial n_{\mathrm{PCM}}} \tag{10}$$

$$-k_{\mathrm{PCM}} \frac{\partial T_{\mathrm{PCM}}}{\partial n_{\mathrm{PCM}}} = -k_{\mathrm{FHP}} \frac{\partial T_{\mathrm{PCM}}}{\partial n_{\mathrm{PCM}}} \tag{11}$$

where n represents the direction along the normal of the outer surface of each component, and the $\partial T/\partial n$ represents the temperature gradient of each component.

$$-k_{\mathrm{PCM}} \frac{\partial T_{\mathrm{PCM}}}{\partial n_{\mathrm{PCM}}} = h\left(T_{\mathrm{PCM}} - T_{\mathrm{amb}}\right) \tag{12}$$

$$-k_{\mathrm{FHP}} \frac{\partial T_{\mathrm{FHP}}}{\partial n_{\mathrm{FHP}}} = h\left(T_{\mathrm{FHP}} - T_{\mathrm{amb}}\right) \tag{13}$$

where h is the convection heat transfer coefficient between PCM or FHP and ambient.

The initial condition is $T = T_0$ and $v = v_0$ at $t = 0$, where T_0 is the ambient temperature, while v_0 is the inlet velocity, set as 0.1 m/s. The inlet temperature of the coolant was 37 °C. Moreover, the coolant inlet was set as the velocity inlet boundary condition, the coolant outlet was set as the pressure outlet boundary condition, and the other surfaces adopted the no-slip boundary condition.

3. Model Verification and Comparison

3.1. Verification of Heat Generation Rate

The experimental data of heat generation rates [26] is used for the verification of the simulation model for this application. Figure 2 displays the comparison of the temperature rise trend using the simulation of dynamic and average heat generation rates with the experimental results at the discharge rates of 0.5C, 1C, 1.5C, and 2C and the environmental temperatures of −5 °C, 10 °C, 25 °C, and 40 °C. The dynamic heat generation rate is the transient value obtained from the discharging experiment. The average heat generation rate is the average value of the dynamic heat generation rate during the discharge process. Based on the dynamic heat generation rate, the calculation accuracy of the battery thermal model is the highest, with the maximum temperature difference within 1 °C. Though the battery temperature values were slightly underestimated, the average heat generation model yielded an acceptable engineering accuracy of 5%.

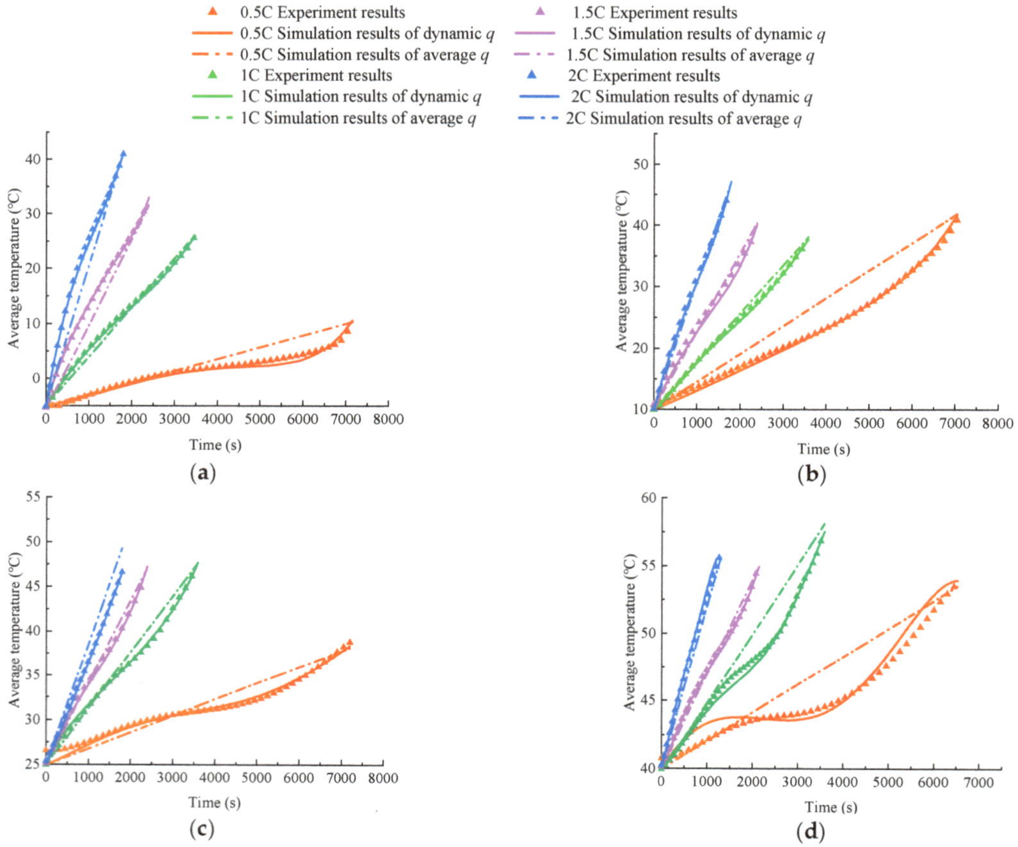

Figure 2. Comparison of battery temperatures obtained using simulation of the dynamic and average heat generation rates with experimental results at environmental temperatures of (**a**) −5°C, (**b**) 10°C, (**c**) 25°C and (**d**) 40°C under discharge rates of 0.5C, 1C, 1.5C and 2C (distinguished in colours).

The simulation result based on the average heat generation rate was lower than the experimental result at the beginning of discharge. There was contact thermal resistance during the experiment, which was ignored during the simulation. Moreover, the reaction heat was not considered in the simulation model. The increase in polarization resistance of the battery led to a rise in polarization heat in the later stage of discharge, resulting in a slightly higher simulation result than the experimental result. There were also specific errors between various devices in the experiment. However, the overall trend between the simulation and experiment results is consistent. The experimental value and the simulated one are almost the same, finally. Therefore, balancing the accuracy and efficiency of calculation, the battery thermal model based on the average heat generation rate will be established in the subsequent work.

3.2. Result Comparison of Battery Modules

Figure 3a,b display the mesh model of Scheme 3 and the independent analysis of the grid number in Scheme 3 under a 1C discharge rate and 25 °C ambient temperature and natural convection environment.

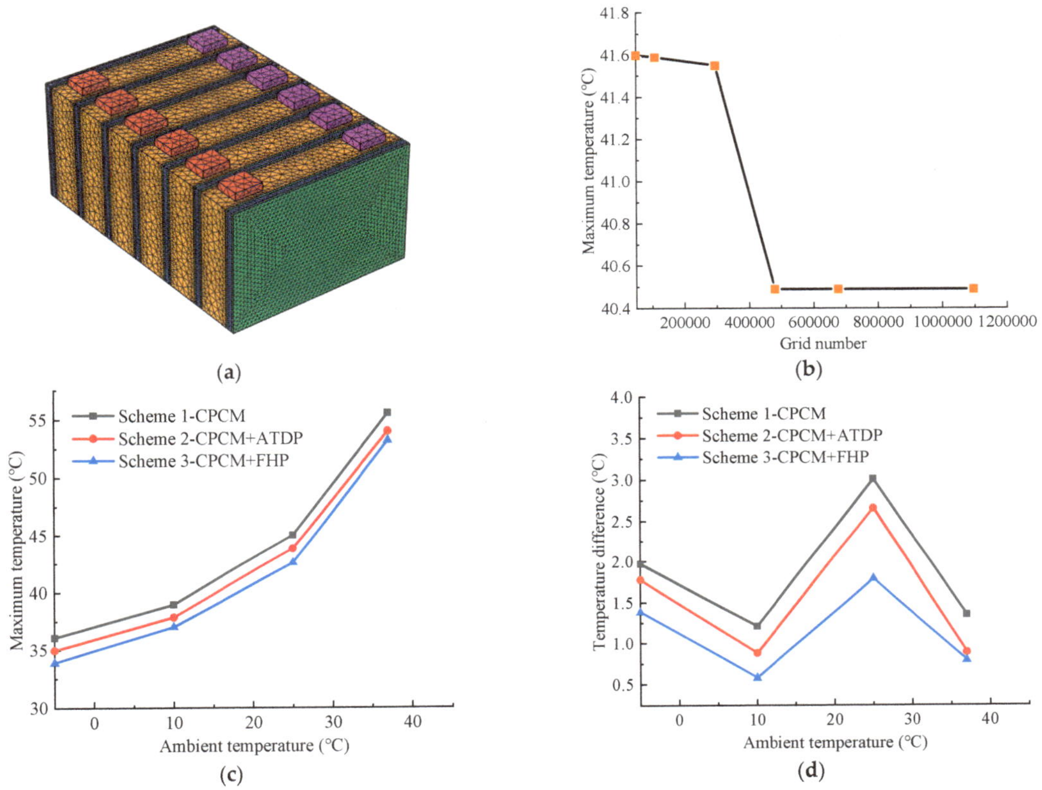

Figure 3. Mesh of Scheme 3 (**a**), independence analysis of grid in Scheme 3 under 1C discharge rate and 25 °C ambient temperature and natural convection (**b**), and comparisons of maximum battery temperatures (**c**) and temperature differences (**d**) in Scheme 1, Scheme 2, and Scheme 3 with 12 FHPs under a 2C discharge rate and ambient temperatures of −5 °C, 10 °C, 25 °C, and 37 °C.

The tetrahedral and hexahedral grids of the battery module were divided using COMSOL Multiphysics. The grid independence study was examined using the number of grids, including 49,423, 108,896, 295,348, 478,898, 675,556, and 1,096,202. It is seen that the maximum temperature decreased with the increase in the grid cells, whose trend was stable at grid number 478,898. Furthermore, the independent analysis of the time step at 25 s was also simulated.

Figure 3c,d are the comparisons of the battery temperatures in Scheme 1, Scheme 2, and Scheme 3 with 12 FHPs under a 2C discharge rate and ambient temperatures of −5 °C, 10 °C, 25 °C, and 37 °C, respectively. It was verified that the maximum temperature and temperature difference of Scheme 3, the combined CPCM with FHPs, were lower than in Schemes 1 and 2. It is ascertained that the BTMS with CPCM and FHPs owned better thermal performance. However, it could not keep the maximum temperature of the batteries within the normal temperature range (50 °C) [4,27,28] under a maximum discharge rate of 2C and a high temperature of 37 °C. Therefore, liquid cooling was assisted in the CPCM–FHP system in the following study to improve thermal performance.

4. Scheme Design and Parameter Discussion

4.1. Scheme Design

From the previous analysis, it can be seen that Scheme 4 has optimal thermal performance. This section proposes three more schemes based on Scheme 4 for comparative verification. The CPCMs in Scheme 4 are set as mica plates, namely, Scheme 4-1. The CPCMs in Scheme 4 are set as air, namely, Scheme 4-2. In addition, all the CPCMs are removed from Scheme 4, namely, Scheme 5, using a combination of seven FHPs and liquid cooling. Figure 4 displays the battery temperatures in the schemes under the 2C discharge rate and 37 °C ambient temperature. It can be seen that the temperatures of the schemes meet the temperature requirements of the BTMS, and the maximum temperature of Scheme 4 is the lowest. In addition, from the perspective of structural simplification, the design of Scheme 5 is also acceptable. To further study the thermal performance of the BTMS, integrating CPCM, FHPs, and liquid cooling, parameter study and optimization will be conducted based on Scheme 4 in the subsequent work.

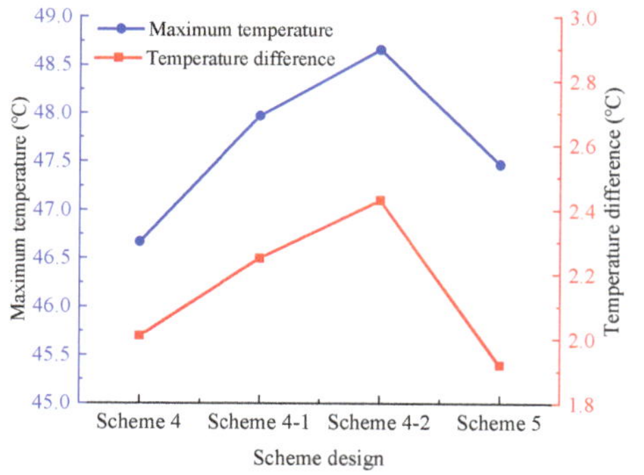

Figure 4. Maximum temperature and temperature difference in the battery module of different schemes under 2C discharge rate and 37 °C ambient temperature.

4.2. Effect of FHP Number

The FHPs are applied between batteries in the BTMS based on CPCM to improve thermal performance. Therefore, the FHP number is set to 6, 8, 12, and 14, respectively, shown in Figure 5, to improve the temperature uniformity of the BTMS. Six FHPs are decorated in Figure 5a, attached to one side of the six batteries. Eight FHPs are set in Figure 5b, which are placed on both sides of the four middle batteries to strengthen the cooling capacity of the battery module. Twelve FHPs are employed in Figure 5c on both sides of the six batteries. Moreover, Figure 5d is a combination of Figure 5c and two FHPs on the two outsides of the battery module.

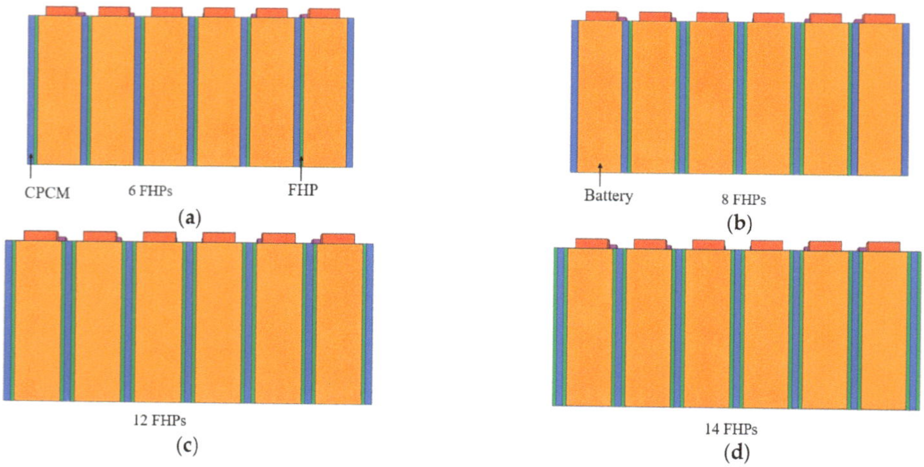

Figure 5. Battery module with different numbers 6, 8, 12 and 14 (**a–d**) of FHPs in Scheme 3.

Air cooling based on forced convection is also a common cooling method of BTMSs. According to Figure 3c, the BMTS using CPCM and FHPs cannot meet the operating temperature requirements. Therefore, the BTMS using CPCM–FHP combined with air cooling and liquid cooling were studied, respectively, in this work.

Figure 6 are the maximum battery temperatures (a,c,e,g) and temperature differences (b,d,f,h) in Scheme 3 versus convective heat transfer coefficient under a 2C discharge rate, different numbers of FHPs and different ambient temperatures of −5 °C (a,b), 10 °C (c,d), 25 °C (e,f), and 37 °C (g,h). It can be seen from Figure 6a,c,e,g that the maximum temperature of the battery module shows a negative correlation with the number of FHPs, decreasing with more FHPs. Moreover, the temperature difference of the battery module with eight FHPs is slightly higher than that with six FHPs. The main reason is that the eight FHPs are arranged on both sides of the middle four batteries, and the outermost two batteries do not have a strong heat dissipation effect. The temperature difference further increases, especially after the CPCM begins to melt. However, the six FHPs are evenly distributed on one side of the six batteries. Hence, the temperature uniformity at the end of the discharge process is good because of the large thermal conductivity of the FHPs. In addition, the temperature difference of the battery modules with six and eight FHPs continues to rise at the convective heat transfer coefficient of 20 ~ 30 $W/(m^2{\cdot}K)$ due to the CPCM beginning to melt to complete liquidity.

When the number of FHPs increases to 12, the maximum temperature of the battery module shows a significant downward trend, which is decreased by 2 °C compared to 6 FHPs, especially in a low-temperature environment of −5 °C and adiabatic conditions. In addition, it shows a similar trend with 14 FHPs but with better temperature uniformity. The main reason is that the 12 FHPs are evenly arranged on both sides of the 6 batteries. The heat released by the batteries is dissipated in a timely manner and stored in the CPCM, leading to a low temperature difference. The 2 additional FHPs in the battery module with 14 FHPs reduce the maximum temperature of the batteries and the minimum temperature of the outermost batteries, resulting in a larger temperature difference.

In addition, it can be seen from Figure 6b,d,f,h that the maximum temperature of the battery module decreases with the convective heat transfer coefficient increasing while the temperature difference increases. Furthermore, the temperature difference of the battery module is significantly impacted by the convective heat transfer coefficient, especially when the FHP number is set to be 6 and 8, and it is decreased when the number of FHPs increases to 12 and 14 because of the higher thermal conductivity of the FHPs. It is indicated that the FHPs can effectively reduce the maximum temperature while significantly improving the temperature uniformity of the BTMS.

Figure 7a–d display the temperature fields of the different FHP numbers on battery temperature under a 2C discharge rate and 37 °C ambient temperature. For the battery modules with six FHPs, the maximum temperature appears near the middle batteries with a slight temperature difference due to the average distribution of FHPs and batteries. For the battery modules with eight FHPs, the temperature of the middle batteries is lower, while that on both sides is higher. The main reason is that the FHPs are set on both sides of the four batteries in the middle. Furthermore, the temperature distribution of the battery module is relatively uniform with the FHPs increased to 12. In addition, the outermost two batteries of the battery module are supercooled when the FHP number is increased to 14, resulting in a larger temperature difference.

All temperature differences of the battery modules with 12 FHPs and 14 FHPs are maintained within 5 °C, meeting the temperature uniformity requirements of a BTMS [29–31] under different ambient temperatures and convective heat transfer coefficients. However, the temperature differences in the battery modules with six FHPs and eight FHPs ultimately exceed 5 °C with the convective heat transfer coefficient increase. Based on the above analysis, the battery module with 12 FHPs has a better maximum temperature and temperature uniformity.

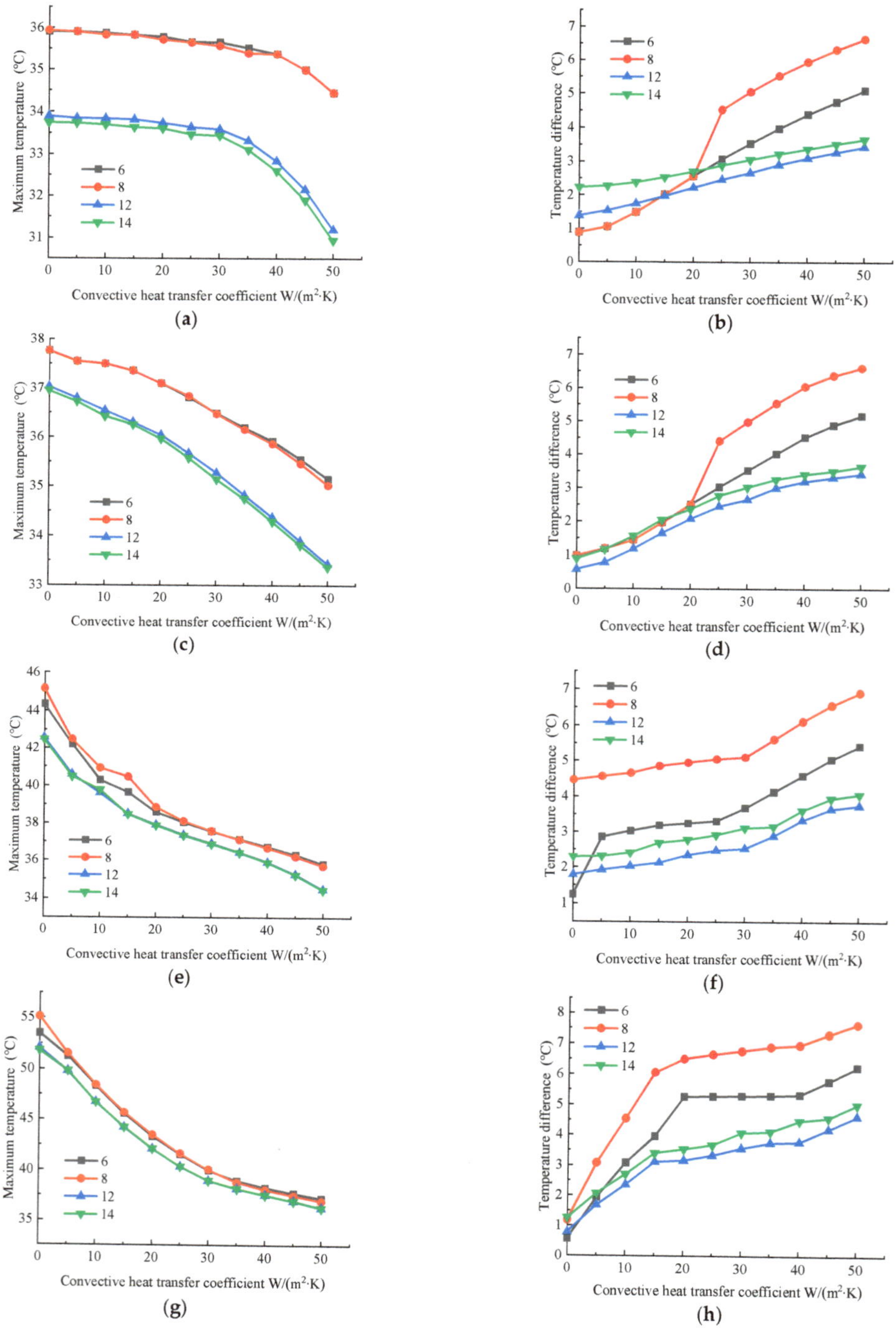

Figure 6. Maximum battery temperatures (**a**,**c**,**e**,**g**) and temperature differences (**b**,**d**,**f**,**h**) in Scheme 3 versus convective heat transfer coefficient under a 2C discharge rate, different numbers of FHPs and different ambient temperatures of −5 °C (**a**,**b**), 10 °C (**c**,**d**), 25 °C (**e**,**f**), and 37 °C (**g**,**h**).

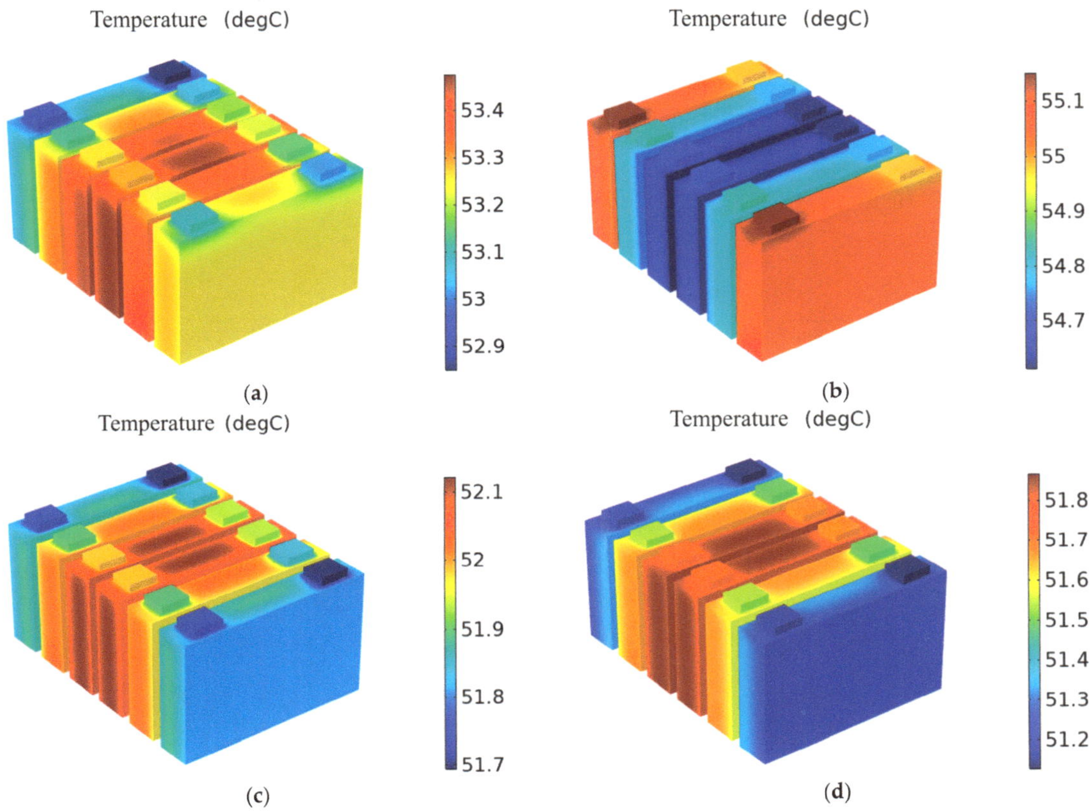

Figure 7. Temperature fields of the battery module with different FHP numbers: 6 (**a**), 8 (**b**), 12 (**c**) and 14 (**d**) under a 2C discharge rate and 37 °C ambient temperature.

4.3. Effect of Coolant Temperature

It is known that Scheme 3, using CPCM and FHPs, has a maximum temperature of over 50 °C under a high-temperature environment of 37 °C and a maximum discharge rate of 2C. Therefore, liquid cooling is added on two sides of the battery module to assist the heat dissipation of Scheme 3 in Figure 1d. Figure 8 shows the change in battery temperature with different coolant temperatures under the 2C discharge rate and 37 °C ambient temperature.

Figure 8a shows that the maximum temperature of the battery module will be affected by the coolant temperature, and the two are positively correlated. Still, the temperature difference has an opposite trend. The maximum temperature reaches 46.67 °C, 6.56 °C lower than that of the BTMS, based on the coupling of CPCM and the FHPs under the coolant temperature of 37 °C, showing that liquid cooling has a good effect on improving the heat dissipation performance of the BTMS. Liquid cooling takes away more and more heat as the coolant temperature decreases. Moreover, the rate of decline is relatively large until the coolant temperature drops to 22 °C. It shows that the heat taken away by the coolant has reached a balance with the heat generated by the battery module. Even if the coolant temperature continues to be lowered, it cannot cool down more effectively. Currently, the maximum temperature of the battery module remains near the initial ambient temperature. In addition, it remains near the phase change temperature of CPCM at 13 °C and 15 °C coolant temperatures due to the melting of the CPCM.

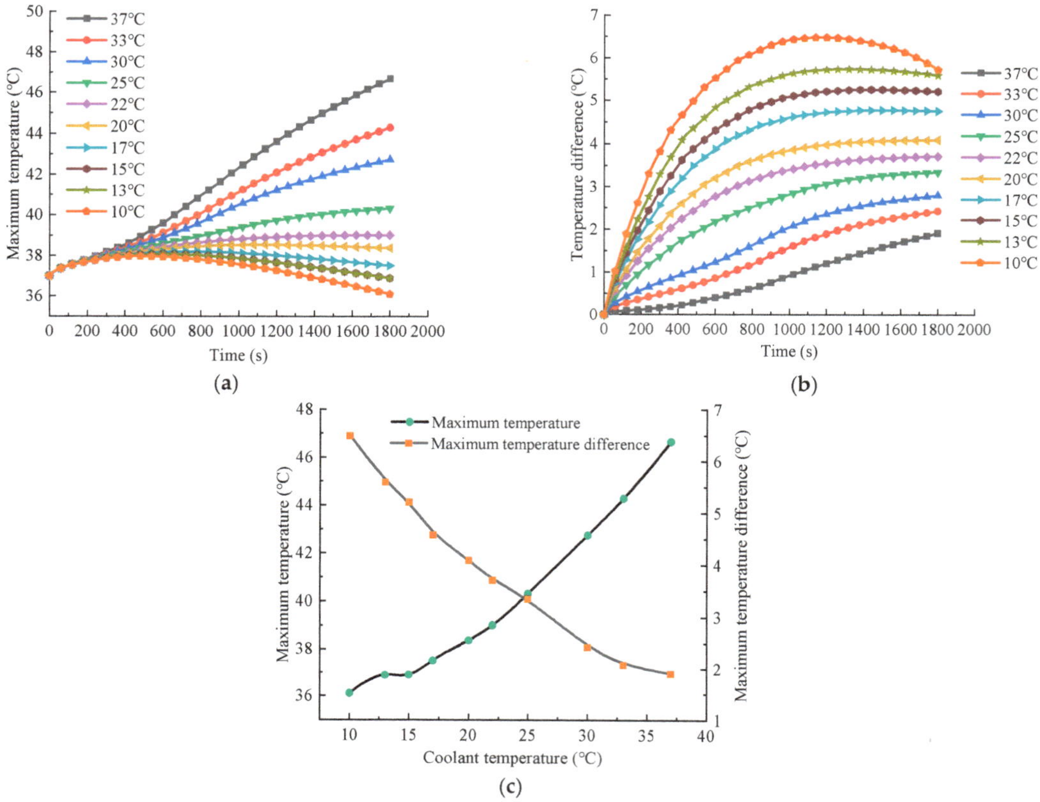

Figure 8. History curves of maximum temperatures (**a**) and temperature differences (**b**) under the conditions of 2C discharge rate, 37 °C ambient temperature, 0.1 m/s velocity and different coolant temperatures of 10 °C, 13 °C, 15 °C, 17 °C, 20 °C, 22 °C, 25 °C, 30 °C, 33 °C and 37 °C, and peak maximum temperature and maximum temperature difference versus coolant temperature (**c**).

In addition, Figure 8b shows that the maximum temperature of the battery module decreases while decreasing the coolant temperature. In contrast, the temperature difference increases, reaching its lowest at 1.91 °C at the coolant temperature of 37 °C, consistent with the ambient temperature. In comparison, it reaches its highest at 4.76 °C at the coolant temperature of 17 °C. Furthermore, the temperature difference of the battery module will exceed 5 °C if we suppose the coolant temperature continues to be lowered, which cannot meet the temperature uniformity requirements of the BTMS during regular operation. Relying solely on reducing the coolant temperature cannot improve the thermal performance of a BTMS. Therefore, the coolant temperature, close to the ambient temperature, should be preferred to balance the maximum temperature and maximum temperature difference of the BTMS.

4.4. *Effect of Coolant Pipe Number*

Figure 9 shows the change in battery temperature with the number of cooling pipes being 0, 2, 4, 6, 8, 10, and 12 under a 2C discharge rate, 37 °C ambient temperature, and a coolant velocity of 0.1 m/s. The maximum temperature of the battery module decreases with the larger number of cooling pipes. It reaches up to 50.32 °C without liquid cooling, which exceeds the working temperature of the BTMS. Furthermore, it continues to decrease with the increasing number of cooling pipes. The main reason is that more cooling pipes can take more battery heat.

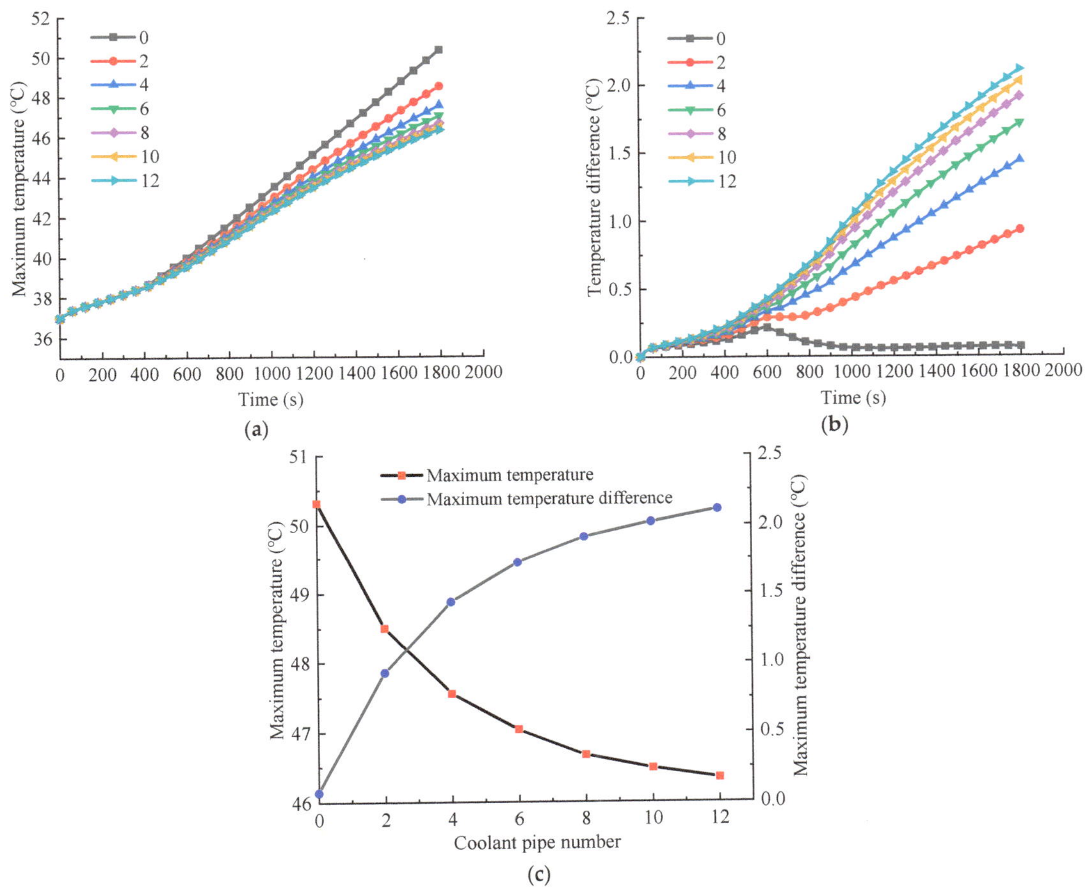

Figure 9. History curves of maximum temperatures (**a**) and temperature differences (**b**) under the conditions of 2C discharge rate, 37 °C ambient temperature, 0.1 m/s velocity and different cooling pipe numbers of 0, 2, 4, 6, 8, 10, and 12, and peak maximum temperature and maximum temperature difference versus cooling pipe number (**c**).

The drop-tends of the battery temperature will be flat when the number of cooling pipes increases to eight or more. The coolant takes the heat away and reaches a balance with the heat from the battery; continuing to increase the flow of the coolant will not further reduce the maximum temperature of the battery module. Figure 9b shows that the battery temperature difference increases with the number of cooling pipes. However, the increasing trend of the temperature difference tends to be stable as the decreasing trend of the maximum temperature tends to be stable. In addition, it is seen from Figure 9c that the maximum temperature and temperature difference exhibit opposite trends with the increased number of cooling pipes. Therefore, selecting the appropriate number of cooling pipes should be considered comprehensively. Based on the above analysis, a BTMS with eight cooling pipes is preferred for the following study under a high discharge rate of 2C, a high-temperature environment of 37 °C, and a 0.1 m/s velocity.

4.5. Effect of Coolant Direction

Figure 10 shows the structure diagram of Scheme 4 in Figure 1d with four coolant flow directions under the 2C discharge rate, 37 °C ambient temperature, and 0.1 m/s coolant velocity. Figure 10a shows the directions of eight cooling pipes set as parallel flow, namely,

Case 1; Figure 10b shows the four cooling pipes on one side as inlets while the four cooling pipes on the other side are outlets, namely, Case 2; Figure 10c is the directions of the four cooling pipes on one side as the counter flow and four cooling pipes on the other side in the horizontal direction, namely, Case 3; Figure 10d is the directions of eight cooling pipes, all as counter flow, namely, Case 4. The difference between Case 3 and Case 4 is that the coolant flow direction on both sides of the horizontal direction is consistent in Case 3 while it is different in Case 4.

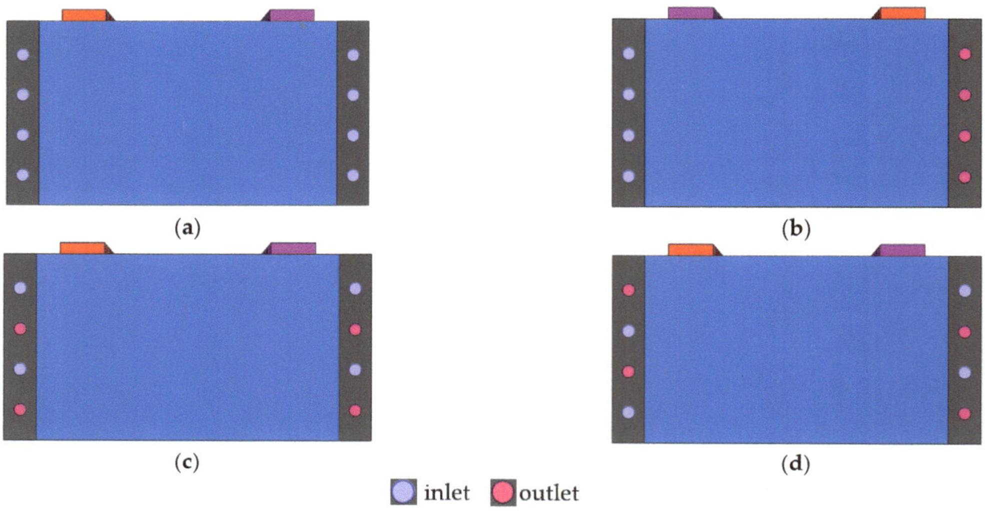

Figure 10. Four cases based on different coolant flow directions of Scheme 4 under a discharge rate of 2C, ambient temperature of 37 °C and coolant flow rate of 0.1 m/s, (**a**) Case 1, eight cooling pipes are a parallel flow; (**b**) Case 2, four cooling pipes on one side as inlets while four are on the other side as outlets; (**c**) Case 3, four cooling pipes on one side as counter flow while the remaining in the direction on the other side are consistent; (**d**) Case 4, eight cooling pipes as counter flow.

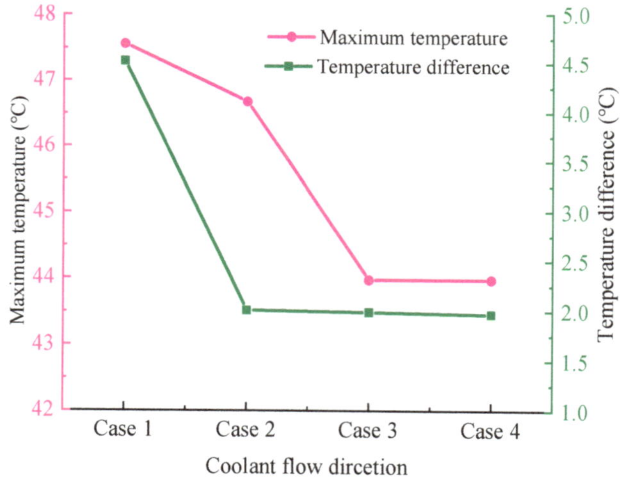

Figure 11. Maximum battery temperature and temperature difference of Scheme 4 in four cases of coolant flow direction, as shown in Figure 10.

Figure 11 displays the battery temperature in the four coolant flow direction cases in Figure 10. It can be seen that the maximum temperature and temperature difference of the battery module is lower with a counter flow of the coolant than with a parallel flow. The main reason is that the coolant inlet temperature is the same as that of the battery module. In contrast, the coolant outlet temperature rises due to the heat absorption from the battery module. The temperature difference between the two ends of the battery module is balanced through the counter flow of the coolant. Therefore, the counter flow direction in Case 3 owns better thermal performance of the BTMS to account for the complexity of the system structure and the light weight of the vehicle and is used for the following work.

5. Optimization Based on Orthogonal Design Method

5.1. Orthogonal Design

The orthogonal design method employs a table to realize multi-factor and multi-variable experiments. A representative test plan is obtained, and further comparisons are made to form the best approach. Therefore, the effects of different factors on the test results can be clarified, which can be used to predict the development trend of related variables. The advantage of the method is that it can obtain the optimal parameter combination that meets the experimental requirements through fewer tests quickly and accurately [32,33].

The methods of range analysis and variance analysis are mainly used in analyzing the test results, which are the numerical tests in orthogonal design, and the basic principles and applicability of the two are significantly different. The range analysis method is simple and intuitive. The importance of factors can be judged according to the size of the range value. In addition, the range analysis method can also quickly obtain the optimal factor combination but cannot accurately determine the effect of various factors on the experimental results. The analysis of variance can not only estimate the experimental error but also judge the impact of the error, so it makes up for the defects of the former method to a certain extent. In addition, it can still obtain the influence of various factors and interactions on the test results, even with many levels and interactions.

The orthogonal test aims to select the optimal parameter combination in the previous parameter study to design a more efficient BTMS structure and better temperature control to prevent thermal safety problems caused by excessive temperatures. The lowest maximum temperature and the lowest temperature difference of the battery module are set as the evaluation indicators in the orthogonal test design under the maximum discharge rate of 2C and a high-temperature environment of 37 °C.

The coolant temperature has a significant impact on the maximum temperature and the temperature difference of the battery module. Therefore, it is necessary to find a suitable coolant temperature under different operating conditions when designing the BTMS. In general, the coolant temperature changes with the ambient temperature, but it can also be controlled with thermoelectric precooling, which is set as an optimized design variable. Therefore, the FHP number, the coolant temperature, and the cooling pipe number are selected for the design. It was concluded that the optimal values of the three factors with four levels appear within a specific range from the previous single-factor analysis. We replaced each factor with a letter: A, FHP number; B, coolant temperature; C, cooling pipe number. There was no interaction among the factors when judging the preliminary experiments, so the influence of interaction is not considered. Therefore, there are three factors and four levels in the orthogonal test. The specific factor levels are shown in Table 3.

The orthogonal table is mainly selected with a small number that can meet the experimental requirements according to the number of factors and levels in an orthogonal test design. This test has three factors and four levels, so the L_{16} (4^5) orthogonal table is employed to determine 16 test schemes. A reasonable header design needs to arrange all factors in the table. There are three factors in this test, so the two extra columns are used as empty columns, designed as error columns, and placed at the end of the orthogonal table. The required orthogonal test schemes can be obtained by combining the previous analyses. The results were obtained using COMSOL Multiphysics and analyzed for range and variance. The orthogonal test schemes and simulation results are shown in Table 4.

Table 3. Values of factors at different levels for the orthogonal test.

Level	A (FHP Number)	B (Coolant Temperature), °C	C (Coolant Pipe Number)
1	6	25	6
2	8	30	8
3	12	33	10
4	14	37	12

Table 4. COMSOL simulation results of all tests in the orthogonal test scheme.

Test	A (FHP Number)	B (Coolant Temperature)	C (Coolant Pipe Number)	T_{max} (°C)	ΔT (°C)
1	6	25	6	37.98	5.50
2	6	30	8	39.11	4.24
3	6	33	10	40.09	3.21
4	6	37	12	43.17	3.36
5	8	25	8	37.60	5.98
6	8	30	6	41.63	5.89
7	8	33	12	40.35	4.13
8	8	37	10	45.15	4.58
9	12	25	10	34.29	3.97
10	12	30	12	37.60	3.01
11	12	33	6	41.79	3.27
12	12	37	8	44.18	2.26
13	14	25	12	33.48	3.62
14	14	30	10	38.23	3.35
15	14	33	8	40.42	2.63
16	14	37	6	45.03	2.77

5.2. Results and Discussion

5.2.1. Range Analysis

The order of significance of the various factors on the maximum temperature of the battery module was B-coolant temperature > C-coolant pipe number > A-FHP number to take the maximum temperature as the evaluation index. The range analysis data for the maximum temperature are shown in Table 5.

Table 5. Range analysis of the maximum temperature of the battery module.

Serial Number	A (FHP Number)	B (Coolant Temperature)	C (Coolant Pipe Number)	e_1	e_2
Mean value 1	40.09	35.84	41.61	40.19	40.29
Mean value 2	41.18	39.14	40.33	39.88	39.69
Mean value 3	39.47	40.66	39.44	40.09	39.85
Mean value 4	39.29	44.38	38.65	39.88	40.20
Range	1.89	8.55	2.96	0.31	0.59

The order of significance of the various factors on the temperature difference was A-FHP number > B-coolant temperature > C-coolant pipe number to take the temperature difference of the battery module as the evaluation index. The range analysis data for the temperature difference are shown in Table 6.

Table 6. Range analysis of the temperature difference of the battery module.

Serial Number	A (FHP Number)	B (Coolant Temperature)	C (Coolant Pipe Number)	e_1	e_2
Mean value 1	4.08	4.77	4.36	3.81	3.93
Mean value 2	5.15	4.12	3.78	3.93	3.78
Mean value 3	3.13	3.31	3.78	3.74	3.75
Mean value 4	3.09	3.32	3.653	3.96	3.99
Range	2.05	1.53	0.83	0.22	0.25

5.2.2. Variance Analysis

According to variance analysis, the coolant temperature had a more significant impact on the experimental results when the maximum temperature was used as the evaluation index, as shown in Table 7. The significance labeled with more stars stands for higher significance. Moreover, the number of FHPs significantly impacted the experimental results when the temperature difference was selected as the evaluation index, as shown in Table 8.

Table 7. Variance analysis uses the maximum temperature as the evaluation index.

Factor	Sum of Squares of Deviations	Degree of Freedom	Ratio of F	Critical Value of F	Significance
A (FHP number)	8.79	3	14.27	4.76	*
B (Coolant temperature)	150.83	3	245.05	4.76	***
C (Coolant pipe number)	19.31	3	31.37	4.76	**
Error	1.23	6			

Note: The number of stars labelled in the last column stands for the significance, whereas the "***" stands for the highest significance, the "**" stands for the moderate significance, and the "*" stands for the low significance.

Table 8. Variance analysis uses the temperature difference as the evaluation index.

Factor	Sum of Squares of Deviations	Degree of Freedom	Ratio of F	Critical Value of F	Significance
A (FHP number)	11.30	3	77.38	4.76	***
B (Coolant temperature)	6.31	3	43.19	4.76	**
C (Coolant pipe number)	1.48	3	10.14	4.76	*
Error	0.29	6			

Note: The number of stars labelled in the last column stands for the significance, whereas the "***" stands for the highest significance, the "**" stands for the moderate significance, and the "*" stands for the low significance.

5.2.3. Optimal Solution Analysis

According to the range and variance analysis results, the optimal solutions determined are different under different test indicators. It is obvious that a high FHP number, low cooling temperature, and a large number of pipes can decrease the maximum battery temperature. In this situation, the $A_4B_1C_4$ condition (Table 3) is obviously optimal without using a range or variance analysis approach. The optimal solution for the minimum temperature difference in the battery is $A_3B_4C_2$, resulting in the lowest temperature difference but, somehow, a higher maximum temperature.

Although a high FHP number and low cooling temperature can decrease the maximum battery temperature, more FHP numbers will increase system weight. A lower coolant temperature will increase the temperature difference of the battery module and the pump power consumption. Therefore, the final scheme is obtained as $A_1B_4C_4$ to target the minimum temperature difference between the maximum temperature of the battery module and the coolant temperature, which meets the requirements of the maximum temperature and temperature difference of the BTMS and greatly reduces the weight and power consumption of the system.

In addition, the original scheme in Figure 11 of Case 3 (number of FHPs: 12, coolant temperature: 37 °C, and number of coolant pipes: 8) is $A_3B_4C_2$. The results of $A_4B_1C_4$, $A_3B_4C_2$, and $A_1B_4C_4$ are shown in Table 9. Figure 12 displays the temperature fields of the original scheme and $A_1B_4C_4$ on battery temperature under the 2C discharge rate and 37 °C ambient temperature. The maximum temperature in $A_1B_4C_4$ is lower than the original scheme, and the temperature distribution of the middle batteries is more uniform. Moreover, the number of FHPs in scheme $A_1B_4C_4$ is fewer than in the original scheme, significantly reducing system weight and improving system economy and safety. Finally, the optimal solution is selected as $A_1B_4C_4$ based on the actual situation of the BTMS and the simulation results.

Table 9. Comparison of simulation results of the four schemes.

Design Schemes	*T*max (°C)	Δ*T* (°C)
A4B1C4	33.48	3.62
A3B4C2	44.18	2.26
A1B4C4	43.17	3.36

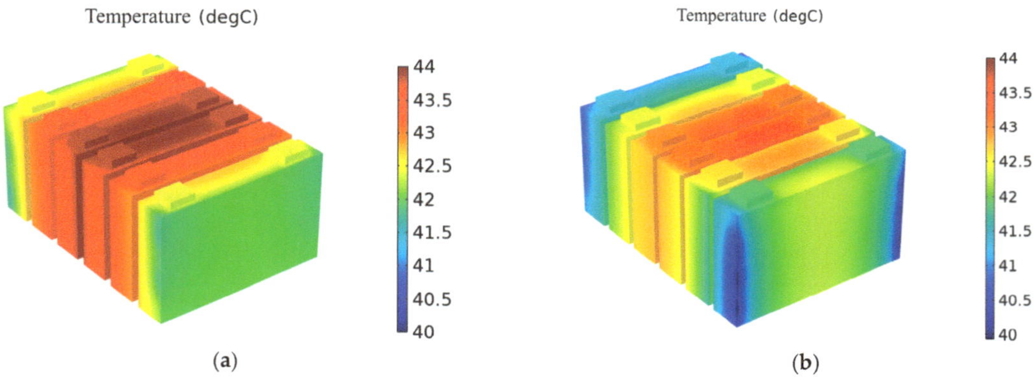

Figure 12. Temperature fields of the battery module in the original scheme ($A_3B_4C_2$) (**a**) and $A_1B_4C_4$ (**b**) under a 2C discharge rate and 37 °C ambient temperature.

The comprehensive balance method was employed to conduct the optimal solution analysis. The formulas for calculating the estimated value of the test indicator are as follows:

$$\begin{cases} \hat{y}_1 = \hat{\mu} + \hat{a}_1 + \hat{b}_1 + \hat{c}_1 \\ \hat{y}_2 = \hat{\mu} + \hat{a}_2 + \hat{b}_2 + \hat{c}_2 \\ \hat{y}_3 = \hat{\mu} + \hat{a}_3 + \hat{b}_3 + \hat{c}_3 \\ \hat{y}_4 = \hat{\mu} + \hat{a}_4 + \hat{b}_4 + \hat{c}_4 \\ \vdots \\ \hat{y}_{13} = \hat{\mu} + \hat{a}_{13} + \hat{b}_{13} + \hat{c}_{13} \\ \hat{y}_{14} = \hat{\mu} + \hat{a}_{14} + \hat{b}_{14} + \hat{c}_{14} \\ \hat{y}_{15} = \hat{\mu} + \hat{a}_{15} + \hat{b}_{15} + \hat{c}_{15} \\ \hat{y}_{16} = \hat{\mu} + \hat{a}_{16} + \hat{b}_{16} + \hat{c}_{16} \end{cases} \quad (14)$$

where $\hat{\mu}$ is the estimated value of mathematical expectation in the experimental data, $\hat{a}_k$ (k = 1, 2, 3, 4) is the effect of factor A (FHP number) corresponding to level k, $\hat{b}_m$ (m = 1, 2, 3, 4) is the effect of factor B (coolant temperature) corresponding to level m, and $\hat{c}_n$ (n = 1, 2, 3, 4) is the effect of factor C (coolant pipe number) corresponding to level n.

The estimated value of the test indicator based on optimal combination factors obtained using the comprehensive balance method is expressed by the following formulas:

$$\begin{cases} \widehat{\mu} = y = \frac{1}{16} \sum_{i=1}^{16} y_i \\ \widehat{a}_k = y_{Ak} - y \\ \widehat{b}_m = y_{Bm} - y \\ \widehat{c}_n = y_{Cn} - y \end{cases} \tag{15}$$

It can be ascertained that the average estimated values of the mathematical expectation of the orthogonal test are 41.51 °C and 4.14 °C, respectively, when the maximum temperature and temperature difference are used as the evaluation indexes, respectively. Moreover, the average value of the experimental indicators corresponding to the levels of each factor can be calculated, as shown in Table 10.

Table 10. Average estimated value $\widehat{y}_\mu$ of indicators (maximum temperature and temperature difference) compared with optimal value.

Parameter	Effect to T_{max}	Effect to ΔT
$\widehat{a}_1$	0.08	0.22
$\widehat{b}_3$	4.37	−0.61
$\widehat{c}_3$	−1.36	−0.33
$\widehat{y}_\mu$ (Estimated)	43.10 °C	3.14 °C
$A_1B_4C_4$ (Optimal)	43.17 °C	3.36 °C

The estimated values obtained through the optimal combination scheme $A_1B_4C_4$ are 43.10 °C and 3.14 °C, respectively, when the evaluation indexes used are the maximum temperature and temperature difference of the battery module. The above results are similar to the data obtained from the previous simulation model $A_1B_4C_4$ (43.17 °C and 3.36 °C), with errors of 0.07 °C and 0.22 °C, respectively. Therefore, the effectiveness and rationality of the selected optimal combination scheme $A_1B_4C_4$ (number of FHPs: 6, coolant temperature: 37 °C, and number of coolant pipes: 12) can be proved.

6. Conclusions

Three schemes of BTMSs using only CPCM (Scheme 1), CPCMs with ATDPs (Scheme 2), and CPCMs with FHPs (Scheme 3) were analyzed for thermal performance in this work. Then, the liquid cooling was added to further assist in the heat dissipation of the CPCM–FHP model. In addition, the effects of the number of FHPs and cooling pipes, the coolant temperature, and the coolant flow direction on the thermal characteristics of battery modules were discussed. Finally, the orthogonal design method was used for optimization tests on multiple factors and levels to improve the light weight of the system while ensuring that the battery module met the requirements of maximum temperature and temperature difference. The conclusions are as follows:

(1) It was found that Scheme 3 could control the maximum temperature and temperature difference of the battery module to a certain extent compared to Scheme 1 and Scheme 2, regardless of the discharge rate or environmental temperature. However, it was unsuitable for high discharge rates and high environmental temperatures.

(2) A BTMS, combined with liquid cooling with CPCM–FHP, can significantly reduce the maximum temperature of the battery module, with a decrease in the coolant temperature and an increase in the number of cooling pipes. Moreover, the maximum temperature and temperature difference of the battery module was lower with a counter flow direction than with a parallel flow.

(3) The number of FHPs, cooling pipes, and the coolant temperature were selected for multi-factor and multi-level analysis in the orthogonal design method. The optimal parameter combinations, with an FHP number of 6, a coolant temperature of 37 °C,

and a cooling pipe number of 12, were obtained to achieve the best thermal performance of the BTMS. The maximum temperature and temperature difference were kept at 43.17 °C and 3.36 °C, respectively, under the maximum discharge rate of 2C and the high-temperature environment of 37 °C. The optimal solution analysis of the optimal scheme was conducted further, proving the effectiveness and rationality of selecting the optimal parameter combination.

Author Contributions: Conceptualization, J.X. and T.Y.; methodology, T.Y. and H.Z.; software, Q.X.; validation, Q.X.; formal analysis, T.Y.; investigation, Q.X.; resources, J.X. and H.Z.; data curation, Q.X. and T.Y.; writing—original draft preparation, Q.X. and T.Y.; writing—review and editing, J.X., H.Z., and J.Z.; visualization, Q.X.; project administration, J.X. and J.Z.; funding acquisition, J.X. and H.Z. All authors have read and agreed to the published version of the manuscript.

Funding: This research was funded by the Research Project of Wuhan University of Technology Chongqing Research Institute (YF 2021-08), the Science and Technology Development Foundation of CMVR from China Merchants Testing Certification Vehicle Technology Research Institute Co., Ltd. (20AKC3), the National Natural Science Foundation of China (51876113), the 111 Project of China (B17034), and the Innovative Research Team Development Program of Ministry of Education of China (IRT_17R83).

Data Availability Statement: The data are contained within the article.

Acknowledgments: Thanks should be given to Yang Jiaxing for his helpful discussions on the experimental validation of the model.

Conflicts of Interest: The authors declare that they have no known competing financial interest or personal relationships that could have appeared to influence the work reported in this paper.

References

1. Li, A.; Yuen, A.C.Y.; Wang, W.; Chen, T.B.Y.; Lai, C.S.; Yang, W.; Wu, W.; Chan, Q.N.; Kook, S.; Yeoh, G.H. Integration of computational fluid dynamics and artificial neural network for optimization design of battery thermal management system. *Batteries* **2022**, *8*, 69. [CrossRef]
2. Yamanaka, T.; Kihara, D.; Takagishi, Y.; Yamaue, T. Multi-Physics equivalent circuit models for a cooling system of a lithium ion battery pack. *Batteries* **2020**, *6*, 44. [CrossRef]
3. Chen, W.; Hou, S.; Shi, J.; Han, P.; Liu, B.; Wu, B.; Lin, X. Numerical analysis of novel air-based li-ion battery thermal management. *Batteries* **2022**, *8*, 128. [CrossRef]
4. Ling, Z.; Wang, F.; Fang, X.; Gao, X.; Zhang, Z. A hybrid thermal management system for lithium ion batteries combining phase change materials with forced-air cooling. *Appl. Energy* **2015**, *148*, 403–409. [CrossRef]
5. Xie, Y.; Tang, J.; Shi, S.; Xing, Y.; Wu, H.; Hu, Z.; Wen, D. Experimental and numerical investigation on integrated thermal management for lithium-ion battery pack with composite phase change materials. *Energy Convers. Manag.* **2017**, *154*, 562–575. [CrossRef]
6. Jiang, G.; Huang, J.; Liu, M.; Cao, M. Experiment and simulation of thermal management for a tube-shell Li-ion battery pack with composite phase change material. *Appl. Therm. Eng.* **2017**, *120*, 1–9. [CrossRef]
7. Anqi, A.; Li, C.; Dhahad, H.; Sharma, K.; Attia, E.-A.; Abdelrahman, A.; Mohammed, A.G.; Alamri, S.; Rajhi, A.A. Effect of combined air cooling and nano enhanced phase change materials on thermal management of lithium-ion batteries. *J. Energy Storage* **2022**, *52*, 104906. [CrossRef]
8. Rao, Z.; Wang, Q.; Huang, C. Investigation of the thermal performance of phase change material/mini-channel coupled battery thermal management system. *Appl. Energy* **2016**, *164*, 659–669. [CrossRef]
9. Bai, F.; Chen, M.; Song, W.; Yu, Q.; Li, Y.; Feng, Z.; Ding, Y. Investigation of thermal management for lithium-ion pouch battery module based on phase change slurry and mini channel cooling plate. *Energy* **2019**, *167*, 561–574. [CrossRef]
10. Ling, Z.; Cao, J.; Zhang, W.; Zhang, Z.; Fang, X.; Gao, X. Compact liquid cooling strategy with phase change materials for Li-ion batteries optimized using response surface methodology. *Appl. Energy* **2018**, *228*, 777–788. [CrossRef]
11. Abdelkareem, M.; Maghrabie, H.; Abo-Khalil, A.; Adhari, O.H.K.; Sayed, E.T.; Radwan, A.; Elsaid, K.; Wilberforce, T.; Olabi, A.G. Battery thermal management systems based on nanofluids for electric vehicles. *J. Energy Storage* **2022**, *50*, 104385. [CrossRef]
12. Abbas, S.; Ramadan, Z.; Park, C. Thermal performance analysis of compact-type simulative battery module with paraffin as phase-change material and flat plate heat pipe. *Int. J. Heat Mass Transf.* **2021**, *173*, 121269. [CrossRef]
13. Wang, Q.; Rao, Z.; Huo, Y.; Wang, S. Thermal performance of phase change material/oscillating heat pipe-based battery thermal management system. *Int. J. Therm. Sci.* **2016**, *102*, 9–16. [CrossRef]
14. Wayan, S.; Muhamad, A.; Made, A. The application of battery thermal management system based on heat pipes and phase change materials in the electric bike. *J. Energy Storage* **2022**, *56*, 106014.

15. Khan, M.; Dhahad, H.; Alamri, S.; Ali, E.A.; Sharma, K. Air cooled lithium-ion battery with cylindrical cell in phase change material filled cavity of different shapes. *J. Energy Storage* **2022**, *50*, 104573. [CrossRef]
16. Faizan, M.; Pati, S.; Randive, P. Implications of novel cold plate design with hybrid cooling on thermal management of fast discharging lithium-ion battery. *J. Energy Storage* **2022**, *53*, 105051. [CrossRef]
17. Yuan, Q.; Xu, X.; Tong, G.; Ding, H. Effect of coupling phase change materials and heat pipe on performance enhancement of Li-ion battery thermal management system. *Int. J. Energy Res.* **2020**, *45*, 5399–5411. [CrossRef]
18. Elnaggar, M.; Abdullah, M.; Munusamy, S. Experimental and numerical studies of finned L-shape heat pipe for notebook PC cooling. *IEEE Trans. Compon. Packag. Manuf. Technol.* **2013**, *3*, 978–988. [CrossRef]
19. Chen, W. Research on Heat Control System for Vehicular Power Battery Pack Based on Heat Pipe Technology. Master's thesis, South China University of Technology, Guangzhou, China, 2014.
20. Wei, K. Structure Design and Themal Analysis of Power Battery Thermal Management Based on Flat Heat Pipes. Master's thesis, Beijing Jiaotong University, Beijing, China, 2019.
21. Li, R. Study on the Heat Dissipation Performance of Power Battery Liquid Cooling Thermal Management System of Electric Vehicle Based on Flat Heat Pipe. Master's thesis, Jiangsu University, Zhenjiang, China, 2019.
22. Xu, X.; Zhang, H.; Cao, L.; Yi, Z.; Li, P.; Guo, H. Heat generation and surrogate model for large-capacity nickel-rich prismatic lithium-ion battery as against 18650 battery. *J. Loss Prev. Process Ind.* **2022**, *77*, 104783. [CrossRef]
23. Zhang, H.; Wu, X.; Wu, Q.; Xu, S. Experimental investigation of thermal performance of large-sized battery module using hybrid PCM and bottom liquid cooling configuration. *Appl. Therm. Eng.* **2019**, *159*, 113968. [CrossRef]
24. Song, L.; Zhang, H.; Yang, C. Thermal analysis of conjugated cooling configurations using phase change material and liquid cooling techniques for a battery module. *Int. J. Heat Mass Transf.* **2019**, *133*, 827–841. [CrossRef]
25. COMSOL Multiphysics®. Available online: https://www.comsol.com (accessed on 16 June 2023).
26. Xin, Q.; Yang, T.; Zhang, H.; Yang, J.; Zeng, J.; Xiao, J. Experimental and numerical study of lithium-ion battery thermal management system using composite phase change material and liquid cooling. *J. Energy Storage* **2023**, *in press*.
27. Liu, H.; Wei, Z.; He, W.; Zhao, J. Thermal issues about Li-ion batteries and recent progress in battery thermal management systems: A review. *Energy Convers. Manag.* **2017**, *150*, 304–330. [CrossRef]
28. Wang, Q.; Ping, P.; Zhao, X.; Chu, G.; Sun, J.; Chen, C. Thermal runaway caused fire and explosion of lithium ion battery. *J. Power Sources* **2012**, *208*, 210–224. [CrossRef]
29. Paul, S.; Diegelmann, C.; Kabza, H.; Tillmetz, W. Analysis of ageing inhomogeneities in lithium-ion battery systems. *J. Power Sources* **2013**, *239*, 642–650. [CrossRef]
30. Yang, X.; Tan, S.; Liu, J. Thermal management of Li-ion battery with liquid metal. *Energy Convers. Manag.* **2016**, *117*, 577–585. [CrossRef]
31. Chen, K.; Song, M.; Wei, W.; Wang, S. Structure optimization of parallel air-cooled battery thermal management system with U-type flow for cooling efficiency improvement. *Energy* **2018**, *145*, 603–613. [CrossRef]
32. Xie, J.; Zang, M.; Wang, S.; Ge, Z. Optimization investigation on the liquid cooling heat dissipation structure for the lithium-ion battery package in electric vehicles. *Proc. Inst. Mech. Eng. Part D J. Automob. Eng.* **2017**, *231*, 1735–1750. [CrossRef]
33. Li, N. Design and Optimization of Heat Dissipation Structure for Thermal Management System of Lithium Ion Batteries. Master's thesis, Southwest Jiaotong University, Chengdou, China, 2019.

batteries

MDPI

Article

Investigating the Effect of Different Bidirectional Pulsed Current Parameters on the Heat Generation of Lithium-Ion Battery at Low Temperatures

Ranjun Huang [1,2], Gang Wei [1,2], Bo Jiang [1,2], Jiangong Zhu [1,2], Xiangmin Pan [3], Xueyuan Wang [1,2,*], Xiangyang Zhou [3], Jiping Ye [3], Xuezhe Wei [1,2] and Haifeng Dai [1,2,*]

1 Clean Energy Automotive Engineering Center, Tongji University, Shanghai 201804, China; ranjun_huangtj@tongji.edu.cn (R.H.); tju_wg@tongji.edu.cn (G.W.); jiangbo15@tongji.edu.cn (B.J.); zhujiangong@tongji.edu.cn (J.Z.); weixzh@tongji.edu.cn (X.W.)
2 School of Automotive Studies, Tongji University, Shanghai 201804, China
3 Shanghai AI NEV Innovative Platform Co., Ltd., Shanghai 201804, China; xiangminp@smvic.com.cn (X.P.); zhouxy@sanevip.com (X.Z.); saiheik@163.com (J.Y.)
* Correspondence: 7wangxueyuan@tongji.edu.cn (X.W.); tongjidai@tongji.edu.cn (H.D.)

Abstract: Bidirectional pulsed current (BPC) heating has proven to be an effective method for internal heating. However, current research has primarily focused on the impact of symmetrical BPC on battery heat generation, while neglecting the influence of different BPC parameters. To address this gap, this paper investigates the effects of various BPC parameters on battery heat generation. Initially, an electro-thermal coupled model of the battery is constructed based on the results of electrochemical impedance spectroscopy (EIS) tests conducted at different temperatures and amplitudes at 20% state of charge (SOC). The validation results of the model demonstrate that the absolute errors of voltage and temperature are generally less than 50 mV and 1.2 °C. Subsequently, the influence of BPC parameters on battery heat generation is examined under different terminal voltage constraints, temperatures, and frequencies. The findings at 20% SOC reveal that symmetrical BPC does not consistently correspond to the maximum heating power. The proportion of charge time and discharge time in one cycle, corresponding to the maximum heating power, varies depending on the charge and discharge cut-off voltages. Moreover, these variations differ across frequencies and temperatures. When the terminal voltage is constrained between 3 V and 4.2 V, the maximum heat power corresponds to a discharge time share of 0.55 in one cycle. In conclusion, the results underscore the complex relationship between BPC parameters and battery heat generation, which can further enhance our understanding of effective heating strategies for batteries.

Keywords: lithium-ion battery; alternating current (AC) heating; heating power; electro-thermal coupled model; low temperatures

Citation: Huang, R.; Wei, G.; Jiang, B.; Zhu, J.; Pan, X.; Wang, X.; Zhou, X.; Ye, J.; Wei, X.; Dai, H. Investigating the Effect of Different Bidirectional Pulsed Current Parameters on the Heat Generation of Lithium-Ion Battery at Low Temperatures. *Batteries* **2023**, *9*, 457. https://doi.org/10.3390/batteries9090457

Academic Editors: Carlos Ziebert and Thomas Wetzel

Received: 27 June 2023
Revised: 11 August 2023
Accepted: 4 September 2023
Published: 5 September 2023

1. Introduction

Due to their high energy and power density, long cycle life, and low self-discharge, lithium-ion batteries (LIBs) have been widely used in electric vehicles (EVs), energy storage systems (ESSs), and electronic products [1,2]. However, at low temperatures, LIBs suffer from problems such as a decrease in discharge capacity [3], a decline in cycle performance [4], and difficulty in charging [5]. These problems seriously affect the performance of EVs, ESSs, and electronic products at low temperatures. Moreover, lithium plating may occur during the low-temperature charging process [5]. On the one hand, lithium plating will reduce the number of available lithium ions in the battery, which will reduce the life of the battery. On the other hand, the deposited lithium dendrites will pierce the separator and lead to an internal short circuit, further threatening the safety of the battery. Therefore, LIBs need to be heated in advance before use at low temperatures.

The heating methods for LIBs can be divided into external heating and internal heating [6]. External heating uses an external heat source to heat the battery by conductive heating or convective heating [7]. Internal heating uses the internal resistance, or the heat-generating device embedded inside the battery to heat the battery. Compared with external heating, internal heating has the advantages of uniform temperature rise, high heating efficiency, and low energy consumption [8]. Therefore, internal heating has gained the attention of many researchers. Wang et al. [9] proposed a self-heating battery design by integrating nickel foil inside the battery. The new battery can be heated from −20 °C to 0 °C within 20 s. Although this method exhibited high heating efficiency, it required substantial structural modifications, resulting in increased costs and potential safety concerns. In contrast to the embedded nickel foil in the battery, the internal resistance heating method relies on the heat generated by the charging or discharging processes. Internal resistance heating can be further categorized into discharge heating and AC heating [10,11]. Discharge heating employs the discharge current to generate heat in the battery, while AC heating utilizes alternating currents of specific frequencies. Since AC heating includes both charging and discharging phases, it generally exhibits higher heating efficiency and lower energy consumption compared to discharge heating methods.

Although AC heating methods encompass a variety of waveforms [12–17], sinusoidal alternating current (SAC) and BPC are the most commonly utilized. Zhang et al. [18] employed an SAC with a frequency of 1 Hz to heat the battery from −20 °C to 5 °C in 900 s. Guo et al. [19] devised an SAC heating method without lithium plating using a three-electrode battery. Ruan et al. [20] optimized the frequency of SAC heating and identified the optimal heating frequency for different temperatures. At this frequency, the battery could be heated from −15.4 °C to 5.6 °C in 338 s. In contrast to frequency optimization, Zhang et al. [21] adjusted the amplitude of SAC heating for various temperatures at the predetermined heating frequency and achieved a temperature increase from −20 °C to 0 °C in 300 s. The advantages of BPC heating over SAC heating have been noticed and studied in recent years. Both Zhu et al. [22] and Zhang et al. [23] conducted comparative analyses of SAC heating and BPC heating. They observed that BPC heating exhibited higher heating power than SAC heating under the same voltage constraint. Wu et al. [24] proposed a BPC heating strategy with temperature-dependent amplitude and heated the battery from −20 °C to 5 °C within 308 s. Li et al. [16] achieved BPC heating by reconfiguring the battery pack structure and traction motor drive system, which can heat the battery from −7 °C to 29.1 °C in 300 s. Qin et al. [25] investigated various types of BPC heating methods and found that the battery can be heated from −8 °C to 25 °C in 250 s.

Despite the evident advantages of BPC heating, the existing studies on BPC heating have primarily focused on using the same charge and discharge time and the same charge and discharge amplitude in one waveform cycle [16,22–25]. Little attention has been given to investigating the impact of different BPC parameters, such as varying charge and discharge times and amplitudes within a waveform cycle, on battery heat generation. In addition, the relationship between BPC parameters and battery heat generation under different charge and discharge cut-off voltages, temperatures, and frequencies remains unclear. To address these knowledge gaps, this study presents an electro-thermal coupled model for a 1 Ah pouch battery at 20% state of charge (SOC), which is constructed based on the fractional-order equivalent circuit model. The electro-thermal coupled model serves two purposes. Firstly, it is employed to calculate the battery heating power. Secondly, it determines the battery's terminal voltage and temperature rise. The parameters of the electro-thermal coupled model are determined through electrochemical impedance spectroscopy (EIS) tests conducted at different temperatures and current amplitudes. To indirectly validate the accuracy of the calculated battery heating power, the temperature rise experiments of the battery are conducted. Finally, the electro-thermal coupled model is utilized to comprehensively analyze the relationship between BPC parameters and battery heating power under different charge and discharge cut-off voltages, temperatures, and frequencies.

2. Experiments

2.1. Experiment Platform

The experimental platform and tests in this paper are shown in Figure 1. The battery test platform includes a temperature acquisition recorder (Hongrun OHR-XH710), a temperature chamber (Binder MK56), a host computer, an electrochemical workstation (EnergyLab XM), and a power booster (Solartron analytical, 24 V/10 A). The sample lithium-ion battery is a nickel–cobalt–manganese (NCM)/graphite battery with a nominal capacity of 1 Ah. The operating voltage range is 3 V–4.2 V. The temperature acquisition recorder is used to obtain the surface temperature of the battery. The EnergyLab XM is used to perform the EIS test as well as the BPC test. Due to the small current range of the EnergyLab XM, the power booster is used to increase the range.

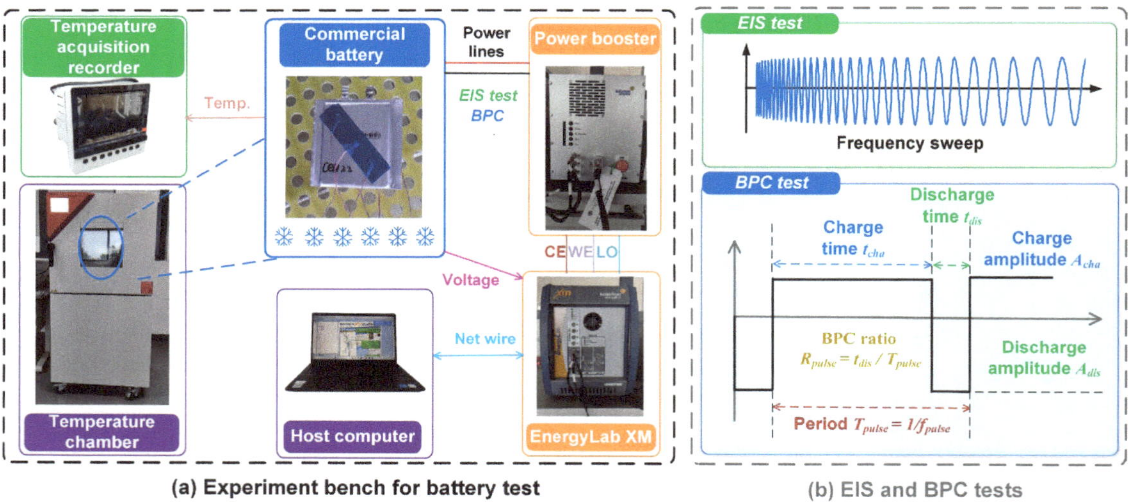

Figure 1. Experiment platform and tests.

2.2. EIS Test at Different Temperatures and AC Amplitude

Given that the impedance of the battery is influenced by both temperature and current amplitude, EIS tests are conducted on the battery at various temperatures and AC amplitudes to determine the parameters of the fractional-order equivalent circuit model. In addition, the SOC of the battery studied is set at 20% before conducting the EIS test. According to [26,27], lithium plating is more likely to occur under large-rate charging and low-temperature conditions. In addition, a high SOC increases the risk of lithium plating [28]. Since the EIS test includes the charging process, lithium plating may occur during the large-current EIS test. To mitigate this risk, a low SOC is chosen here. However, an excessively low SOC may expose the battery to the danger of over-discharge during the EIS test. Consequently, an SOC of 20% is chosen as a balanced compromise. The specific procedure involved initially charging the battery to full capacity using Constant Current Constant Voltage (CCCV) charging, with a charge current of 1 A and a charge cut-off current of 0.02 A. The constant voltage is 4.2 V. Subsequently, the battery is discharged with a constant current of 1 A after a resting period of 30 min. Discharging is terminated once the discharged capacity reached 0.8 Ah. At this time, the SOC of the battery is 20%. Considering the actual capacity of the battery is 1.05 A, the actual SOC of the battery is 23.8%. The effect of SOC is not discussed in this paper, and the error in SOC is only 3.8%. Therefore, the error is ignored. The impedance test is in galvanostatic mode, and the frequency range is 0.1 Hz to 10^6 Hz. Numerous studies [19,22,23,29,30] have focused on low-temperature AC heating strategies. In these investigations, the battery's maximum heating temperature typically hovers around 10 °C. The minimum temperature is usually determined based on the temperature range in which the battery will be used. In addition, the literature calibrates the parameters of the battery every 5 °C. The minimum discharge temperature of the battery used in this paper is −15 °C. Therefore, the test temperatures chosen for this paper are 10 °C, 5 °C, 0 °C, −5 °C, −10 °C, and −15 °C. After each temperature adjustment, the battery needs to be left for 2 h before the impedance test. The impedance tests are performed at different amplitudes of 0.25 A, 0.65 A, 1 A, 1.5 A, 2 A, and 2.5 A. The time interval between successive impedance tests with varying amplitudes at the same temperature is 1 h.

2.3. BPC Test

Before conducting the BPC test, it is necessary to predefine the relevant parameters. Figure 1 illustrates the key parameters of the BPC, which encompass charge time t_{cha}, discharge time t_{dis}, charge amplitude A_{cha}, and discharge amplitude A_{dis}, period T_{pulse}, frequency f_{pulse}, and ratio R_{pulse}. T_{pulse} and f_{pulse} have straightforward interpretations. t_{cha} and t_{dis} signify the duration of the charging and discharging within one cycle of the waveform, respectively. A_{cha} and A_{dis} indicate the magnitude of charging and discharging within one cycle of the waveform, respectively. In addition, R_{pulse} represents the ratio of t_{dis} to T_{pulse} within one cycle of the waveform. An accurate electro-thermal coupled model is crucial for precise calculations of heating power in the battery. The accuracy of the equivalent circuit model can be assessed by comparing the discrepancy between the terminal voltage calculated by the model and the terminal voltage obtained through experimental measurement. Furthermore, the accuracy of the heat generation calculation is evaluated by comparing the temperature rise calculated by the electro-thermal coupled model with the temperature rise measured experimentally. Consequently, the BPC test consists of two types: voltage testing and temperature testing. For voltage testing, the test temperatures include 10 °C, −5 °C, and −15 °C. The specific parameters of different BPCs at each temperature are presented in Table 1. The frequencies of the BPCs include three kinds of 1 Hz, 50 Hz, and 100 Hz. Each frequency includes two BPCs with different parameters, denoted as BPC I and BPC II, respectively. BPC I has a discharge amplitude of 3.5 A and a charge amplitude of 1.5 A. The ratio is 0.5. For BPC II, both discharge amplitude and charge amplitude are 2.5 A. The ratio is 0.7. During the voltage testing, both current and voltage are collected by the EnergyLab XM. The sampling frequency is 100 times the corresponding BPC frequency. Moreover, a 2 h stabilization period is set after each temperature adjustment, and the time interval between BPC tests conducted at the same temperature is set to 30 min. As an example of the test procedure at 10 °C, the battery is initially placed in a 10 °C environment for 2 h to reach the desired temperature. Subsequently, the BPC I@1 Hz test is performed using the EnergyLab XM and the power booster. Since the EnergyLab XM has a high sampling frequency, a sufficient amount of current and voltage data can be obtained in just a few seconds. After the BPC I@1 Hz test, the battery is allowed to stand for 30 min to stabilize. The BPC II@1 Hz test is then conducted following the same methodology. Similar procedures are followed for the other BPCs listed in Table 1. Upon completing the BPC II@100 Hz test, it signifies the completion of the voltage testing at 10 °C. Subsequently, the temperature is adjusted to other specified values to continue the testing. Once all the measurements at −5 °C and −15 °C are completed, the voltage testing is considered finished. For temperature testing, unlike voltage testing, it is only necessary to obtain the temperature rise data of the battery under different BPCs in Table 2 without the need for high-frequency sampling of current and voltage. The BPCs for temperature testing consist of three different frequencies. There are two types of BPCs at each frequency, BPC III and BPC IV. The ratios of BPC III and BPC IV are 0.5 and 0.7, respectively. For BPC III, both the discharge amplitude and charge amplitude are 2.5 A. The discharge amplitude and charge amplitude of BPC IV are 1.5 A and 3.5 A, respectively. The temperature testing is performed according to the following procedure. The battery is first placed at −13 °C for 2 h. Subsequently, the BPC III@1

Hz is applied to the battery using the EnergyLab XM and the power booster, while the temperature of the battery surface is measured simultaneously. After a 30 min duration, the heating of the battery is discontinued. The battery is left for 1 h and then subjected to BPC IV@1 Hz. The other BPCs are conducted following the same procedure. Upon completion of the BPC IV@100 Hz, it indicates the conclusion of the temperature testing.

Table 1. The BPCs for voltage testing.

Frequency f_{pulse} (Hz)	Ratio R_{pulse}	Discharge Amplitude A_{dis} (A)	Charge Amplitude A_{cha} (A)	Abbreviation
1	0.5	3.5	1.5	I@1 Hz
	0.7	2.5	2.5	II@1 Hz
50	0.5	3.5	1.5	I@50 Hz
	0.7	2.5	2.5	II@50 Hz
100	0.5	3.5	1.5	I@100 Hz
	0.7	2.5	2.5	II@100 Hz

Table 2. The BPCs for temperature testing.

Frequency f_{pulse} (Hz)	Ratio R_{pulse}	Discharge Amplitude A_{dis} (A)	Charge Amplitude A_{cha} (A)	Abbreviation
1	0.5	2.5	2.5	III@1 Hz
	0.7	1.5	3.5	IV@1 Hz
10	0.5	2.5	2.5	III@10 Hz
	0.7	1.5	3.5	IV@10 Hz
100	0.5	2.5	2.5	III@100 Hz
	0.7	1.5	3.5	IV@100 Hz

3. Mathematic Model

3.1. Calculation of Heating Power

The heat generation of the battery can be quantified by the heating power. Without considering the heat of mixing and side reactions, the heat generation of the battery consists mainly of reaction heat and Joule heat [31]. The reaction heat is determined by the effective entropy potential of the battery and is reversible heat for the electrochemical reaction [32]. For AC heating, the reaction heat generated by the charging and discharging process can cancel each other. Therefore, AC heating can be used to consider only Joule heat [33]. Joule heat includes ohmic internal resistance heat generation and polarization internal resistance heat generation [25]. Therefore, based on the parameter definitions of the BPC in Figure 1, the heating power q of the battery under BPC conditions can be expressed as

$$q = q_{dis} + q_{cha} = \frac{A_{dis}^2 Z'_{dis} t_{dis}}{T_{pulse}} + \frac{A_{cha}^2 Z'_{cha} t_{cha}}{T_{pulse}} \tag{1}$$

where q_{dis} and q_{cha} denote the discharge heating power and charge heating power, respectively. Z'_{dis} and Z'_{cha} represent the real part of the impedance corresponding to the discharge and the real part of the impedance corresponding to the charge, respectively. They can be calculated by the fractional-order equivalent circuit model. In addition, the accuracy of the heating power calculation cannot be directly verified. Indirect validation is performed by comparing experimental measurements of cell temperature with calculated values from the thermal model. Therefore, an accurate electro-thermal coupled model is necessary for the calculation and verification of the heating power.

3.2. Electro-Thermal Coupled Model

The fractional-order equivalent circuit model can accurately describe the impedance variation of the battery at different frequencies [34]. Hence, the electrical model in the electro-thermal coupled model is constructed using a fractional-order equivalent circuit model. The fractional-order equivalent circuit model is shown in Figure 2. The constant voltage source U_{oc} represents the open circuit voltage (OCV) of the battery. The resistance R_0 represents the ohmic resistance of the battery. The remaining polarization process inside the battery is described by two parallel structures of resistance and constant phase element (CPE). In addition, the battery in this paper is a 1 Ah pouch cell. Considering the relatively small size of the battery, the temperature distribution gradient of the battery is neglected. Therefore, the lumped thermal model is used to describe the temperature rise variation of the battery.

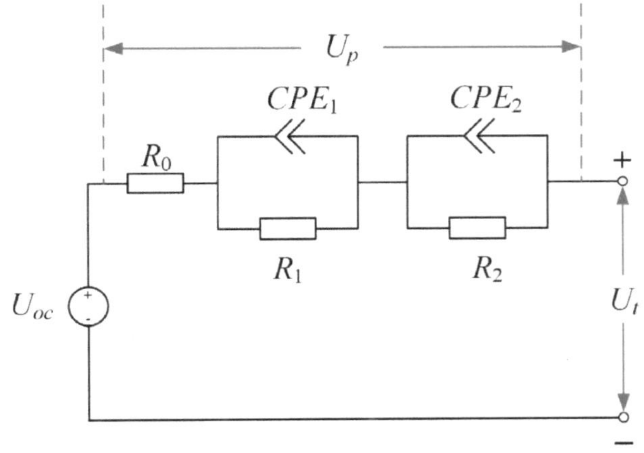

Figure 2. The fractional-order equivalent circuit model.

For the fractional-order equivalent circuit model, the real part of the impedance Z' can be calculated as [21]

$$Z' = R_0 + \frac{R_1 + \omega^{n_1} R_1^2 Q_1 \cos\frac{n_1\pi}{2}}{1 + 2\omega^{n_1} R_1 Q_1 \cos\frac{n_1\pi}{2} + \omega^{2n_1} R_1^2 Q_1} + \frac{R_2 + \omega^{n_2} R_2^2 Q_2 \cos\frac{n_2\pi}{2}}{1 + 2\omega^{n_2} R_2 Q_2 \cos\frac{n_2\pi}{2} + \omega^{2n_2} R_2^2 Q_2} \quad (2)$$

where ω is the angular frequency of the current. Q and n are the generalized capacitor and depression factors of CPE. The terminal voltage U_t of the battery can be expressed as

$$U_t = U_{oc} + U_p \quad (3)$$

where U_p is the polarization voltage, which can be expressed as [34]

$$U_p(s) = Z(s)I(s) = \left(\frac{\sum_{k=1}^{4} a_k s^{\alpha_k}}{\sum_{k=1}^{4} b_k s^{\beta_k}} + R_0 \right) I(s) = \left(\frac{R_1}{R_1 Q_1 s^{n_1} + 1} + \frac{R_2}{R_2 Q_2 s^{n_2} + 1} + R_0 \right) I(s) \quad (4)$$

the coefficients are expressed as

$$\begin{cases} [\, a_1 \;\; a_2 \;\; a_3 \;\; a_4 \,] = [\, R_1R_2Q_2 \;\; R_1 \;\; R_1R_2Q_1 \;\; R_2 \,] \\ [\, \alpha_1 \;\; \alpha_2 \;\; \alpha_3 \;\; \alpha_4 \,] = [\, n_2 \;\; 0 \;\; n_1 \;\; 0 \,] \\ [\, b_1 \;\; b_2 \;\; b_3 \;\; b_4 \,] = [\, R_1R_2Q_1Q_2 \;\; R_1Q_1 \;\; R_2Q_2 \;\; 1 \,] \\ [\, \beta_1 \;\; \beta_2 \;\; \beta_3 \;\; \beta_4 \,] = [\, n_1+n_2 \;\; n_1 \;\; n_2 \;\; 0 \,] \end{cases} \quad (5)$$

Fractional calculus is difficult to solve directly. The Grünwald–Letnikov (GL) definition can discretize fractional calculus and realize the solution of fractional-order equations [35,36]. Then, the solution for the polarization voltage is

$$U_p(t) = \frac{1}{\sum_{k=1}^{4}\frac{b_k}{\Delta t^{\beta_k}}}\left[\sum_{k=1}^{4}\sum_{g=0}^{[t/\Delta t]}\frac{a_k}{\Delta t^{\alpha_k}}(-1)^g\binom{\alpha_k}{g}I(t-g\Delta t)+\right.$$
$$\left.\sum_{k=1}^{4}\sum_{g=0}^{[t/\Delta t]}\frac{b_k}{\Delta t^{\beta_k}}(-1)^g\binom{\beta_k}{g}I(t-g\Delta t)R_0-\sum_{k=1}^{4}\sum_{g=0}^{[t/\Delta t]}\frac{b_k}{\Delta t^{\beta_k}}(-1)^g\binom{\beta_k}{g}U_p(t-g\Delta t)\right] \tag{6}$$

where Δt is the sampling time interval. $[t/\Delta t]$ is the integer part of $t/\Delta t$. $\binom{\mu}{g}$ is the Newton binomial term and can be expressed as

$$\binom{\mu}{g} = \begin{cases} \frac{\mu(\mu-1)\ldots(\mu-g+1)}{g!} & (g>0) \\ 1 & (g=0) \end{cases} \tag{7}$$

where μ represents the non-integer order of fractional calculus, representing α_k and β_k.

The lumped thermal model is expressed as [18]

$$mc\frac{dT}{dt} = q - hS(T - T_{am}) \tag{8}$$

where m is the battery mass, which is 22 g. c is the specific heat capacity. The value provided by the battery manufacturer is 1.1 $Jg^{-1}K^{-1}$. T and T_{am} represent battery temperature and ambient temperature, respectively. S is the surface area of the battery, and the value is 73.05 cm^2. h is the equivalent heat transfer coefficient, which needs to be identified by experiment. The procedure begins by placing the battery in a −15 °C environment for 2 h. Subsequently, the battery is heated to approximately 10 °C using the BPC heating method. The battery is then left to rest for 1.5 h. The temperature change during the resting phase of the battery is used to calculate h.

3.3. Determinations of Model Parameters

3.3.1. Electrical Model Parameters

Figure 3 displays the EIS test results conducted at different temperatures and current amplitudes. It is evident from the figure that the high-frequency impedance arc is minimally influenced by the current amplitude. However, the low- and medium-frequency impedance arcs gradually diminish in size as the current amplitude increases. Moreover, these characteristic frequencies are 8 Hz, 6.3 Hz, 5 Hz, 20 Hz, 25 Hz, and 63 Hz with decreasing temperature. These characteristic frequencies remain relatively stable around 6 Hz at temperatures above 0 °C. However, as the temperature drops below 0 °C, this characteristic frequency increases. These characteristic frequencies are determined based on the degree of impedance variation with current amplitude. Furthermore, as the temperature decreases, the reduction in impedance arc becomes more pronounced. Firstly, at the same temperature, the impedance arcs contract with increasing current amplitude, which is related to the solid electrolyte interphase (SEI)/cathode electrolyte interphase (CEI) processes and the charge transfer process. According to our previous research [37], the resistances of the SEI/CEI processes and the charge transfer process decrease with increasing current amplitude at low temperatures, which leads to the contraction of the impedance arc. Secondly, as the temperature decreases, the SEI/CEI processes and the charge transfer process slow down, causing the corresponding frequencies to decrease. However, in Figure 3, the characteristic frequency decreases with decreasing temperature only at temperatures above 0 °C. For temperatures below 0 °C, the characteristic frequency increases. This phenomenon may be related to the conduction process. The conduction process is the contact impedance of particle/particle and particle/current-collector, and it usually occurs more quickly than SEI/CEI processes and the charge transfer process [38]. At relatively high temperatures, the conduction process is less affected by current, with only the SEI/CEI processes and the charge transfer processes being influenced. However, once the temperature drops to a certain level, the conduction process may also be affected by the current. Furthermore, the conduction process exhibits a relatively broad frequency distribution. As the temperature decreases, the influence of the current on this process becomes increasingly pronounced. Consequently, the characteristic frequency progressively increases when the temperature drops below 0 °C.

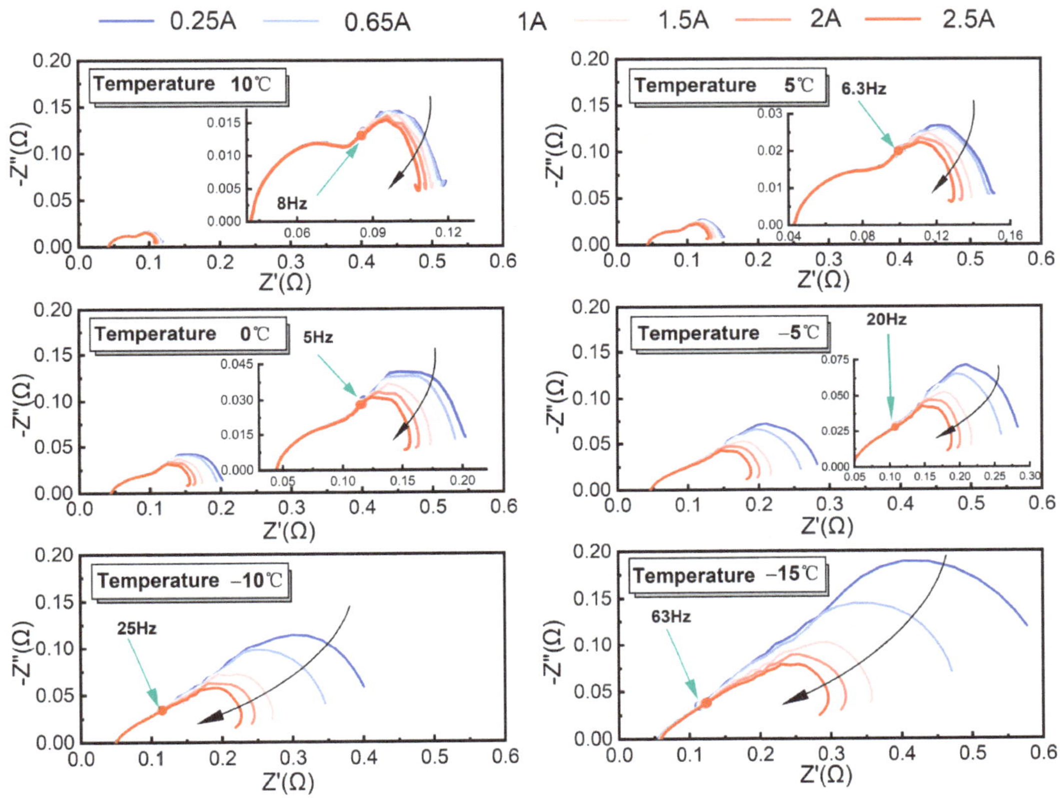

Figure 3. EIS test results at different temperatures and amplitudes.

Figure 4a illustrates the correlation between each parameter of the fractional-order equivalent circuit model and the impedance arc. The calculation of each parameter is fitted using Zview software [39]. The calculated results correspond to the experimental values as depicted in Figure 4b–h. These results demonstrate that the parameters of the fractional-order model are influenced not only by temperature but also by current amplitude. It should be noted that the effect of current amplitude on R_0 seems to be negligible in Figure 3. This is due to R_0 being relatively small and its influence being negligible compared to temperature. Despite this, both temperature and current are taken into consideration when considering the fractional-order equivalent circuit model parameters to ensure accuracy.

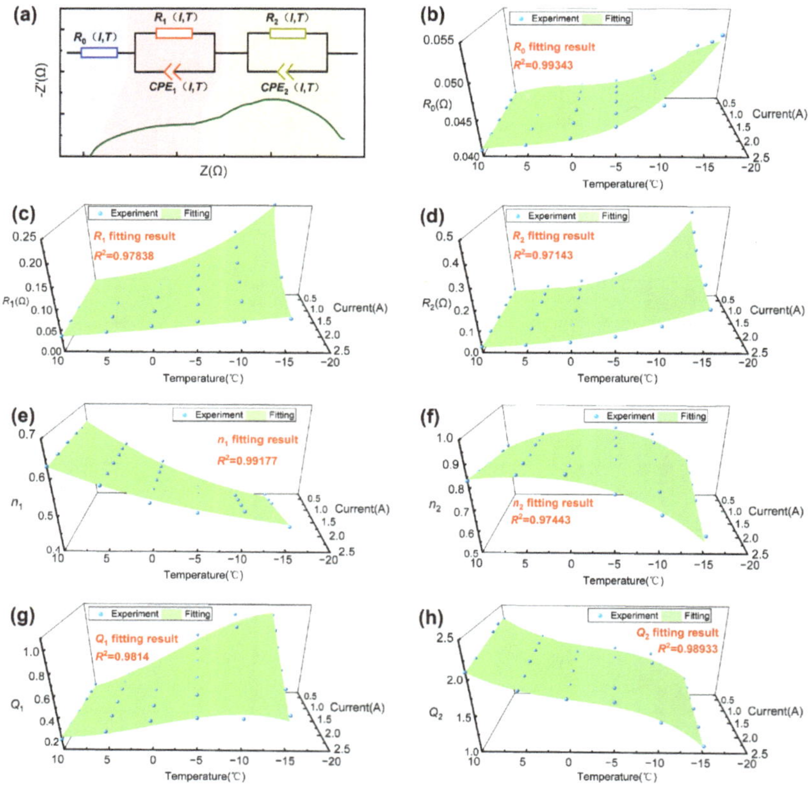

Figure 4. Electrical model parameters and the fitting results under different temperatures and amplitudes. (**a**) The relationship between the equivalent circuit model and the EIS test. (**b**) The fitting result of R_0. (**c**) The fitting result of R_1. (**d**) The fitting result of R_2. (**e**,**g**) The fitting results of CPE_1. (**f**,**h**) The fitting results of CPE_2.

Referring to [22,37], relationship between R_1 and the current and temperature and relationship between R_2 and the current and temperature can be expressed as

$$R(I,T) = \frac{\ln\left\{ \frac{1}{2M}e^{KT}I + \left[\left(\frac{1}{2M}e^{KT}I \right)^2 + 1 \right]^{0.5} \right\}}{\frac{1}{2M}XI} \quad (9)$$

where R represents R_1 or R_2. M, X, and K denote the coefficients of the equation. For R_0 and the parameters of CPE, the polynomial is used to represent the relationship between them and the current and temperature in this paper, which can be expressed as

$$A(I,T) = d_0 + d_1T + d_2I + d_3T^2 + d_4TI + d_6T^3 + d_7T^2I \quad (10)$$

where A refers to R_0 and the parameters of CPE. d denotes the coefficients of the polynomial. The fitting results for each parameter of the fractional-order equivalent circuit model are presented in Figure 4b–h. The coefficient of determination R^2 reflects the goodness of fit. The closer R^2 is to 1, the better the fit is demonstrated. The coefficient of determination R^2 for R_1 and R_2 are 0.97838 and 0.97143, respectively, indicating the accuracy of Equation (9). Furthermore, the coefficient of determination R^2 for R_0, n_1, Q_1, n_2, and Q_2 are 0.99343, 0.99177, 0.9814, 0.97443, and 0.98933. Among the R_0 and the parameters of CPE, n_2 exhibits the smallest R^2 value of 0.97443, supporting the accuracy of Equation (10). Consequently,

Equations (9) and (10) effectively capture the relationship between the parameters of the fractional-order equivalent circuit model and the temperature and current, which provides a reliable foundation for the calculation of battery heating power.

3.3.2. Calculation of Equivalent Heat Transfer Coefficient h

Another important parameter is the equivalent heat transfer coefficient h of the thermal model. h can be calculated by the temperature change of the battery during resting. The solution of Equation (8) is

$$T = \frac{q + hST_{am}}{hS}\left(1 - e^{-\frac{hS}{mc}t}\right) + T_{t=0}e^{-\frac{hS}{mc}t} \tag{11}$$

where $T_{t=0}$ indicates the temperature of the battery at the start moment. When q is 0, Equation (11) becomes

$$T = T_{am} + (T_{t=0} - T_{am})e^{-\frac{hS}{mc}t} \tag{12}$$

Figure 5 shows the fitting result of the battery cooling curve. The maximum absolute temperature error is less than 0.5 °C. The calculated value for h is 7.0993 $Wm^{-2}K^{-1}$.

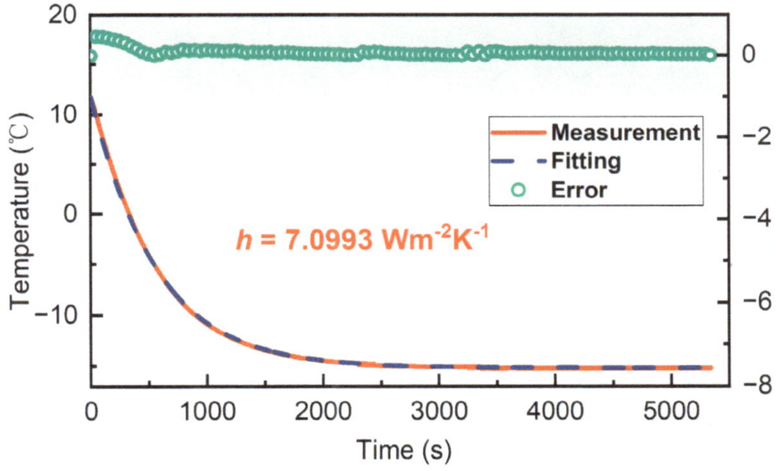

Figure 5. Battery cooling curve fitting result.

4. Results and Discussions

4.1. Validation of the Electro-Thermal Coupled Model

Figure 6 shows the measured voltages and the calculated voltages of BPCs I and II at different frequencies (1 Hz, 50 Hz, 100 Hz) and temperatures (10 °C, −5 °C, −10 °C). In the error plots, the gray areas indicate the absolute value of error is less than 0.05 V. The error of the terminal voltage calculated by the electro-thermal coupled model is larger at the frequency of 1 Hz and the temperature of −15 °C. Some voltage errors are even greater than 0.1 V. Nevertheless, the absolute value of the voltage error is in general within 0.05 V in others. Only a few errors are outside the gray area. Therefore, the results indicate that the electro-thermal coupled model can calculate the voltages of BPCs at different temperatures and frequencies relatively accurately. This ensures the accuracy of the BPC heating power calculation under various voltage constraints in the subsequent analysis.

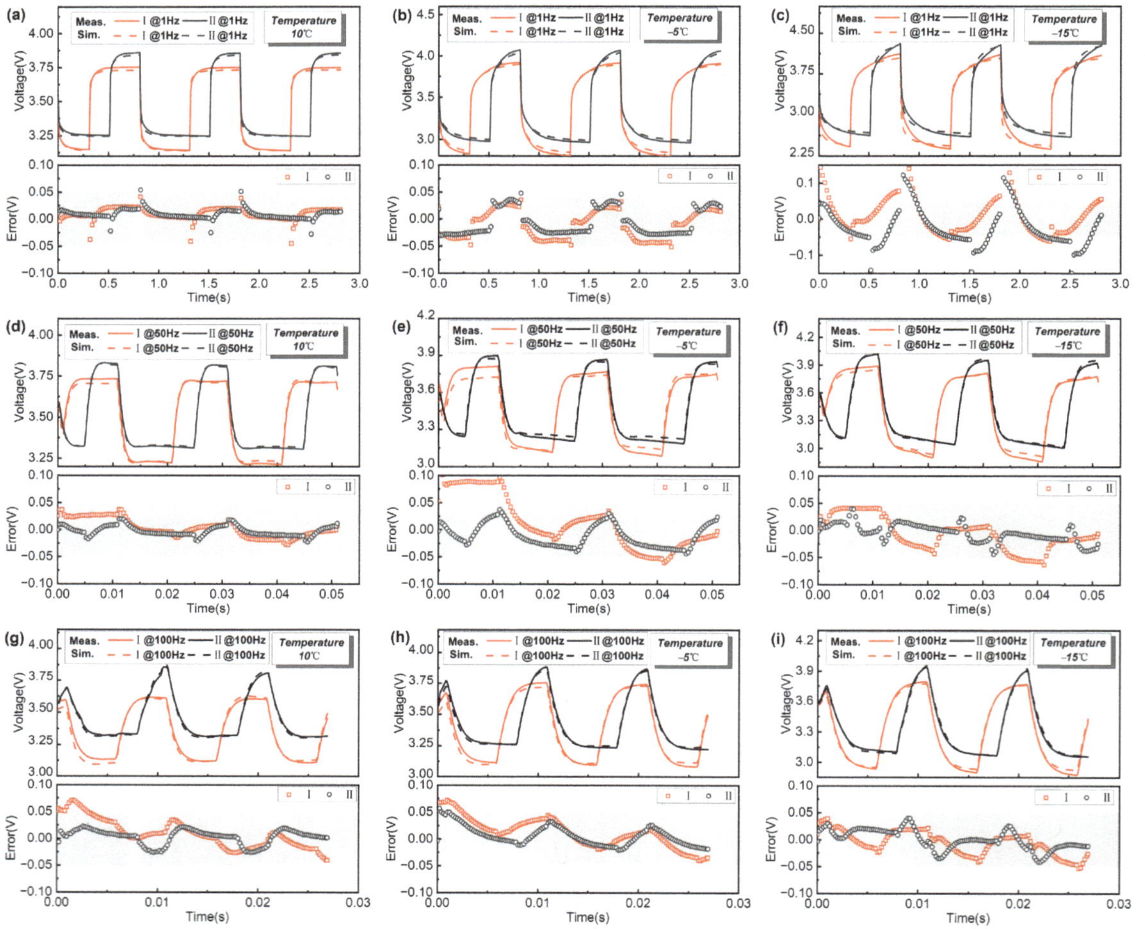

Figure 6. Battery voltage validation at different temperatures and frequencies. (**a**,**d**,**g**) The results at 10 °C. (**b**,**e**,**h**) The results at −5 °C. (**c**,**f**,**i**) The results at −15 °C.

The accuracy of the battery heating power calculation is validated through battery temperature rise experiments. Figure 7 illustrates the temperature variations of the battery under BPCs III and IV at different frequencies (1 Hz, 10 Hz, 100 Hz) and the temperature values computed by the electro-thermal coupled model. For BPC III, the absolute temperature error remains within 1.2 °C, while for BPC IV, it is within 1 °C. These results demonstrate that the developed electro-thermal coupled model accurately predicts the temperature changes in the battery. Furthermore, it confirms the capability of the electro-thermal coupled model to accurately calculate the heating power of BPC with different parameters at various temperatures. Additionally, it is observed that for both BPC III and BPC IV, a lower frequency corresponds to a higher balanced temperature. This can be attributed to the larger real part of the impedance at a lower frequency, resulting in higher heating power.

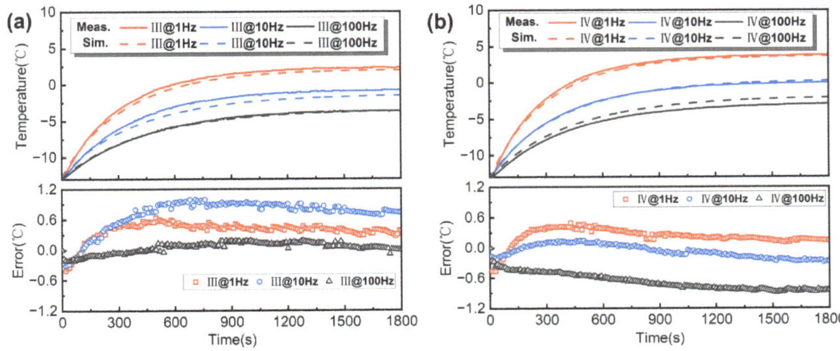

Figure 7. Battery temperature validation at different frequencies. (**a**) The results of BPC III. (**b**) The results of BPC IV.

4.2. Determination of BPC Parameters

The definition of BPC in Figure 1 contains several parameters, which makes it difficult to study the heating power variation of the BPC with different parameters. In practical scenarios of AC heating applications, there are two primary types of AC heating schemes. One category encompasses the AC heating scheme with energy injection, as seen in instances where battery packs are heated through charge devices [40–43]. The second type involves the AC heating scheme without energy injection. In the latter approach, the design of the AC heating scheme commonly leverages circuit elements to configure heating systems, effectively employing controlled switching elements to induce AC heating in batteries [13,14,16,44]. For the AC heating scheme with energy injection, adhering to an equilibrium between discharge and charge ampere-hours in a single cycle may not be theoretically essential. In contrast, for the AC heating schemes without energy injection, it is preferable to uphold parity between discharge and charge ampere-hours in a single cycle, thereby mitigating energy consumption. Given the context of EV applications, the AC heating scheme without energy injection proves more expedient for implementation. Furthermore, the AC heating scheme with energy injection can also embrace the strategy of equalizing the charge and discharge ampere-hours. Consequently, this paper delves into an analysis of the influence exerted by varying BPC parameters on heating power under the stipulation of maintaining equal charge and discharge ampere-hours. This can be expressed as

$$A_{dis}t_{dis} = A_{cha}t_{cha} \tag{13}$$

Subject to the constraints outlined in Equation (13), the BPC can be uniquely determined by the three parameters: frequency f_{pulse}, ratio R_{pulse}, and discharge amplitude A_{dis}. In addition, the failure boundary of the battery also needs to be considered during the design of AC heating strategies. Therefore, in this study, the heating power corresponding to different BPC parameters must satisfy the failure boundary constraints. Although various approaches have been proposed by researchers for the development of AC heating strategies, the determination of constraint boundaries remains a topic of controversy. Currently, recognized failure modes of low-temperature AC heating include lithium plating, overcharge, and over-discharge [25,30,45], which have been the focal points of recent research. Controlling the polarization voltage of the battery, specifically the charge and discharge cut-off voltages, is one way to mitigate these failure modes. However, the relationships between lithium plating and terminal voltage, overcharge and terminal voltage, and over-discharge and terminal voltage under high-frequency AC conditions remain unclear. Thus, to encompass a wide range of terminal voltage, this paper considers charge cut-off voltage constraints within the range of 4.2 V–5 V and discharge cut-off voltage constraints within the range of 2 V–3 V. In addition, although the parameters determining the BPC include frequency f_{pulse}, ratio R_{pulse}, and discharge amplitude A_{dis}, A_{dis} is determined by the voltage constraint. When the frequency f_{pulse} and ratio R_{pulse} are determined, there is a maximum amplitude A_{dis} so that the terminal voltage is exactly at the constraint boundary and the corresponding heating power is maximum. This heating power is the object of the subsequent study in this paper.

In addition to the discharge amplitude A_{dis}, two other parameters that require attention are frequency f_{pulse} and ratio R_{pulse}. The ratio R_{pulse} takes values between 0 and 1. The values of 0 and 1 represent complete charging and complete discharging, respectively. Therefore, 0 and 1 cannot be taken. For this research, the ratio analysis ranges from 0.05 to 0.95 and the calculation of the heating power is conducted at an interval of 0.05. Moreover, the selection of frequency range is based on the EIS test results of the battery. Four frequencies (0.5 Hz, 5 Hz, 100 Hz, and 10,000 Hz) are chosen to ensure the coverage of the entire impedance distribution as extensively as possible. Figure 8 illustrates the distribution of these four frequencies over the EIS at different temperatures. The current amplitude corresponding to each impedance is 0.25 A.

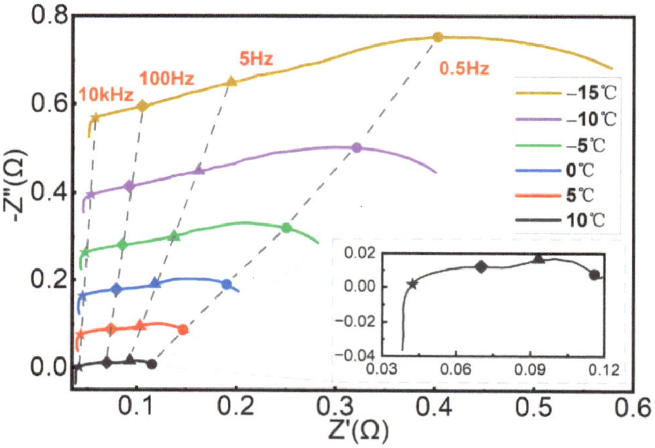

Figure 8. Distribution of the studied frequencies over the impedances.

4.3. *Heating Power of the Battery with Different BPC Parameters*

Figure 9 shows the variation of the relationship between R_{pulse} and heating power with frequency and temperature at different discharge cut-off voltages. The legends for the 0.5 Hz, 5 Hz, and 100 Hz results in the figure range from 0 W to 150 W, while the legends for the 10,000 Hz results range from 0 W to 300 W. The black areas in the plots indicate that the heating power exceeds the maximum value in the legend. For further analysis of the heating power distribution, Figure A1 provides detailed information for each case. The heating power falling in the black area can be examined and analyzed in Figure A1. In each subplot of Figures 9 and A1, it is evident that the heating power gradually increases as the discharge cut-off voltage decreases at the same R_{pulse}. When the discharge cut-off voltage is set to a smaller value, a higher discharge current is required to reach the discharge cut-off voltage. Additionally, based on Equation (13), the charge current also increases accordingly. Consequently, the battery heating power increases. Then, under the same discharge cut-off voltage constraint, the battery heating power becomes larger first and then smaller as R_{pulse} increases in most cases. The heating power variation can be attributed to the combined effects of currents and time of discharge and charge. To illustrate this phenomenon, the results for 100 Hz@−5 °C with the discharge cut-off voltage of 2.7 V as shown in Table 3. Combined with Equation (1), currents and the time of discharge and charge are relatively large in terms of magnitude. (In calculating the heating power, the discharge time is t_{dis}/T_{pulse} and the charge time is t_{cha}/T_{pulse} according to Equation (1)) Although Z'_{dis} and Z'_{cha} exhibit variation, their impact on heating power is comparatively minor. Therefore, currents and the time of discharge and charge are the main factors affecting the battery heating power. When R_{pulse} increases, the discharge current diminishes while the discharge heating power initially rises and subsequently declines. Notably, the discharge heating power reaches its peak when R_{pulse} is 0.8. When R_{pulse} is less than 0.8, although the discharge current decreases, the degree of decrease is not large. In addition, the discharge time in one cycle increases. Therefore, the discharge heating power increases. When R_{pulse} exceeds 0.8, although the discharge time increases, the discharge current is significantly smaller. Therefore, the discharge heating power decreases. Combined with the above analysis, the reason for the increase in the discharge heating power is attributed to the elevation in the discharge time, while the reason for the decline in the discharge heating power is the reduction in the discharge current. In addition, when R_{pulse} increases, it can be seen from Table 3 that the charge current increases, and the charge heating power initially rises and subsequently declines. When R_{pulse} is 0.85, the charge heating power has the maximum value. In the same way as the analysis of the discharge heating power, the reason for the increase in the charge heating power can be attributed to the elevation in the charge current. The decrease in the charge heating power can be attributed to the reduction in charge time. Finally, the changes in the discharge heating power and the charge heating power together determine the changes in the battery heating power.

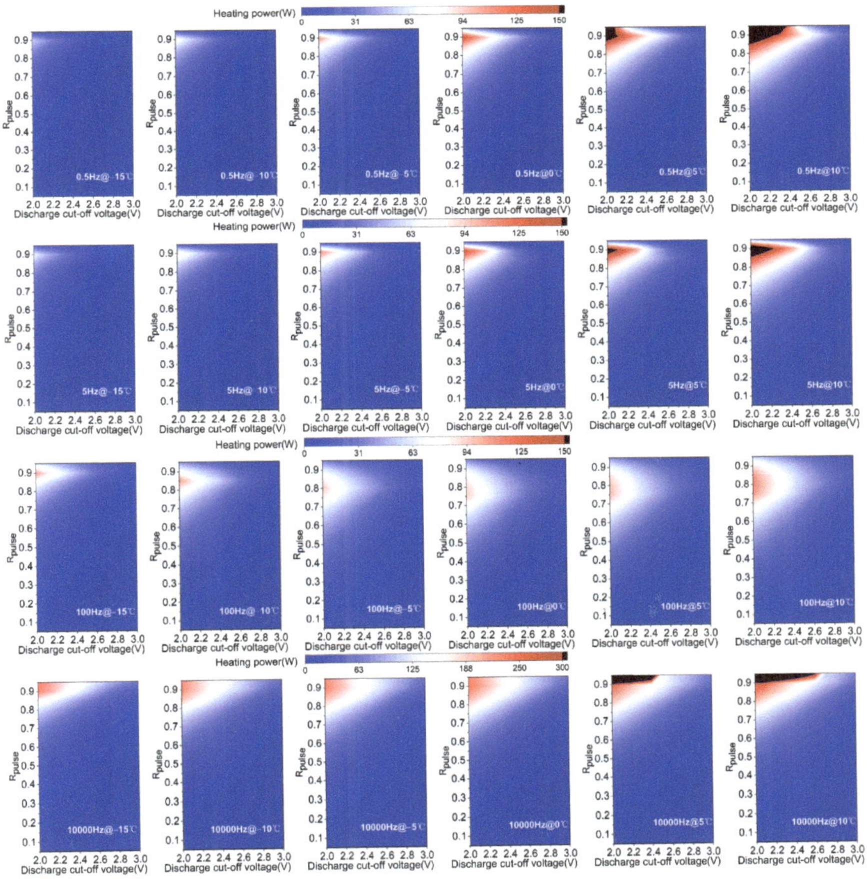

Figure 9. Variation of the relationship between R_{pulse} and heating power with frequency and temperature at different discharge cut-off voltages.

Table 3. Parameter variations in Equation (1) at −5 °C with different R_{pulse} values under the discharge cut-off voltage of 2.7 V when the BPC frequency is 100 Hz.

R_{pulse}	**0.05**	**0.10**	**0.15**	**0.20**	**0.25**	**0.30**	**0.35**	**0.40**	**0.45**	**0.50**	**0.55**	**0.60**	**0.65**	**0.70**	**0.75**	**0.80**	**0.85**	**0.90**	**0.95**
A_{dis} (A)	14.49	13.80	13.89	13.07	12.81	12.58	12.38	12.20	12.04	11.88	11.74	11.61	11.48	11.36	11.25	11.14	9.21	6.49	3.44
Z'_{dis} (Ω)	0.0619	0.0624	0.0628	0.0631	0.0633	0.0635	0.0637	0.0638	0.0640	0.0641	0.0643	0.0644	0.0645	0.0647	0.0648	0.0649	0.0672	0.0715	0.079
q_{dis} (W)	0.65	1.19	1.69	2.15	2.60	3.02	3.42	3.80	4.17	4.53	4.87	5.21	5.53	5.84	6.15	6.45	4.84	2.71	0.89
A_{cha} (A)	0.76	1.53	2.36	3.27	4.27	5.39	6.67	8.13	9.84	11.88	14.35	17.41	21.32	26.51	33.75	44.57	52.18	58.43	65.33
Z'_{cha} (Ω)	0.0841	0.0827	0.0823	0.0796	0.0766	0.0738	0.0711	0.0687	0.0664	0.0641	0.0620	0.0600	0.0580	0.0560	0.0541	0.0522	0.0513	0.0506	0.0501
q_{cha} (W)	0.047	0.18	0.39	0.68	1.05	1.50	2.06	2.73	3.54	4.53	5.75	7.27	9.22	11.82	15.41	20.74	20.93	17.29	10.68
q (W)	0.70	1.36	2.08	2.83	3.64	4.52	5.47	6.53	7.71	9.06	10.62	12.48	14.75	17.66	21.56	27.19	25.77	20.00	11.57

Furthermore, the relationship between heating power and frequency and the relationship between heating power and temperature exhibit variability across different R_{pulse} values and discharge cutoff voltages. Figure 9 demonstrates that the area covered in white and red becomes larger as the temperature increases at the same frequency, which shows that the area covered by the high heating power is increasing. Therefore, at the same discharge cut-off voltage and R_{pulse}, the battery heating power may become larger with the increase in temperature. This phenomenon can be explained by the variation of battery resistance with temperature. In Figure 8, the resistance becomes small as the temperature increases. Thus, at the same discharge cut-off voltage, the battery needs a larger current to reach the constraint. From the above analysis, it is clear that the battery heating power is mainly affected by the current. Therefore, the battery heating power becomes large as the current increases. In addition, Figure A1 shows that the relationship between the maximum heating power and temperature is not certain. For example, at a frequency of 0.5 Hz, the R_{pulse} value corresponding to the maximum heating power remains at 0.9 within the temperature range of −15 °C to 0 °C. However, when temperatures are 5 °C and 10 °C, the R_{pulse} value associated with the maximum heating power increases to 0.95. Moreover, the maximum heating power gradually increases with rising temperature. Conversely, when the frequency is 100 Hz, the R_{pulse} value linked to the maximum heating power tends to decrease with the increase in temperature. The difference in the maximum heating power is not significant. Additionally, the variations of R_{pulse} corresponding to the maximum heating power with frequency at the same discharge cut-off voltage are not the same at temperatures of −15 °C and 5 °C. When the temperature is below 0 °C, Figure 9 demonstrates that the area covered in white and red becomes large as the frequency increases at the same temperature, which shows that the area covered by the high heating power is increasing. However, when the temperature exceeds 0 °C, this phenomenon is not obvious. At the same temperature, the area covered in white and red seems to first diminish and then grow large with increasing frequency. At a frequency of 100 Hz, the region is the smallest and the heating power in this region is significantly small. The reason for the difference can be attributed to variations in battery resistance. For example, Table 4 demonstrates the parameter variations in Equation (1) with different BPC frequencies at −15 °C and 10 °C when the discharge cut-off voltage is 2.2 V and R_{pulse} is 0.85. Both Z'_{dis} and Z'_{cha} decrease gradually with increasing frequency at −15 °C and 10 °C. However, compared to the results at 10 °C, Z'_{dis} and Z'_{cha} decrease more significantly at −15 °C, which also leads to a greater increase in discharge and charge currents at −15 °C. Thus, the battery heating power at −15 °C becomes large with the increase in frequency. When the temperature is 10 °C, the increase in discharge and charge currents from 0.5 Hz to 100 Hz is not obvious and Z'_{dis} and Z'_{cha} decrease, thus causing the battery heating power to reduce. When the frequency is 10,000 Hz, Z'_{dis} and Z'_{cha} are significantly smaller, which makes a significant increase in the discharge and charge currents. Consequently, the battery heating power becomes large again. Comprehensive analysis of the above shows that, when the temperature is less than 0 °C, because the battery resistance reduces more obviously with the increase in frequency, the battery heating power is mainly dominated by the discharge and charge currents. Therefore, the battery heating power gradually becomes large with the increase in frequency. When the temperature is equal to or greater than 0 °C, as the battery resistance decreases less with the increase of frequency, the battery heating power is dominated by the discharge and charge currents, in conjunction with Z'_{dis} and Z'_{cha}. Therefore, the battery heating power changes uncertainly with the increase in frequency. In addition, the variation of the maximum heating power with different frequencies is similar to the variation of the maximum heat production power with different temperatures. There is no obvious pattern as can be seen in Figure A1.

Figure 10 shows the variation of the relationship between R_{pulse} and heating power with frequency and temperature at different charge cut-off voltages. The legends in the figure range from 0 W to 130 W. The black areas in the plots indicate that the heating power exceeds the maximum value in the legend. Figure A2 demonstrates detailed information for each case. In each subplot of Figures 10 and A2, the battery heating power at the same R_{pulse} increases gradually as the charge cut-off voltage becomes larger. This is similar to the results for the discharge cut-off voltage. A larger charge cut-off voltage needs a higher current to reach the constraint boundary. The heating power will also become larger. Similar to the discharge cut-off voltage, the battery heating power becomes large first and then smaller as R_{pulse} increases in most cases under the same charge cut-off voltage. Unlike the discharge cut-off voltage, the large heating power is mainly concentrated in the interval where R_{pulse} is small. To illustrate this phenomenon, Table 5 shows the results for 100 Hz@−5 °C with the charge cut-off voltage of 4.8 V. It can be seen that the battery heating power first increases and then decreases with the increase of R_{pulse}, which is the same as the changes of discharge heating power and charge heating power. Moreover, the maximum values of the battery heating power, discharge heating power, and charge heating power correspond to R_{pulse} of 0.25, 0.2, and 0.45, respectively. Similar to the analysis of the discharge cut-off voltage, due to the small influence of Z'_{dis} and Z'_{cha} on the battery heating power, the discussion will exclusively focus on the currents and time of charge and discharge. For the discharge heating power, with the increase of R_{pulse}, the variation of the discharge heating power is the same as the variation of the discharge current. Therefore, the change in the discharge heating power is mainly dominated by the discharge current. For the charge heating power, with the increase of R_{pulse}, the charge current gradually increases, while the charge time gradually decreases. The initial increase in charge heating power can be attributed to the increase in charge current. As the charge time decreases sharply afterward, the dominant factor of the charge heating power changes from the charge current to the charge time. As a result, the charge heating power becomes small with the increase in R_{pulse}.

Table 4. Parameter variations in Equation (1) at −15 °C and 10 °C with different BPC frequencies under the discharge cut-off voltage of 2.2 V when the R_{pulse} is 0.85.

Temperature (°C)	**−15**				**10**			
f_{pulse} **(Hz)**	**0.5**	**5**	**100**	**10,000**	**0.5**	**5**	**100**	**10,000**
A_{dis} (A)	6.13	7.12	10.91	18.89	17.74	18.14	18.26	28.37
Z'_{dis} (Ω)	0.2085	0.1578	0.0956	0.0637	0.0765	0.0652	0.0522	0.0435
q_{dis} (W)	6.66	6.79	9.67	19.31	20.47	18.24	14.79	29.74
A_{cha} (A)	34.73	40.34	61.81	107.04	100.54	102.81	103.49	160.79
Z'_{cha} (Ω)	0.1062	0.0960	0.0742	0.0589	0.0587	0.0536	0.0433	0.0414
q_{cha} (W)	19.21	23.44	42.52	101.30	89.03	85.05	69.54	160.71
q (W)	25.87	30.24	52.19	120.61	109.51	103.29	84.33	190.45

Similar to the discharge cut-off voltage, the effect of temperature and frequency on the heating power is also diverse. First, it can be seen from Figure 10 that at the same frequency, the area covered by high heating power becomes larger as the temperature increases. Therefore, at the same charge cut-off voltage and R_{pulse}, the battery heating power may become large with the increase in temperature. The reason is the same as that for the discharge cut-off voltage, which is caused by the decrease in battery resistance due to the increase in temperature. When the battery resistance decreases, the current increases at the same charge cut-off voltage. Thus, the battery heating power becomes large. In addition, the relationship between the maximum heating power and temperature is not clear in Figure A2. For example, at 0.5 Hz, the R_{pulse} corresponding to the maximum heating power at the same charge cut-off voltage is 0.1 when the temperature is less than 5 °C. When the temperature is 10 °C, the R_{pulse} becomes 0.05 and the maximum heating power becomes larger as the temperature increases. At 10 kHz, the R_{pulse} of the maximum heating power at the same charge cut-off voltage becomes larger as the temperature increases. Then, at the same temperature, the result of high heating power distribution with frequency is similar to that of discharge cut-off voltage. When the temperature is less than 0 °C, the region of high heating power distribution increases with frequency. When the temperature is equal to or greater than 0 °C, the region of high heating power distribution decreases and then increases with frequency. To explain this phenomenon, Table 6 shows the parameter variations in Equation (1) with different BPC frequencies at −15 °C and 10 °C when the charge cut-off voltage is 4.8 V and R_{pulse} is 0.2. Whether the temperature is −15 °C or 10 °C, the battery resistance gradually gets small as the frequency increases. However, at −15 °C, the resistance decreases more significantly, which makes the increase in current at −15 °C more pronounced. Therefore, the battery heating power becomes progressively larger with increasing frequency at −15 °C. When the temperature is 10 °C, the current does not increase much from 0.5 Hz to 100 Hz, while the resistance decreases. Therefore, the battery heating power decreases. When the frequency is 10,000 Hz, the current is a little larger. Therefore, the battery heating power becomes large again. Comprehensive analysis of the above shows that at −15 °C, the battery heating power becomes larger with the increase of frequency, which is determined by the current. The change of battery heating power at 10 °C is determined by the current and resistance together. In addition, the variation of the maximum heating power with different frequencies in Figure A2 is similar to the result of discharge cut-off voltage. Neither of them has a clear pattern.

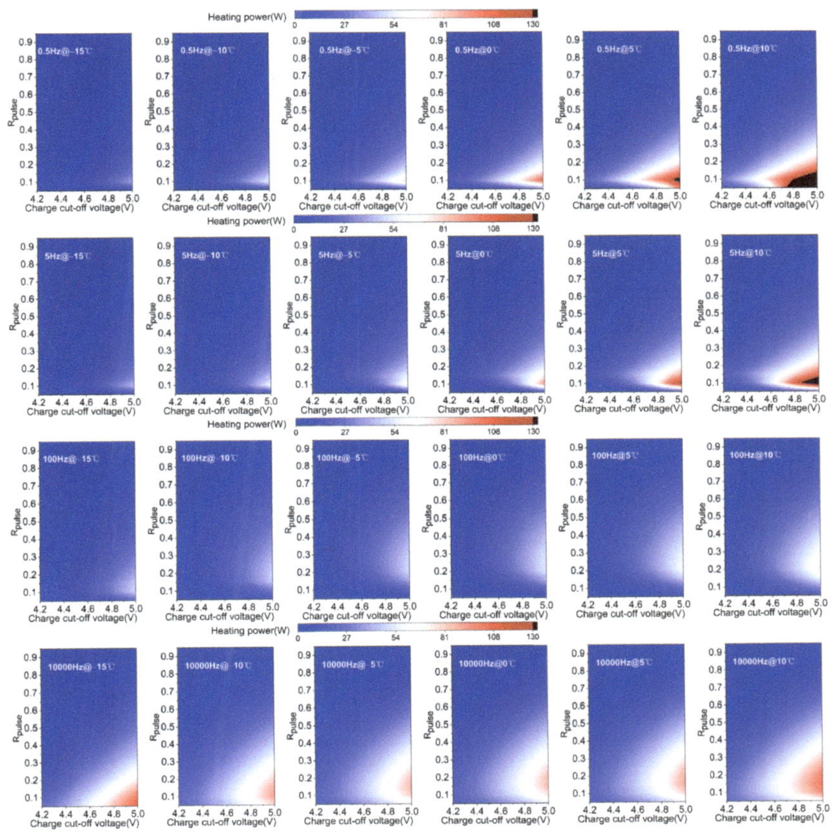

Figure 10. Variation of the relationship between R_{pulse} and heating power with frequency and temperature at different charge cut-off voltages.

Table 5. Parameter variations in Equation (1) at −5 °C with different R_{pulse} values under the charge cut-off voltage of 4.8 V when the BPC frequency is 100 Hz.

R_{pulse}	0.05	0.10	0.15	0.20	0.25	0.30	0.35	0.40	0.45	0.50	0.55	0.60	0.65	0.70	0.75	0.80	0.85	0.90	0.95
A_{dis} (A)	43.79	48.18	51.54	48.93	42.83	37.02	31.83	27.30	23.35	19.88	16.83	14.12	11.69	9.51	7.53	5.73	4.08	2.58	1.21
Z'_{dis} (Ω)	0.0523	0.0517	0.0513	0.0516	0.0525	0.0534	0.0546	0.0558	0.0571	0.0586	0.0603	0.0622	0.0643	0.0668	0.0696	0.0730	0.0771	0.0820	0.0832
q_{dis} (W)	5.02	12.01	20.45	24.72	24.06	21.97	19.35	16.63	14.02	11.59	9.40	7.44	5.72	4.23	2.96	1.92	1.09	0.49	0.12
A_{cha} (A)	2.31	5.35	9.10	12.23	14.28	15.86	17.14	18.20	19.10	19.88	20.57	21.18	21.72	22.19	22.59	22.91	23.14	23.22	22.93
Z'_{cha} (Ω)	0.0823	0.0739	0.0673	0.0638	0.0621	0.0609	0.0601	0.0595	0.0590	0.0586	0.0583	0.0580	0.0578	0.0576	0.0574	0.0573	0.0572	0.0572	0.0573
q_{cha} (W)	0.42	1.91	4.73	7.64	9.49	10.74	11.48	11.83	11.85	11.59	11.10	10.41	9.54	8.51	7.33	6.02	4.60	3.08	1.51
q (W)	5.43	13.91	25.19	32.36	33.55	32.71	30.83	28.46	25.87	23.19	20.50	17.85	15.26	12.73	10.29	7.94	5.69	3.58	1.62

Table 6. Parameter variations in Equation (1) at −15 °C and 10 °C with different BPC frequencies under the charge cut-off voltage of 4.8 V when the R_{pulse} is 0.2.

Temperature (°C)			**−15**				**10**	
f_{pulse} **(Hz)**	**0.5**	**5**	**100**	**10,000**	**0.5**	**5**	**100**	**10,000**
A_{dis} (A)	19.59	23.44	35.59	59.85	61.52	63.33	64.69	84.00
Z'_{dis} (Ω)	0.1294	0.1110	0.0795	0.0608	0.0626	0.0560	0.0449	0.0426
q_{dis} (W)	9.93	12.20	20.13	43.52	47.37	44.93	37.55	60.15
A_{cha} (A)	4.90	5.86	8.90	14.96	15.38	15.83	16.17	21.00
Z'_{cha} (Ω)	0.2292	0.1669	0.0991	0.064	0.0784	0.0665	0.0530	0.0435
q_{cha} (W)	4.40	4.59	6.27	11.47	14.84	13.34	11.10	15.35
q (W)	14.32	16.78	26.41	54.99	62.21	58.27	48.65	75.50

The variations of the relationship between R_{pulse} and heating power with frequency and temperature for different discharge and charge cut-off voltages are analyzed in Figures 9 and 10, respectively. However, when the AC heating strategy in practice is designed, the charge and discharge cut-off voltages should be considered simultaneously. Figure 11 shows the variation of the relationship between R_{pulse} and heating power with charge and discharge cut-off voltages at −5 °C and 100 Hz. The curves in the figure represent the maximum heating power that can be achieved at the corresponding charge and discharge cut-off voltages. When both charge and discharge cut-off voltages are considered, the battery heating power is determined by the smaller one. Therefore, the intersection of the heating power curves for the charge and discharge cut-off voltages is the maximum heating power under the constraints of charge and discharge terminal voltages, such as the green dots in Figure 11. If the value of R_{pulse} is smaller than the R_{pulse} value of the intersection, the maximum heating power is determined by the discharge cut-off voltage. Conversely, if the value of R_{pulse} is larger than the R_{pulse} value of the intersection, the maximum heating power is determined by the charge cut-off voltage. The two figures on the right in Figure 11 demonstrate the influences of the discharge cut-off voltage and the charge cut-off voltage on the position of the intersection, respectively. As the discharge cut-off voltage increases, the intersection gradually shifts towards a larger R_{pulse} value. Conversely, as the charge cut-off voltage decreases, the intersection gradually shifts towards a smaller R_{pulse} value. Moreover, the heating power corresponding to R_{pulse} of 0.5 is not the maximum heating power. For example, when the charge cut-off voltage and the discharge cut-off voltage are 5 V and 3 V, the maximum heating power corresponds to R_{pulse} of about 0.8.

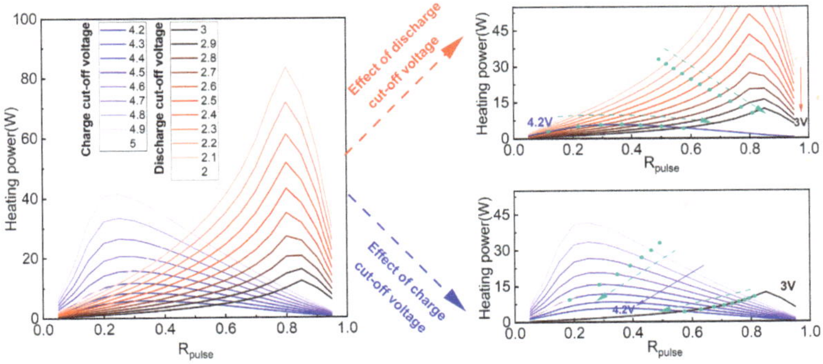

Figure 11. Variation of the relationship between R_{pulse} and heating power with charge/discharge cut-off voltages at −5 °C and 100 Hz.

The intersection of the heating power curves for the charge and discharge cut-off voltages corresponds to the maximum heating power. Thus, the impact of different frequencies and temperatures on the heating power under the constraints of both charge and discharge cut-off voltage can be assessed through the intersection. In this study, the charge and discharge cut-off voltages of the battery are 4.2 V and 3 V, respectively. Therefore, the maximum heating power at these voltage constraints serves as the basis for analyzing the effects of frequency and temperature, and the results are presented in Figure 12. Figure 12a,b depict the variations in maximum heating power with frequency and temperature and the distribution of the corresponding R_{pulse} value, respectively. Notably, the maximum heating power increases with temperature for a given frequency. Similarly, at a constant temperature, the maximum heating power exhibits continuous growth with the increase in frequency. Figure 12b reveals that the value of R_{pulse} corresponding to the maximum heating power is predominantly 0.55. The value of R_{pulse} corresponding to the maximum heating power is 0.5 only when the frequency is 0.5 Hz and the temperatures are 5 °C and 10 °C.

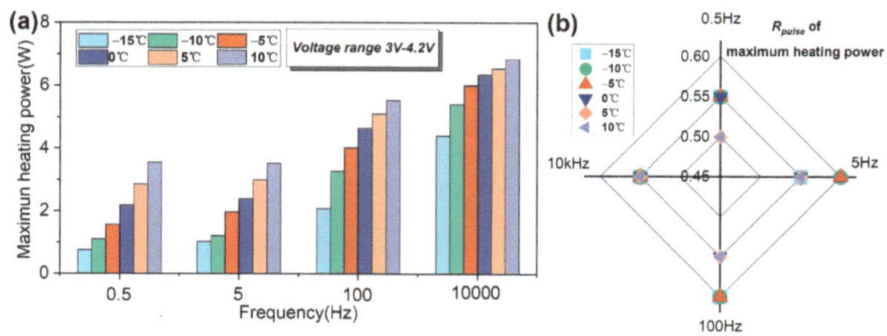

Figure 12. Maximum heating power and the corresponding R_{pulse} at different frequencies and temperatures when the voltage range is 3 V–4.2 V. (**a**) Maximum heating power. (**b**) R_{pulse} of maximum heating power.

5. Conclusions

This paper investigates the impact of various BPC parameters on the heat generation of the battery by the electro-thermal coupled model under the premise that the charge capacity and discharge capacity are equal. The main conclusions of this study can be summarized as follows:

(1) The heating power of the battery increases with a decrease in the discharge cut-off voltage and an increase in the charge cut-off voltage. Under the discharge cut-off voltage constraint, higher heating powers are observed in the range of larger R_{pulse} values. Conversely, the higher heating powers are mainly concentrated in the range of smaller R_{pulse} values for the charge cut-off voltage constraint.

(2) The symmetrical BPC approach does not result in maximum heating power. The relationship between the maximum heating power and R_{pulse} is influenced by both the charge cut-off voltage and the discharge cut-off voltage. Specifically, the R_{pulse} of the maximum heating power increases with an increase in the discharge cut-off voltage, while it decreases with a decrease in the charge cut-off voltage.

(3) When the voltage constraint range is set between 3 V and 4.2 V, the R_{pulse} of the maximum heating power is primarily 0.55. Moreover, under the same frequency, the maximum heating power gradually increases with rising temperatures. At frequencies of 0.5 Hz, 5 Hz, 100 Hz, and 10,000 Hz, the maximum heating power exhibits a progressive increase with higher frequencies at a constant temperature.

BPC heating offers higher heating efficiency compared to SAC heating. However, BPC involves multiple parameters, making it challenging to optimize the BPC heating strategy. The research approach in this paper provides the possibility of optimizing the BPC heating strategy. Moreover, the findings of this study demonstrate that the widely used symmetrical BPC heating is not the most efficient strategy. To enhance heating efficiency when designing BPC heating, intervals with higher frequencies should be considered. Additionally, the optimization of the BPC heating strategy should not only consider amplitude and frequency, but also the BPC ratio. In future work, the terminal voltages corresponding to lithium plating, overcharge, and over-discharge of the battery will be determined. Under these constraints, the frequency, the ratio, and the amplitude of BPC heating will be optimized.

Author Contributions: Conceptualization, R.H.; Methodology, R.H., J.Z., X.W. (Xueyuan Wang) and X.Z.; Software, R.H.; Validation, G.W.; Formal analysis, R.H., B.J. and X.W. (Xueyuan Wang); Investigation, G.W.; Resources, X.P., X.Z., J.Y. and H.D.; Writing—original draft, R.H.; Writing—review & editing, R.H., B.J., J.Z., X.P., X.W. (Xueyuan Wang), X.Z. and J.Y.; Supervision, H.D.; Project administration, X.W. (Xuezhe Wei) and H.D.; Funding acquisition, H.D. All authors have read and agreed to the published version of the manuscript.

Funding: This research was funded by the National Natural Science Foundation of China (NSFC, Grant number U20A20310) and sponsored by Program of Shanghai Academic/Technology Research Leader (22XD1423800).

Data Availability Statement: The authors are unable or have chosen not to specify which data have been used.

Acknowledgments: Ranjun Huang would like to thank the assistance of Dongdong Qiao (School of Automotive Studies, Tongji University, Shanghai 201804, China) and Guangxu Zhang (School of Automotive Studies, Tongji University, Shanghai 201804, China) for their help in drawing the pictures in this paper.

Conflicts of Interest: The authors declare no conflict of interest.

Appendix A

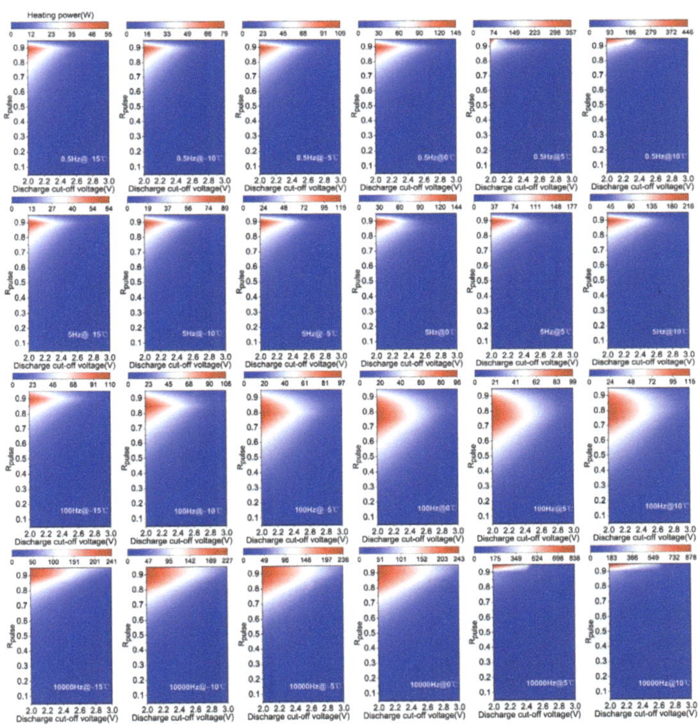

Figure A1. Variation of the relationship between R_{pulse} and heating power with frequency and temperature at different discharge cut-off voltages.

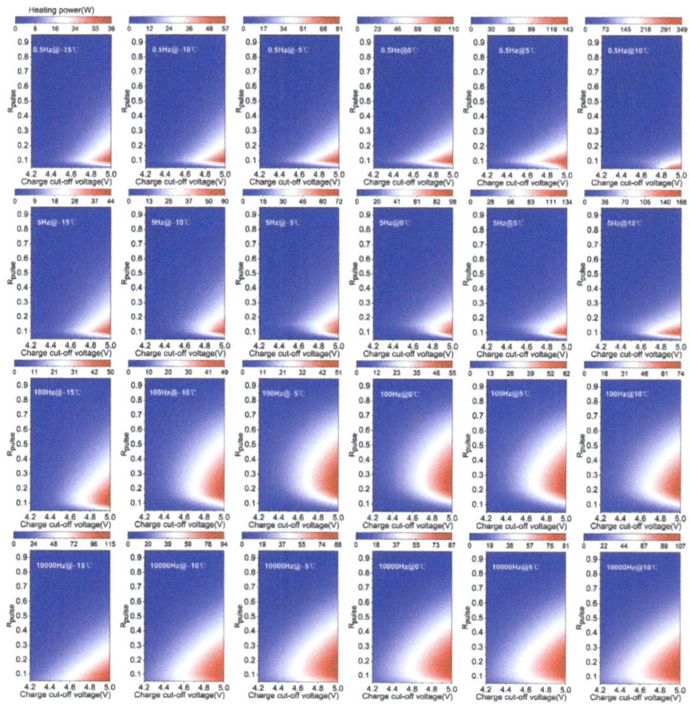

Figure A2. Variation of the relationship between R_{pulse} and heating power with frequency and temperature at different charge cut-off voltages.

References

1. Wei, G.; Huang, R.; Zhang, G.; Jiang, B.; Zhu, J.; Guo, Y.; Han, G.; Wei, X.; Dai, H. A comprehensive insight into the thermal runaway issues in the view of lithium-ion battery intrinsic safety performance and venting gas explosion hazards. *Appl. Energy* **2023**, *349*, 121651. [CrossRef]
2. Zhou, W.; Huang, R.J.; Liu, K.; Zhang, W.G. A novel interval-based approach for quantifying practical parameter identifiability of a lithium-ion battery model. *Int. J. Energy Res.* **2020**, *44*, 3558–3573. [CrossRef]
3. Rodrigues, M.-T.F.; Babu, G.; Gullapalli, H.; Kalaga, K.; Sayed, F.N.; Kato, K.; Joyner, J.; Ajayan, P.M. A materials perspective on Li-ion batteries at extreme temperatures. *Nat. Energy* **2017**, *2*, 17108. [CrossRef]
4. Ouyang, M.; Chu, Z.; Lu, L.; Li, J.; Han, X.; Feng, X.; Liu, G. Low temperature aging mechanism identification and lithium deposition in a large format lithium iron phosphate battery for different charge profiles. *J. Power Sources* **2015**, *286*, 309–320. [CrossRef]
5. You, H.; Dai, H.; Li, L.; Wei, X.; Han, G. Charging Strategy Optimization at Low Temperatures for Li-Ion Batteries Based on Multi-Factor Coupling Aging Model. *IEEE Trans. Veh. Technol.* **2021**, *70*, 11433–11445. [CrossRef]
6. Wang, Y.; Zhang, X.; Chen, Z. Low temperature preheating techniques for Lithium-ion batteries: Recent advances and future challenges. *Appl. Energy* **2022**, *313*, 118832. [CrossRef]
7. Hu, X.; Zheng, Y.; Howey, D.A.; Perez, H.; Foley, A.; Pecht, M. Battery warm-up methodologies at subzero temperatures for automotive applications: Recent advances and perspectives. *Prog. Energy Combust. Sci.* **2020**, *77*, 100806. [CrossRef]
8. Lin, C.; Kong, W.; Tian, Y.; Wang, W.; Zhao, M. Heating Lithium-Ion Batteries at Low Temperatures for Onboard Applications: Recent Progress, Challenges and Prospects. *Automot. Innov.* **2022**, *5*, 3–17. [CrossRef]
9. Wang, C.Y.; Zhang, G.; Ge, S.; Xu, T.; Ji, Y.; Yang, X.G.; Leng, Y. Lithium-ion battery structure that self-heats at low temperatures. *Nature* **2016**, *529*, 515–518. [CrossRef] [PubMed]
10. Wu, S.; Xiong, R.; Li, H.; Nian, V.; Ma, S. The state of the art on preheating lithium-ion batteries in cold weather. *J. Energy Storag.* **2020**, *27*, 101059. [CrossRef]
11. Qu, Z.G.; Jiang, Z.Y.; Wang, Q. Experimental study on pulse self–heating of lithium–ion battery at low temperature. *Int. J. Heat Mass Transf.* **2019**, *135*, 696–705. [CrossRef]
12. Stuart, T.A.; Hande, A. HEV battery heating using AC currents. *J. Power Sources* **2004**, *129*, 368–378. [CrossRef]
13. Shang, Y.; Liu, K.; Cui, N.; Wang, N.; Li, K.; Zhang, C. A Compact Resonant Switched-Capacitor Heater for Lithium-Ion Battery Self-Heating at Low Temperatures. *IEEE Trans. Power Electron.* **2020**, *35*, 7134–7144. [CrossRef]

14. Hu, Z.; Li, Y.; Liu, F.; Zhao, B.; Li, W.; Yang, R.; Xie, C.; Shi, Y. Thermal Characteristics Investigation of Lithium-Ion Battery Under High-Frequency AC Excitation in Low-Temperature Environment. *IEEE Trans. Transp. Electrif.* **2022**, *8*, 407–419. [CrossRef]
15. Zhu, C.; Cao, Y.; Zhang, H.; Lu, F.; Zhang, X. Comprehensive Design and Optimization of an Onboard Resonant Self-Heater for EV Battery. *IEEE Trans. Transp. Electrif.* **2021**, *7*, 452–463. [CrossRef]
16. Li, Y.; Gao, X.; Qin, Y.; Du, J.; Guo, D.; Feng, X.; Lu, L.; Han, X.; Ouyang, M. Drive circuitry of an electric vehicle enabling rapid heating of the battery pack at low temperatures. *iScience* **2021**, *24*, 101921. [CrossRef]
17. Du, C.; Peng, Q.; Chen, F.; Deng, K.; Chen, J.; Deng, C.; Hu, M. Investigation on the method of battery self-heating using motor pulse current. *Proc. Inst. Mech. Eng. Part D J. Automob. Eng.* **2021**, *236*, 2399–2409. [CrossRef]
18. Zhang, J.; Ge, H.; Li, Z.; Ding, Z. Internal heating of lithium-ion batteries using alternating current based on the heat generation model in frequency domain. *J. Power Sources* **2015**, *273*, 1030–1037. [CrossRef]
19. Guo, S.; Xiong, R.; Shen, W.; Sun, F. Aging investigation of an echelon internal heating method on a three-electrode lithium ion cell at low temperatures. *J. Energy Storage* **2019**, *25*, 100878. [CrossRef]
20. Ruan, H.; Jiang, J.; Sun, B.; Zhang, W.; Gao, W.; Wang, L.Y.; Ma, Z. A rapid low-temperature internal heating strategy with optimal frequency based on constant polarization voltage for lithium-ion batteries. *Appl. Energy* **2016**, *177*, 771–782. [CrossRef]
21. Zhang, L.; Fan, W.; Wang, Z.; Li, W.; Sauer, D.U. Battery heating for lithium-ion batteries based on multi-stage alternative currents. *J. Energy Storage* **2020**, *32*, 101885. [CrossRef]
22. Zhu, J.; Sun, Z.; Wei, X.; Dai, H. An alternating current heating method for lithium-ion batteries from subzero temperatures. *Int. J. Energy Res.* **2016**, *40*, 1869–1883. [CrossRef]
23. Zhang, Z.; Yang, X.; Jian, J.; Wang, S.; Gong, J.; Zhang, J. Rapid and deposition-free preheating of lithium-ion cell with square wave current. *Int. J. Energy Res.* **2022**, *46*, 18. [CrossRef]
24. Wu, X.; Cui, Z.; Chen, E.; Du, J. Capacity degradation minimization oriented optimization for the pulse preheating of lithium-ion batteries under low temperature. *J. Energy Storage* **2020**, *31*, 101746. [CrossRef]
25. Qin, Y.; Du, J.; Lu, L.; Gao, M.; Haase, F.; Li, J.; Ouyang, M. A rapid lithium-ion battery heating method based on bidirectional pulsed current: Heating effect and impact on battery life. *Appl. Energy* **2020**, *280*, 115957. [CrossRef]
26. Waldmann, T.; Hogg, B.-I.; Wohlfahrt-Mehrens, M. Li plating as unwanted side reaction in commercial Li-ion cells—A review. *J. Power Sources* **2018**, *384*, 107–124. [CrossRef]
27. Tomaszewska, A.; Chu, Z.; Feng, X.; O'Kane, S.; Liu, X.; Chen, J.; Ji, C.; Endler, E.; Li, R.; Liu, L.; et al. Lithium-ion battery fast charging: A review. *eTransportation* **2019**, *1*, 100011. [CrossRef]
28. Konz, Z.M.; McShane, E.J.; McCloskey, B.D. Detecting the Onset of Lithium Plating and Monitoring Fast Charging Performance with Voltage Relaxation. *ACS Energy Lett.* **2020**, *5*, 1750–1757. [CrossRef]
29. Jiang, J.; Ruan, H.; Sun, B.; Wang, L.; Gao, W.; Zhang, W. A low-temperature internal heating strategy without lifetime reduction for large-size automotive lithium-ion battery pack. *Appl. Energy* **2018**, *230*, 257–266. [CrossRef]
30. Ge, H.; Huang, J.; Zhang, J.; Li, Z. Temperature-Adaptive Alternating Current Preheating of Lithium-Ion Batteries with Lithium Deposition Prevention. *J. Electrochem. Soc.* **2015**, *163*, A290–A299. [CrossRef]
31. Forgez, C.; Vinh Do, D.; Friedrich, G.; Morcrette, M.; Delacourt, C. Thermal modeling of a cylindrical $LiFePO_4$/graphite lithium-ion battery. *J. Power Sources* **2010**, *195*, 2961–2968. [CrossRef]
32. Onda, K.; Kameyama, H.; Hanamoto, T.; Ito, K. Experimental Study on Heat Generation Behavior of Small Lithium-Ion Secondary Batteries. *J. Electrochem. Soc.* **2003**, *150*, A285–A291. [CrossRef]
33. Liu, G.; Zhang, Z.; Gong, J.; Li, Q.; Zhou, Y.; Zou, H. A Square Wave Alternating Current Preheating with High Applicability and Effectiveness of Preventing Lithium Plating. *Processes* **2023**, *11*, 1089. [CrossRef]
34. Ruan, H.; Sun, B.; Jiang, J.; Zhang, W.; He, X.; Su, X.; Bian, J.; Gao, W. A modified-electrochemical impedance spectroscopy-based multi-time-scale fractional-order model for lithium-ion batteries. *Electrochim. Acta* **2021**, *394*, 139066. [CrossRef]
35. Zou, C.; Hu, X.; Dey, S.; Zhang, L.; Tang, X. Nonlinear Fractional-Order Estimator with Guaranteed Robustness and Stability for Lithium-Ion Batteries. *IEEE Trans. Ind. Electron.* **2017**, *65*, 5951–5961. [CrossRef]
36. Wang, B.; Li, S.E.; Peng, H.; Liu, Z. Fractional-order modeling and parameter identification for lithium-ion batteries. *J. Power Sources* **2015**, *293*, 151–161. [CrossRef]
37. Huang, R.; Wang, X.; Jiang, B.; Chen, S.; Zhang, G.; Zhu, J.; Wei, X.; Dai, H. Revealing the electrochemical impedance characteristics of lithium-ion battery (nickel-cobalt-aluminum vs. graphite) under various alternating current amplitudes. *J. Power Sources* **2023**, *566*, 232929. [CrossRef]
38. Shafiei Sabet, P.; Sauer, D.U. Separation of predominant processes in electrochemical impedance spectra of lithium-ion batteries with nickel-manganese-cobalt cathodes. *J. Power Sources* **2019**, *425*, 121–129. [CrossRef]
39. Available online: https://www.ameteksi.com/products/software/zview-software-en (accessed on 31 July 2023).
40. Guillen Montenegro, L.E.; Villegas Pico, H.N. Ac-Heating and Fast-Charging Power Requirements of EV Battery Packs in Subzero Temperature. *IEEE Trans. Transp. Electrif.* **2023**. [CrossRef]
41. Cao, W.; Xu, X.; Wei, Z.; Wang, W.; Li, J.; He, H. Synergized Heating and Optimal Charging of Lithium-ion Batteries at Low Temperature. *IEEE Trans. Transp. Electrif.* **2022**. [CrossRef]
42. Suh, I.-S.; Lee, M.; Kim, J.; Oh, S.T.; Won, J.-P. Design and experimental analysis of an efficient HVAC (heating, ventilation, air-conditioning) system on an electric bus with dynamic on-road wireless charging. *Energy* **2015**, *81*, 262–273. [CrossRef]

43. Soares, R.; Djekanovic, N.; Wallmark, O.; Loh, P.C. Integration of Magnified Alternating Current in Battery Fast Chargers Based on DC–DC Converters Using Transformerless Resonant Filter Design. *IEEE Trans. Transp. Electrif.* **2019**, *5*, 925–933. [CrossRef]
44. Zhu, C.; Han, J.; Zhang, H.; Lu, F.; Liu, K.; Zhang, X. Modeling and Control of An Integrated Self-Heater for Automotive Batteries Based on Traction Motor Drive Reconfiguration. *IEEE J. Emerg. Sel. Top. Power Electron.* **2023**, *11*, 384–395. [CrossRef]
45. Ghassemi, A.; Hollenkamp, A.F.; Chakraborty Banerjee, P.; Bahrani, B. Impact of high-amplitude alternating current on LiFePO4 battery life performance: Investigation of AC-preheating and microcycling effects. *Appl. Energy* **2022**, *314*, 12. [CrossRef]

batteries

MDPI

Article

Performance Analysis of the Liquid Cooling System for Lithium-Ion Batteries According to Cooling Plate Parameters

Nayoung You [1], Jeonggyun Ham [2], Donghyeon Shin [3] and Honghyun Cho [2,*]

1 Department of Mechanical Engineering, Graduate School of Chosun University, 309 Pilmundaero, Gwangju 61452, Republic of Korea; you_ny0326@chosun.kr
2 Department of Mechanical Engineering, Chosun University, 309 Pilmundaero, Gwangju 61452, Republic of Korea; orchiders@chosun.kr
3 Korea Automotive Technology Institute, 303 Pungse-ro, Cheonan 31214, Republic of Korea; dhshin@katech.re.kr
* Correspondence: hhcho@chosun.ac.kr; Tel.: +82-62-230-7050; Fax: +82-62-230-7055

Abstract: In this study, the effects of battery thermal management (BTM), pumping power, and heat transfer rate were compared and analyzed under different operating conditions and cooling configurations for the liquid cooling plate of a lithium-ion battery. The results elucidated that when the flow rate in the cooling plate increased from 2 to 6 L/min, the average temperature of the battery module decreased from 53.8 to 50.7 °C, but the pumping power increased from 0.036 to 0.808 W. In addition, an increase in the width of the cooling channel and number of channels resulted in a decrease in the average temperature of the battery module and a reduction in the pumping power. The most influential variable for the temperature control of the battery was an increase in the flow rate. In addition, according to the results of the orthogonal analysis, an increase in the number of cooling plate channels resulted in the best cooling performance and reduced pumping power. Based on this, a cooling plate with six channels was applied to both the top and bottom parts, and the top and bottom cooling showed sufficient cooling performance in maintaining the average temperature of the battery module below 45 °C.

Keywords: lithium-ion battery; cylindrical battery; liquid cooling; battery thermal management; pumping power; cooling performance

Citation: You, N.; Ham, J.; Shin, D.; Cho, H. Performance Analysis of the Liquid Cooling System for Lithium-Ion Batteries According to Cooling Plate Parameters. *Batteries* **2023**, *9*, 538. https://doi.org/10.3390/batteries9110538

Academic Editors: Jinsheng Xiao, Hengyun Zhang and Sousso Kelouwani

Received: 18 September 2023
Revised: 25 October 2023
Accepted: 25 October 2023
Published: 30 October 2023

1. Introduction

Massive consumption of fossil fuels depletes non-environmentally friendly energy sources and exposes the earth to severe problems, including climate change and global warming [1]. Therefore, as concerns about environmental issues and resource depletion have grown worldwide, considerable focus is placed on eco-friendly energy sources for replacing fossil fuels. In light of the current situation, interest in electric vehicles that do not release pollutants like carbon dioxide and nitrogen oxide has increased in many nations, even industrialized nations. Due to its high energy density, extended lifespan, environmental friendliness, and cost-effectiveness, lithium-ion batteries (LIBs) have become popular in electric cars (EVs) and hybrid electric vehicles (HEVs) [2,3]. An important issue to be solved for the stable operation of LIBs is the excessive heat generation owing to rapid charging and discharging. An increase in the generated heat of the battery raises the battery temperature, which boosts the production of more battery heat. Generally, the safe operating temperature of a battery pack is a maximum temperature of 40 °C and a temperature difference between batteries of 5 °C or less. For safe and normal battery operation, maintaining these temperatures below the safe operating temperatures is important [4]. Excessive heat generation in battery packs causes stability problems and decreases the overall performance and lifespan of the battery system. In particular, problems such as the formation of SEI (Solid Electrolyte Interphase) inside the battery occur, gradually reducing battery capacity [5]. Thus, a more efficient battery thermal management system (BTMS) is needed [6–8]. Especially as the capacity and current of batteries have rapidly increased, battery safety has become a major concern.

There are three different categories of cooling systems utilized in battery thermal management systems: air cooling, liquid cooling, and phase change (phase change material (PCM) and heat pipe) cooling. First, the air cooling method has a disadvantage because air has a lower heat capacity and thermal conductivity than liquids. However, it was widely used in early BTMS owing to the simplicity of its structure, its light weight, and its low cost in the cycle configuration [9,10]. It was confirmed by Fan et al. [11] and Yu et al. [12] that an air cooling system employing the 18650 battery maintains the temperature of the battery below the safe operating temperature at 0.5 and 1 C-rate and is an appropriate BTMS at a low C-rate. Zhou et al. [13] conducted a study using an air distribution pipe model in an air cooled system. At a rate of 3 C-rate or higher, the air cooling system could maintain the battery below the safe operating temperature. However, as the C-rate increased, the pumping power increased rapidly to control the temperature of the battery. As a result, the air cooling system is not practical for high C-rates because the rapid rise in pumping power impacts the power capacity of the electric vehicle negatively.

Due to its large heat capacity and thermal conductivity, the liquid cooling system is more effective than the air cooling technique. It also has benefits like high-temperature uniformity, high efficiency, and economic viability [14]. The liquid cooling system is divided into a direct cooling method, in which batteries are cooled by directly contacted thermal fluid and an indirect cooling method, in which batteries are indirectly cooled by a cooling plate cooled by thermal fluid. The liquid cooling system is classified into two types: a direct cooling method, where batteries are cooled by thermal fluid that comes into direct contact with them, and an indirect cooling method, where batteries are cooled indirectly by thermal fluid through cooling plates. By reducing contact heat resistance, the direct cooling technique maximizes battery cooling performance [15]. Liu et al. [16] evaluated the thermal behavior of dynamically cycling batteries exposed to static and flowing mineral oil (MO) in a series of tests using oil-immersed battery cooling systems. Even at a 4 C-rate discharge, the battery temperature can be kept below 35 °C at a flow rate of 5 mL/min below 30 °C when the flow rate exceeds 15 mL/min. Kim et al. [17] examined the cooling performance in relation to the heat transfer fluid (HTF), the number of battery cooling block ports, and the header width, and optimized the model by using the liquid immersion cooling method on a 1S16P battery module. The cooling model performed best when there were three ports and a 5 mm header width, which was consistent with low-pressure drop and good cooling performance. When the flow rate of R134a was 3 and 5 L/min, the maximum temperature of the battery and the temperature difference between the batteries were, respectively, 30.5 °C and 4.91 °C and 28.7 °C and 3.28 °C at 25 °C outside settings. The results reached were appropriate for preserving a consistent battery temperature range. However, immersion cooling is not widely used commercially owing to its fluid cost, weight, and performance uncertainties [18].

Despite having less cooling effectiveness than the direct cooling method, indirect cooling has been the subject of several studies since it is an effective method when considering factors like electrical stability, fluid cost, weight, etc. The importance of a range of cold plate channel sizes and designs in battery cooling was numerically supported by Pulugundla et al. [19]. Lai et al. [20] and Zhou et al. [21] applied structures such as a half helical duct and thermally conductive structure (TCS) to the outside of the cell in the battery liquid cooling system. These structures improve the cooling performance by increasing the contact area between the cells and the cooling device. It was confirmed that the half helical duct and TCS kept the battery below the safe operating temperature at 5 C-rate with high heat generation. Zhao et al. [22] and Qian et al. [23] studied improving cooling performance using a liquid cooling system. Their results revealed that increasing the flow rate reduced the temperature of the battery module effectively and improved the temperature uniformity.

The PCM cooling method can maintain the temperature in the battery and release it as needed while using little to no fan power. The benefits of adopting this strategy approach are more obvious when the ambient temperature is low, and the energy utilization efficiency can be significantly increased [4]. However, when the PCM completely changes its phase, achieving accurate control of the battery temperature is difficult. Accordingly, cooling using PCMs requires additional research and is studied along with other cooling methods. Lv et al. [24] employed liquid cooling and PCM to manage the heat generation of a 18650 battery. They reported that at a 2 C-rate, the composite system reduced the maximum temperature of the battery by 15 °C more than the single system. Wu et al. [25] conducted an experimental study that combined a PCM, a heat pipe, and liquid cooling for a 18650 battery. As the discharge cycle of the battery increased at a 3 C-rate, the composite system reduced the maximum temperature by 28 °C or more than the single system. Accordingly, it was confirmed that the combined PCM and liquid cooling system is an appropriate BTMS to maintain the battery below a safe operating temperature. The liquid cooling system is a basic component of battery thermal management in the combined system.

The liquid cooling system can manage the produced heat of the battery at a high C-rate, and it is a basic component for maintaining high efficiency even in a phase-change cooling system. Currently, cylindrical, prismatic, and pouch-type batteries are used in EVs. In particular, the liquid cooling BTMS for a cylindrical battery has a curved surface compared with prismatic pouch-type batteries; thus, it is more difficult to mount a cooling plate on the surface. However, cylindrical batteries have the advantages of better consistency, more material processing, and longer lifetimes, and are more adaptable to situations where multiple batteries must be combined in series and parallel [26]. Much information is available on the cooling capacity according to the structure of the cooling plate in a rectangular battery module. However, due to the structural characteristics of cylindrical batteries, many studies have applied cooling plates to the sides, but few studies have applied cooling plates to the top and bottom. Using a cold plate to the side is complicated because the cold plate must be designed to contact between battery cells. Therefore, in this study, a numerical analysis of cooling performance and pump output was performed to apply cooling plates to the top and bottom of a cylindrical battery. In the 24S16P module structure, the heating characteristics of the 21700 cylindrical battery module and the heating environment of the battery were numerically analyzed. In addition, to effectively control the heating of the cylindrical battery module, the average temperature, cooling capacity, and pumping power of the battery module according to changes in the structure of the cooling plate and operating conditions were compared and analyzed. Additionally, analysis findings for the cooling system, which maintained a temperature that could guarantee stable battery performance, were also provided. The obtained results will help improve the cooling performance of a cylindrical battery using liquid cooling and optimize the parameters of the cooling system. In addition, it can enhance the safety and performance of EVs through battery temperature management.

2. Methods and Simulation

2.1. Simulation Model

The cooling performance of the liquid cooling system using a 21700 cylindrical battery cell was investigated using ANSYS Fluent 2022 R2. The battery model in this study used the battery parameters from the anode and the cathode material of $Li(Ni_{0.8}Co_{0.1}Mn_{0.1})O_2$, and the battery specifications used in the simulation are presented in Table 1. A schematic and 3D model of the battery module and BTMS used in the simulation are shown in Figure 1 [27]. As shown in Figure 1a, to uniformly dissipate heat from the battery, a thermal pad was positioned between the battery module and cooling plate. A cooling plate was placed at the bottom to control heat generation from the battery. In Figure 1b, the battery module is depicted. The 384 cylindrical battery cells were composed of 24S16P. Figure 1c depicts the internal shape of the cooling plate, which was composed of five U-type channels with dimensions of 440 × 580 × 6 mm^3. The cooling water flow path of the cooling plate had a U-shape, and the cooling water entered the inlet branches and flowed in the cooling plate. The gap between the channels was maintained at 10 mm. In this study, the widths of the channels are expressed as L_1 and L_2, and the height is expressed as h_1. Water is the HTF that circulates inside the cooling plate to cool the battery. The correlations of the thermal properties of the water are shown in Table 2.

Table 1. Specifications of a 21700 battery.

Parameter	Specification
Diameter (mm)	21
Height (mm)	70
Weight (g)	67.5
Nominal capacity (Ah)	4.95
Nominal voltage (V)	3.69
Charge cut-off voltage (V)	4.25

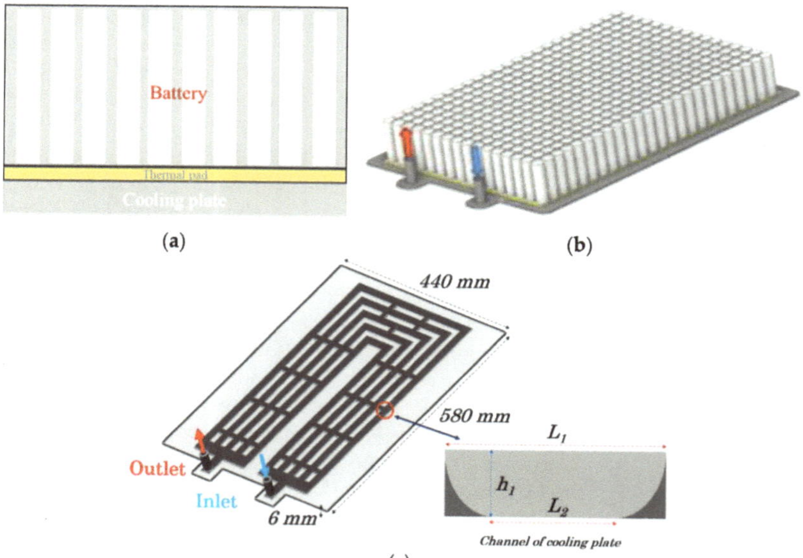

Figure 1. Battery module and BTMS with a cooling plate for liquid cooling: (**a**) schematic diagram of the battery module; (**b**) 3D model of the battery module; (**c**) structure of a cooling plate.

Table 2. Thermal properties of HTF [28].

Parameter	Specification
Density (kg/m^3)	$765.33 + 1.8142T - 0.0035T^2$
Specific heat (J/kg·K)	$28070 - 281.7T + 1.25T^2 - 2.48 \times 10^{-3}T^3 + 1.857 \times 10^{-6}T^4$
Thermal conductivity (W/m·K)	$-0.5752 + 6.397 \times 10^{-3}T - 8.151 \times 10^{-6}T^2$
Viscosity (Pa·s)	$9.67 \times 10^{-2} - 8.207 \times 10^{-4}T + 2.344 \times 10^{-6}T^2 - 2.244 \times 10^{-9}T^3$

2.2. Governing Equations

This study was carried out under the following presumptions: first, the fluid flow was a single-phase flow; second, the flow field was completely filled with HTF; third, thermal resistance and radiation emission were neglected; fourth, the HTF was an incompressible fluid. The mass, energy, and momentum conservation equations were used to analyze a cylindrical battery cooling system. The turbulence model of the realizable k–ε model was used because it can be applied to a wide flow range and is highly accurate in stream and heat transfer analysis [29]. It is the same when cold plates are used at the top and bottom of the battery module.

Bernari et al. [28] calculated the heating value of the battery. The proposed internal resistance was used to estimate the heat generation quantity using the heat generation correlation Equation (1).

$$Q_{gen} = Q_{ir} + Q_{re} = I^2R - IT_{cell}\frac{\partial U_{OCV}}{\partial T}, \quad (1)$$

where Q_{gen} is the total heating value generated by the battery cell, and Q_{ir} and Q_{re} are the irreversible and reversible heats generated during charging and discharging, respectively. I and R are the charging/discharging current and resistance, T is the temperature of the battery cell, and $\partial U_{OCV}/\partial T$ is the temperature coefficient of open-circuit voltage. The equation used by Lai et al. [13] for the temperature coefficient of the open-circuit voltage by Lai et al. [20] is given in Equation (2).

$$\frac{\partial U_{OCV}}{\partial T} = -0.355 + 2.154 \times 2.154 \times SOC - 2.869 \times SOC^2 + 1.028 \times SOC^3, \quad (2)$$

where SOC represents the state of charge of the battery.

Battery heat generation was calculated by averaging the heat generated during the discharge process. The pumping power used for liquid cooling was calculated using Equation (3).

$$P_{pump} = \Delta P Q_w, \quad (3)$$

where P_{pump} is the pumping power, ΔP is the pressure drop between the inlet and outlet of the cooling plate, and Q_w is the flow rate of HTF.

2.3. Initial and Boundary Conditions

Table 3 shows the test conditions for the simulation. The ambient and initial temperature of the battery module and HTF was set to 25 °C. The number of channels was determined considering the size of the cooling plate, and the channel width was determined by considering the ratio of the top and bottom widths based on standard conditions. The standard condition is marked with a (*) based on the cooling plate of Kim [27]. To investigate the influence on the flow rate (Q_1), channel width (L_1/L_2), and number of channels (n_1), the average temperature of the battery module and the average temperature difference between the upper and lower parts were calculated and compared. To examine the cooling performance of the BTMS, the pressure drops at the inlet and outlet of the cooling plate as well as the pumping power were calculated.

Table 3. Simulation conditions.

	Parameters	Range
Boundary condition and flow rate	Ambient temperature (°C)	25
	Initial temperature (°C)	25
	Battery volumetric heat generation (W/m^3)	225,000
	Heat transfer coefficient (W/m^2·K)	10
	Flow rate (L/min)	2, 3 *, 4, 5, 6
Channel shape	Top (L_1)/Bottom width (L_2) (mm)	15.3/8.1, 17.3/10.1, 19.3/12.1 *, 21.3/14.1, 23.3/16.1
	Height (h_1) (mm)	3.6
	Number of the channel (n_1)	4, 5 *, 6

* Standard condition.

2.4. Validation of Simulation Model

In this study, the open-circuit voltage of the temperature coefficient and internal resistance of a 21700 cylindrical battery was used to predict the heat generated from the cylindrical battery, and a heating value prediction model for the cylindrical battery was developed. The results were also contrasted with those of Sheng et al. [30], who tested the heating value of 21700 cylindrical battery cells at rates of 1–2.5 C. Based on this, the heating value of the cylindrical battery cell was predicted at a rate of 3 C-rate conditions, and the cooling performance was predicted according to the change in the cooling plate structure and operating conditions during the cylindrical battery module cooling.

Because computational fluid dynamics analyze physical phenomena using a divided grid, the number of grids dramatically influences the overall analysis results. Therefore, ensuring the reliability of the analysis results through a grid-independent test based on the number of grids is necessary. Figure 2 shows the average temperature of the battery module for each mesh node under the heating value condition of the 3 C-rate. As the number of mesh nodes increased, the temperature of the battery module decreased, and it was 52.17 °C when the number of mesh nodes reached 2,809,041. As mesh nodes increased to 4,675,798, it was 52.17 °C, and the temperature did not change significantly as the number of mesh nodes increased. Therefore, in this study, 2,809,041 mesh nodes were used to analyze the battery modules.

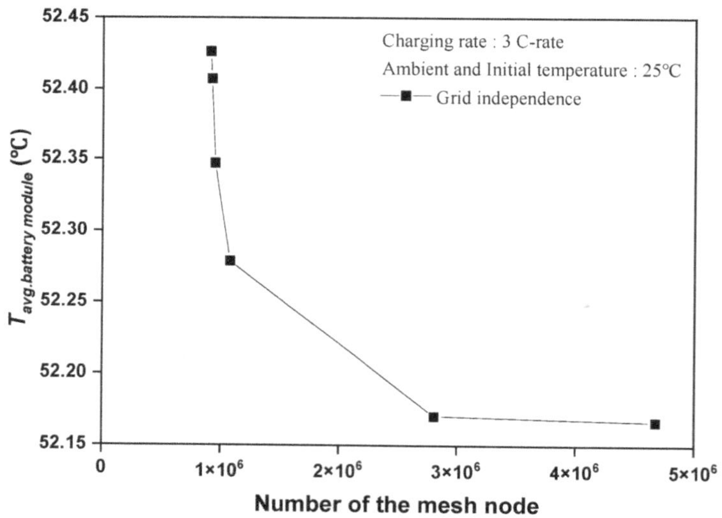

Figure 2. Grid independence test of the mesh node.

Figure 3 shows a comparison of the results of the heating value prediction model of this study with the experimental results of the 1–2.5 C rate of Sheng et al. [30]. The change over time in the average temperature of the battery predicted in the simulation exhibited a trend similar to that of the experimental values at all C-rates. The average battery temperature, as reported by Sheng et al. [30] under the 1 C-rate discharging process, was 46.7 °C, and the predicted temperature of this study was 34.8 °C. The maximum and average error rates were 27% and 15%, respectively. The battery's average temperature during the 2 C-rate discharging condition was 67.5 °C and 57.1 °C, respectively; the maximum error rate was 15%, and the average error rate was 5.6%. The average temperature of the battery at 2.5 C-rate was 80.8 and 75.1 °C, respectively. Compared to high C-rate discharge conditions, low-rate discharge conditions require longer discharge times and are more affected by the charging environment. In the case of low-rate discharge conditions, because the discharge time is extended, the total heat loss of the cell during discharge is more significant than that under high-rate discharge conditions. Therefore, under low-rate discharge conditions, the uncertainty due to heat loss is considerable, which causes greater errors in experimental and analysis results. Additionally, the initial temperature is 30 °C in the experiment of Sheng et al. [29], but it is 25 °C in this study; therefore, it is 5 °C higher than the initial temperature of this study. It affected the battery heat generation and loss, leading to a rather significant inaccuracy. The simulation error, however, is within the reliable error range and is a maximum of 9.8%.

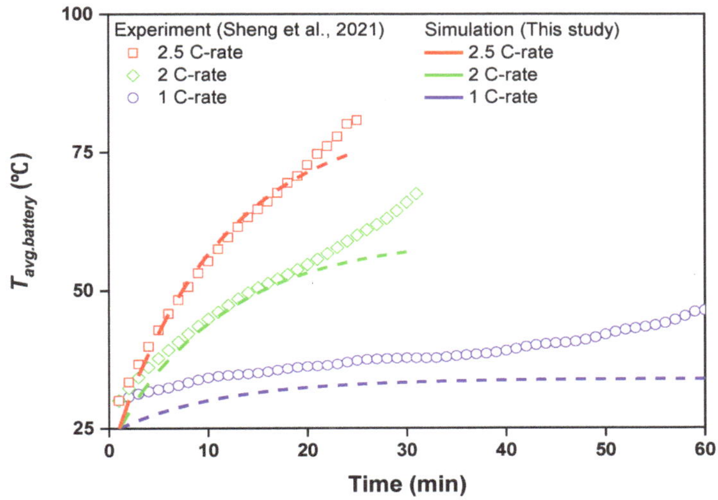

Figure 3. Comparison of the results between this study and those reported by Sheng et al. [30].

3. Results and Discussion

3.1. Effect of the Inlet Flow Rate of Cooling Fluid

Figure 4 depicts the impact on the cylindrical battery's cooling performance at a flow rate of 2 to 6 L per minute under a three-cycle discharging process. After 3 C-rate discharges, the battery module's average temperature was 85.4 °C; when the flow rate increased, this temperature reduced to 79.4 °C on average. The average battery temperature decreased by 3.12 °C, from 53.8 °C to 50.7 °C, as the flow rate rose from 2 to 6 L/min. The average temperature difference between the top and bottom of the battery increased with the flow rate; however, this difference was small. Figure 4b depicts the pressure drop and pumping power of the cylindrical battery cooling according to the flow rate. The flow rate increases with an increase in the pressure drop. The pressure drop increased from 1.09 to 8.08 kPa when the flow rate was raised from 2 to 6 L/min. As the flow rate rose, the pressure drop rose. Accordingly, the pumping power increased from 0.04 to 0.81 W. The pumping power increased 22 times when the flow rate rose from 2 to 6 L/min, although the battery temperature drop was just 3.1 °C. Raising the pump power for cooling in an electric vehicle can reduce the driving range due to increased electric power consumption. Despite a higher flow rate, battery cooling capability has a limit. Because of the cylindrical shape of the 21700 battery, installing the cooling plate at the bottom does not provide a sufficient heat transfer area for cooling. Thus, a wide heat transfer area needs to be designed to manage local excessive heat when a battery is used in a fast-charging and discharging environment.

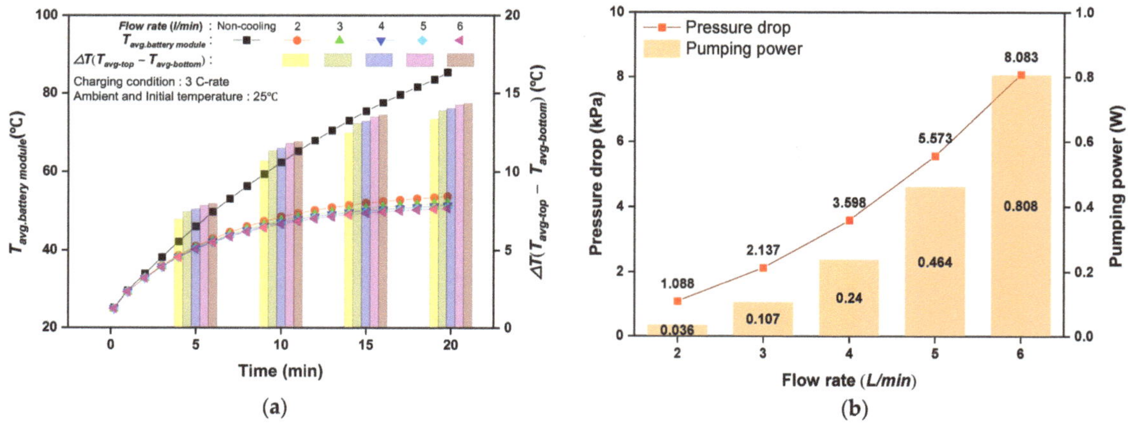

Figure 4. Cooling performance according to the flow rate of the HTF: (**a**) average temperature variation of the battery module; (**b**) pressure drop and pumping power.

3.2. Effect of Channel Width and Number

Figure 5 shows the change in the cooling performance according to the change in the channel width of the cylindrical battery cooling plate. The average temperature of the battery module varies with channel width, as shown in Figure 5a. The width of the channel was varied, while the gap between the channels was maintained at 10 mm. When cooling was not performed during the 3 C-rate discharge process, the average temperature of the battery module was 85.4 °C. As the channel width increased, the average temperature of the battery module decreased. The average temperature of the battery module was 52.7 °C when the cooling plate's width was 15.3 mm and fell to 51.9 °C when it was 23.3 mm—a 0.81 °C difference. Compared with the case without cooling, the highest battery temperature reduction rate was 39.2% at a channel width of 23.3 mm. In addition, the average temperature difference between the upper and bottom regions of the battery increased by 0.27 °C, from 13.7 °C to 14.0 °C, while the width of the cooling plate channel increased from 15.3 to 23.3 mm. This was due to an increase in the contact area between the lower portion of the battery module and the area through which the cooling fluid flowed when the cooling plate channel's width widened. As such, the temperature of the battery module decreased, but the upper part of the battery module was not significantly affected. The pressure drop and pumping power in regard to channel width are shown in Figure 5b. The pressure drop on the cooling plate decreased from 2.53 to 1.89 kPa as the width rose from 15.3 to 23.3 mm. As a result, the pumping power dropped from 0.13 to 0.09 W by 26%. This is due to the fact that the pressure drop was minimized by widening the cooling path, while maintaining the same flow rate. As the channel width of the cooling plate increased, the average temperature of the battery module decreased; however, the temperature difference between the upper and lower parts of the module increased.

Figure 6 depicts the cylindrical battery's cooling performance according to the quantity of cooling plate channels. The average temperature of the battery module is shown in Figure 6a as a function of the quantity of cooling plate channels. As the number of cooling plate channels increased from four to six under the 3 C-rate discharging condition, the average temperature of the battery module decreased by 1.55 °C from 53.3 °C to 51.7 °C. Consequently, the average rate of temperature decrease reached a maximum of 39.4%. In addition, as the number of cooling plate channels increased from four to six, the average temperature difference between the upper and lower parts of the battery module increased from 13.5 °C to 14.0 °C. Thus, it was established that improving the cooling performance of the cylindrical battery could be accomplished by increasing the number of cooling plate channels. This was due to an increased contact area between the battery module and the cooling route caused by the increased number of cooling plate channels. The differences in pressure drop and pumping power based on the quantity of cooling plate channels are depicted in Figure 6b. As the number of cooling plate channels increased from four to six, the pressure drop decreased from 2.34 to 1.86 kPa, and the pumping power decreased by 20.5% from 0.12 to 0.09 W. The fluid flow rate through each channel on the cooling plate reduced as the number of channels rose, minimizing the pressure drop in the cooling plate. Additionally, as the number of cooling plate channels increased, the pumping power was reduced and the cooling plate's performance was enhanced. It was confirmed that the design of the optimal channel number in the cooling plate is an essential factor in battery thermal management and the pumping power of electric vehicles.

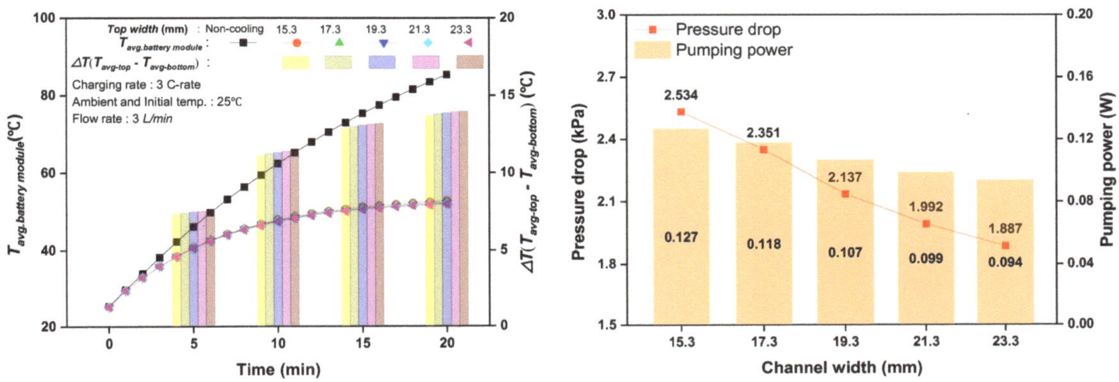

(a) (b)

Figure 5. Cooling performance according to channel width: (**a**) average temperature variation of the battery module; (**b**) pressure drop and pumping power.

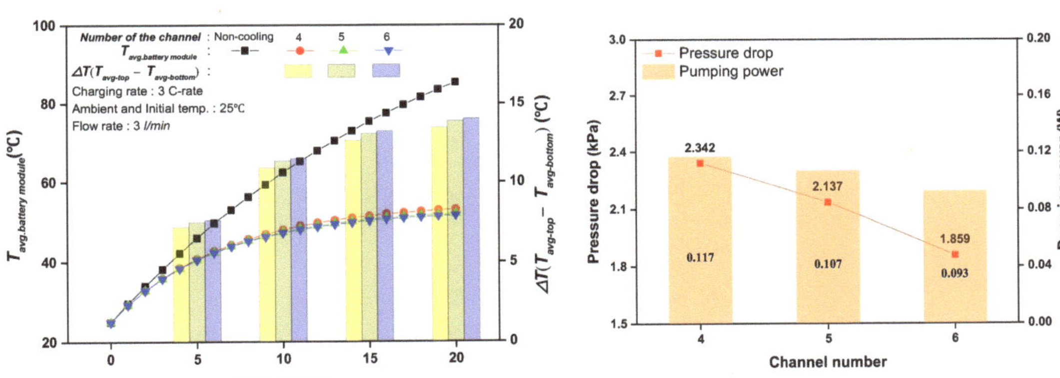

(a) (b)

Figure 6. Result of cooling performance according to channel number: (**a**) average temperature variation of the battery module; (**b**) pressure drop and pumping power.

3.3. Result of the Orthogonal Analysis

The orthogonal analysis method reflects the degree of influence of a variable on the evaluation index within the influence range of the variable. The larger the range of influence (R_i), the greater the degree of influence of the variable [31]. The influence of flow rate, channel width, and channel number on battery thermal management performance and pumping power were assessed in this study utilizing the results of the orthogonal analysis approach. R_i is the influence range, and Equation (4) can be used to assess it.

$$R_i = \max(k_1, k_2, k_3) - \min(k_1, k_2, k_3), \quad (4)$$

where k_1 is the result obtained under conditions smaller than the standard condition, k_2 is the result obtained under the standard conditions, and k_3 is the result obtained under conditions larger than the standard condition. The results of the orthogonal analysis are shown in Table 4.

Table 4. Results of the orthogonal analysis.

		T_{avg} [°C]	ΔT_{avg} [°C]	W_p [W]
Q_1	k_1	53.77	13.37	0.04
	k_2	52.17	13.88	0.11
	k_3	51.43	14.02	0.24
	R_i	2.34	0.65	0.20
L_1	k_1	52.36	13.82	0.12
	k_2	52.17	13.88	0.11
	k_3	52.00	13.94	0.10
	R_i	0.36	0.12	0.02
n_1	k_1	53.27	13.48	0.12
	k_2	52.17	13.88	0.11
	k_3	51.72	14.04	0.09
	R_i	1.55	0.56	0.03

In the average temperature (T_{avg}) and average temperature difference (ΔT_{avg}) of the battery module, the largest R was obtained in the flow rate change and the lowest R in the channel width change. Therefore, the degree of influence on the average temperature and average temperature difference was in the order of $Q_1 > n_1 > L_1$. In terms of the pumping power, the largest R was in the flow rate change, and the smallest R was in the channel width change. Therefore, the degree of influence of the pumping power was similar to that of the average temperature. When considering the average battery temperature, the larger the influence range, the more advantageous it is for temperature control performance. On the other hand, the smaller the pumping power influence range, the smaller the electric vehicle's power consumption. In terms of battery thermal management, the influence range of the number of channels has the second largest value. Additionally, its pumping power value is the lowest. Therefore, the quantity of channels in the cooling plate is the variable that affect BTM most significantly.

3.4. Result of Cooling Performance by Adapting an Improved Cooling System

Based on the optimized model from the orthogonal analysis results, considering the cooling performance under the 3 C-rate discharging condition, it was confirmed that the optimal cooling performance could be obtained in the model with Q = 3 L/min, L = 19.3 mm, and n = 6. However, when the cooling plate was applied only to the lower part, the temperature of the battery module was higher than the safe operating temperature of 40 °C. In particular, the temperature was slightly higher in the upper part where the cooling plate was not in contact. To reduce the heat generation of the battery module effectively, the cooling performance was investigated by applying a cooling plate to the top and bottom of the battery module. Figure 7 depicts the variances in the average temperatures of three C-rate discharge situations. The temperature contours of the battery modules with bottom cooling are shown in Figure 7a,b, respectively. When only bottom cooling was applied, the highest temperature of the battery module was 62.3 °C at the outlet side. In addition, the average temperature of the battery module was 51.7 °C. When top and bottom cooling were applied, the highest temperature of the battery module was 41.5 °C at the outlet side of the battery module, and the average temperature was 35.8 °C.

As a result, the battery module's maximum temperature dropped by 20.7 °C in comparison to when only bottom cooling was used.

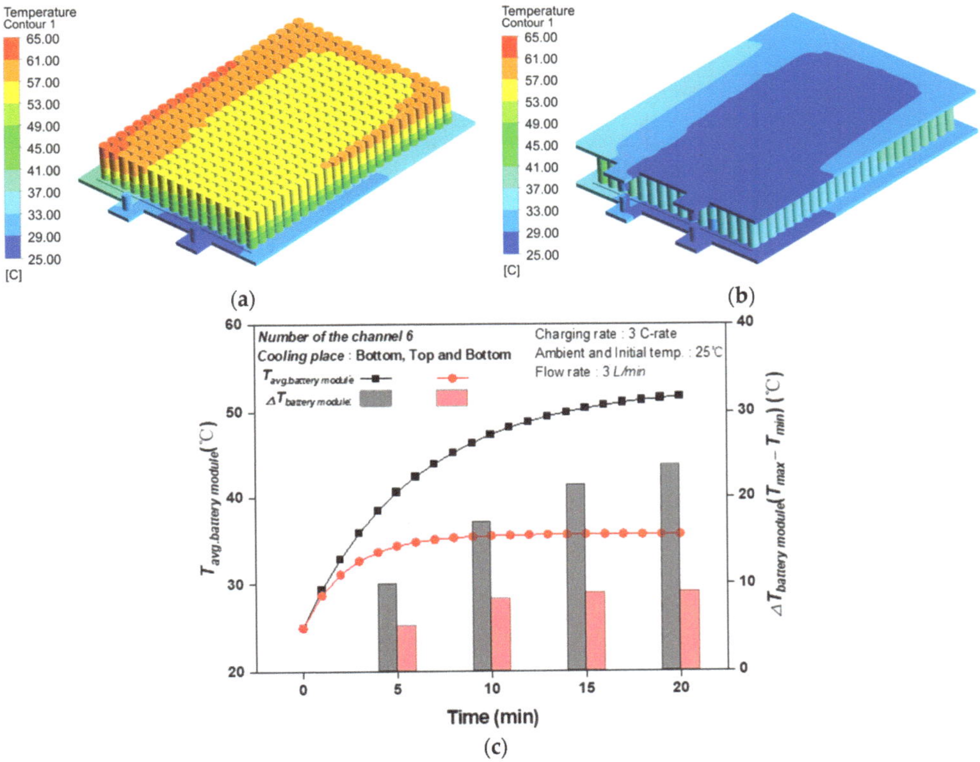

Figure 7. Comparison of bottom cooling and top and bottom cooling; (**a**) bottom cooling of the battery module; (**b**) top and bottom cooling of the battery module; (**c**) temperature variation of the bottom cooling and top and bottom cooling.

The average temperature and temperature differential of the battery module for bot-tom cooling and top and bottom cooling are compared in Figure 7c. The average temperatures of the battery module for only bottom cooling and top and bottom cooling were 51.7 °C and 35.8 °C, respectively. By applying top and bottom cooling, the average temperature of the battery module could be reduced by 16 °C compared to only bottom cooling. The maximum and minimum temperature differences of the battery module in bottom cooling and top and bottom cooling were 22.8 °C and 7.1 °C, respectively. The cooling plate's pressure drop rose by around 2.03 times, from 1.86 kPa for bottom cooling to 3.78 kPa for top and bottom cooling. As top and bottom cooling was applied, the pumping power was slightly higher, but the average temperature of the battery module could be maintained below 40 °C. Battery cell temperatures above 40 °C during charging and discharging have a negative impact on the formation of SEI inside the battery [32], but the top and bottom cooling in this study was robust enough to withstand this, and performance degradation and capacity reduction were minimized. Therefore, it was confirmed that maintaining a safe operating temperature could be accomplished when cooling plates were simultaneously applied to the top and bottom of the battery module. The temperature difference of the battery module was relatively uniform compared to bottom cooling, even though it did not reach the recommended level (5 °C). It is believed that the recommendations for temperature difference within the module can be achieved through improvements in the battery cooling plate.

4. Conclusions

This study investigated the influence of a number of variables, including the flow rate, channel width, and channel number, on the liquid cooling performance of a 21700 cylindrical battery module. The results showed that the average temperature of the battery module decreased from 53.8 °C to 50.7 °C when the flow rate in the cooling plate increased from 2 to 6 L/min, but that the pumping power increased from 0.04 to 0.81 W due to the higher pressure drop. Therefore, increasing the cooling flow rate improves the cooling performance; however, selecting an appropriate flow rate is important because the pumping power increases. In addition, as the width increased from 15.3 to 23.3 mm, the average temperature of the battery module decreased from 52.7 °C to 51.9 °C. Pumping power dropped from 0.13 to 0.09 W by 26%. Under the 3 C-rate discharging condition, the average battery module temperature was reduced by 1.55 °C and the pumping power by 20.5% as the number of cooling plate channels rose from four to six. Nevertheless, it was difficult to achieve the safe operating temperature of the battery module under 3 C-rate discharge conditions by installing a cooling plate at the bottom. To achieve a safe operating temperature, it is necessary to increase the heat transfer area of the battery.

Orthogonal analysis was conducted to investigate the influence of each variable on the cooling performance of the battery module. It was confirmed that increasing the number of channels was the most effective method for improving the cooling performance and reducing the pumping power. Based on the optimized model from the orthogonal analysis results, the cooling performance was investigated by applying top and bottom cooling to enhance cooling. As a result, when compared to the case where just bottom cooling was used, the average temperature and maximum temperature of the battery module dropped by 16.0 °C and 20.7 °C, respectively. Moreover, the average temperature could be maintained below 40 °C for safe operation. Therefore, a safe operating temperature could be effectively maintained when cooling plates were simultaneously applied to the top and bottom of the battery module, and the reliability and lifetime of the battery system could be improved by significantly reducing the temperature difference in the battery module.

Author Contributions: Conceptualization, H.C. and D.S.; methodology N.Y. and J.H.; validation, H.C. and D.S.; formal analysis, N.Y.; investigation, N.Y. and J.H.; resources, N.Y. and J.H.; writing—original draft preparation, N.Y.; writing—review and editing, J.H. and H.C.; supervision, H.C. All authors have read and agreed to the published version of the manuscript.

Funding: This work was supported by "Eco-friendly Car Sector in Development Technology" of the Korea Institute of Industrial Technology Evaluation and Management (KEIT), granted financial resource from the Ministry of Trade, Industry & Energy, Republic of Korea. (No. 20011906) and "Regional Innovation Strategy (RIS)" through the National Research Foundation of Korea (NRF) funded by the Ministry of Education (MOE) (2021RIS-002).

Data Availability Statement: Not applicable.

Conflicts of Interest: The authors declare no conflict of interest.

Nomenclature

LIB	Lithium-ion battery
EV	Electric vehicle
HEV	Hybrid electric vehicle
BTMS	Battery management system
PCM	Phase change material
TCS	Thermally conductive structure
L_1	Top width of cooling channel (mm)
L_2	Bottom width of cooling channel (mm)
h_1	Height of cooling channel (mm)
Q_{gen}	Total heating value of battery cells (W)
Q_{ir}	Charge/discharge irreversible heat (W)
Q_{re}	Charge/discharge reversible heat

I	Current (A)
R	Resistance (Ω)
T_{cell}	Battery cell temperature (°C)
U_{OCV}	Open circuit voltage (V)
W_{pump}	Pumping power (W)
ΔP	Inlet and outlet pressure difference (kPa)
Q_w	Flow rate of water (L/min)
Q_1	Flow rate (L/min)
T_{avg}	Average temperature of battery module ((°C)
ΔT_{avg}	Average temperature difference of battery module ((°C)
K_1	Smaller condition result
K_2	Standard condition result
K_3	Larger condition result
R_i	Influence range
*	Standard condition

References

1. Falih, H.; Hamed, A.J.; Khalifa, A.H.N. Techno-Economic Assessment of a Hybrid Connected PV Solar System. *Int. J. Air-Cond. Refrig.* **2022**, *30*, 1–15. [CrossRef]
2. Liu, X.; Chen, Z.; Zhang, C.; Wu, J. A Novel Temperature-Compensated Model for Power Li-Ion Batteries with Dual-Particle-Filter State of Charge Estimation. *Appl. Energy* **2014**, *123*, 263–272. [CrossRef]
3. Zhang, K.; An, Y.; Wei, C.; Qian, Y.; Zhang, Y.; Feng, J. High-Safety and Dendrite-Free Lithium Metal Batteries Enabled by Building a Stable Interface in a Nonflammable Medium-Concentration Phosphate Electrolyte. *ACS Appl. Mater. Interfaces* **2021**, *13*, 50869–50877. [CrossRef]
4. Fang, Y.; Ye, F.; Zhu, Y.; Li, K.; Shen, J.; Su, L. Experimental Investigation on System Performances and Transient Response of a Pumped Two-Phase Battery Cooling System Using R1233zd. *Energy Rep.* **2020**, *6*, 238–247. [CrossRef]
5. Pinson, M.B.; Bazant, M.Z. Theory of SEI Formation in Rechargeable Batteries: Capacity Fade, Accelerated Aging and Lifetime Prediction. *J. Electrochem. Soc.* **2013**, *160*, A243–A250. [CrossRef]
6. Ramadass, P.; Haran, B.; White, R.; Popov, B.N. Capacity Fade of Sony 18650 Cells Cycled at Elevated Temperatures: Part, I. Cycling Performance. *J. Power Sources* **2002**, *112*, 606–613. [CrossRef]
7. Wu, M.S.; Chiang, P.C.J. High-Rate Capability of Lithium-Ion Batteries after Storing at Elevated Temperature. *Electrochim. Acta* **2007**, *52*, 3719–3725. [CrossRef]
8. Ping, P.; Wang, Q.; Huang, P.; Sun, J.; Chen, C. Thermal Behaviour Analysis of Lithium-Ion Battery at Elevated Temperature Using Deconvolution Method. *Appl. Energy* **2014**, *129*, 261–273. [CrossRef]
9. Giuliano, M.R.; Prasad, A.K.; Advani, S.G. Experimental Study of an Air-Cooled Thermal Management System for High Capacity Lithium–Titanate Batteries. *J. Power Sources* **2012**, *216*, 345–352. [CrossRef]
10. Wang, H.; Ma, L. Thermal Management of a Large Prismatic Battery Pack Based on Reciprocating Flow and Active Control. *Int. J. Heat. Mass. Transf.* **2017**, *115*, 296–303. [CrossRef]
11. Fan, Y.; Bao, Y.; Ling, C.; Chu, Y.; Tan, X.; Yang, S. Experimental Study on the Thermal Management Performance of Air Cooling for High Energy Density Cylindrical Lithium-Ion Batteries. *Appl. Therm. Eng.* **2019**, *155*, 96–109. [CrossRef]
12. Yu, X.; Lu, Z.; Zhang, L.; Wei, L.; Cui, X.; Jin, L. Experimental Study on Transient Thermal Characteristics of Stagger-Arranged Lithium-Ion Battery Pack with Air Cooling Strategy. *Int. J. Heat. Mass. Transf.* **2019**, *143*, 118576. [CrossRef]
13. Zhou, H.; Zhou, F.; Xu, L.; Kong, J. QingxinYang Thermal Performance of Cylindrical Lithium-Ion Battery Thermal Management System Based on Air Distribution Pipe. *Int. J. Heat. Mass. Transf.* **2019**, *131*, 984–998. [CrossRef]
14. Liu, H.; Wei, Z.; He, W.; Zhao, J. Thermal Issues about Li-Ion Batteries and Recent Progress in Battery Thermal Management Systems: A Review. *Energy Convers. Manag.* **2017**, *150*, 304–330. [CrossRef]
15. Dubey, P.; Pulugundla, G.; Srouji, A.K. Direct Comparison of Immersion and Cold-Plate Based Cooling for Automotive Li-Ion Battery Modules. *Energies* **2021**, *14*, 1259. [CrossRef]
16. Liu, J.; Fan, Y.; Wang, J.; Tao, C.; Chen, M. A Model-Scale Experimental and Theoretical Study on a Mineral Oil-Immersed Battery Cooling System. *Renew. Energy* **2022**, *201*, 712–723. [CrossRef]
17. Kim, M.; Ham, J.; Shin, D.; Cho, H. Comparison of Cooling Performance in a Cylindrical Battery with Single-Phase Direct Contact Cooling under Various Operating Conditions. *Batteries* **2022**, *8*, 195. [CrossRef]
18. Roe, C.; Feng, X.; White, G.; Li, R.; Wang, H.; Rui, X.; Li, C.; Zhang, F.; Null, V.; Parkes, M.; et al. Immersion Cooling for Lithium-Ion Batteries—A Review. *J. Power Sources* **2022**, *525*, 231094. [CrossRef]
19. Pulugundla, G.; Dubey, P.; Srouji, A. Time-Accurate CFD Analysis of Liquid Cold Plates for Efficient Thermal Performance of Electric Vehicle Li-Ion Battery Modules. *SAE Tech. Pap.* **2019**, *2019*, 01-0500. [CrossRef]
20. Lai, Y.; Wu, W.; Chen, K.; Wang, S.; Xin, C. A Compact and Lightweight Liquid-Cooled Thermal Management Solution for Cylindrical Lithium-Ion Power Battery Pack. *Int. J. Heat. Mass. Transf.* **2019**, *144*, 118581. [CrossRef]

21. Zhou, H.; Zhou, F.; Zhang, Q.; Wang, Q.; Song, Z. Thermal Management of Cylindrical Lithium-Ion Battery Based on a Liquid Cooling Method with Half-Helical Duct. *Appl. Therm. Eng.* **2019**, *162*, 114257. [CrossRef]
22. Zhao, C.; Cao, W.; Dong, T.; Jiang, F. Thermal Behavior Study of Discharging/Charging Cylindrical Lithium-Ion Battery Module Cooled by Channeled Liquid Flow. *Int. J. Heat. Mass. Transf.* **2018**, *120*, 751–762. [CrossRef]
23. Qian, Z.; Li, Y.; Rao, Z. Thermal Performance of Lithium-Ion Battery Thermal Management System by Using Mini-Channel Cooling. *Energy Convers. Manag.* **2016**, *126*, 622–631. [CrossRef]
24. Lv, Y.; Zhou, D.; Yang, X.; Liu, X.; Li, X.; Zhang, G. Experimental Investigation on a Novel Liquid-Cooling Strategy by Coupling with Graphene-Modified Silica Gel for the Thermal Management of Cylindrical Battery. *Appl. Therm. Eng.* **2019**, *159*, 113885. [CrossRef]
25. Wu, W.; Yang, X.; Zhang, G.; Chen, K.; Wang, S. Experimental Investigation on the Thermal Performance of Heat Pipe-Assisted Phase Change Material Based Battery Thermal Management System. *Energy Convers. Manag.* **2017**, *138*, 486–492. [CrossRef]
26. Nelson, P.; Dees, D.; Amine, K.; Henriksen, G. Modeling Thermal Management of Lithium-Ion PNGV Batteries. *J. Power Sources* **2002**, *110*, 349–356. [CrossRef]
27. Kim, M. Performance Analysis of a Direct Contact Battery Cooling System Using R134a. Master's Thesis, Chosun University, Gwangju, Republic of Korea, 2023.
28. Yadav, V.; Baghel, K.; Kumar, R.; Kadam, S.T. Numerical Investigation of Heat Transfer in Extended Surface Microchannels. *Int. J. Heat. Mass. Transf.* **2016**, *93*, 612–622. [CrossRef]
29. Panchal, S.; Khasow, R.; Dincer, I.; Agelin-Chaab, M.; Fraser, R.; Fowler, M. Numerical Modeling and Experimental Investigation of a Prismatic Battery Subjected to Water Cooling. *Numer. Heat. Transf. Part A Appl.* **2017**, *71*, 626–637. [CrossRef]
30. Sheng, L.; Zhang, Z.; Su, L.; Zhang, H.; Zhang, H.; Li, K.; Fang, Y.; Ye, W. A Calibration Calorimetry Method to Investigate the Thermal Characteristics of a Cylindrical Lithium-Ion Battery. *Int. J. Therm. Sci.* **2021**, *165*, 106891. [CrossRef]
31. Wang, J.; Liu, X.; Liu, F.; Liu, Y.; Wang, F.; Yang, N. Numerical Optimization of the Cooling Effect of the Bionic Spider-Web Channel Cold Plate on a Pouch Lithium-Ion Battery. *Case Stud. Therm. Eng.* **2021**, *26*, 101124. [CrossRef]
32. An, S.J.; Li, J.; Daniel, C.; Mohanty, D.; Nagpure, S.; Wood, D.L. The State of Understanding of the Lithium-Ion-Battery Graphite Solid Electrolyte Interphase (SEI) and Its Relationship to Formation Cycling. *Carbon* **2016**, *105*, 52–76. [CrossRef]

batteries

MDPI

Article

Investigation of Heat Transfer Enhancement Techniques on a Scalable Novel Hybrid Thermal Management Strategy for Lithium-Ion Battery Packs

Seham Shahid and Martin Agelin-Chaab *

Faculty of Engineering and Applied Science, Ontario Tech University, 2000 Simcoe Street North, Oshawa, ON L1H 7K4, Canada; seham.shahid@ontariotechu.ca
* Correspondence: martin.agelin-chaab@ontariotechu.ca

Abstract: This paper introduces a novel hybrid thermal management strategy, which uses secondary coolants (air and fluid) to extract heat from a phase change material (paraffin), resulting in an increase in the phase change material's heat extraction capability and the battery module's overall thermal performance. A novel cold plate design is developed and placed between the rows and columns of the cells. The cold plate contains a single fluid body to improve the thermal performance of the battery module. Experimental studies were conducted to obtain the temperature and heat flux profiles of the battery module. Moreover, a numerical model is developed and validated using the experimental data obtained. The numerical data stayed within ±2% of the experimental data. In addition, the ability of nanoparticles to increase the thermal conductivity of water is examined and it is found that the cooling from the liquid cooling component is not sensitive enough to capture the 0.32 W/m K increase in the thermal conductivity of the fluid. Furthermore, in order to enhance the air cooling, fins were added within the air duct to the cold plate. However, this is not feasible, as the pressure drop through the addition of the fins increased by ~245%, whereas the maximum temperature of the battery module reduced by only 0.6 K. Finally, when scaled up to an entire battery pack at a high discharge rate of 7 C, the numerical results showed that the overall temperature uniformity across the pack was 1.14 K, with a maximum temperature of 302.6 K, which was within the optimal operating temperature and uniformity ranges. Therefore, the developed thermal management strategy eliminates the requirement of a pump and reservoir and can be scaled up or down according to the energy and power requirements.

Keywords: phase change material; Li-ion cylindrical cells; computational fluid dynamics (CFD) analysis; battery thermal management; hybrid cooling; liquid channels

Citation: Shahid, S.; Agelin-Chaab, M. Investigation of Heat Transfer Enhancement Techniques on a Scalable Novel Hybrid Thermal Management Strategy for Lithium-Ion Battery Packs. *Batteries* **2024**, *10*, 32. https://doi.org/10.3390/batteries10010032

Academic Editors: Hengyun Zhang, Jinsheng Xiao and Sousso Kelouwani

Received: 1 December 2023
Revised: 12 January 2024
Accepted: 16 January 2024
Published: 18 January 2024

1. Introduction

A major issue encountered by the transportation industry is the reduction in greenhouse gas emissions, as they are responsible for 29% of the overall emissions [1]. To overcome this issue, transportation electrification was introduced, and lithium-ion (Li-ion) batteries are currently being used to develop electric vehicle (EV) packs. This is mainly due to the low rate of the self-discharge of the Li-ion cells, the high energy density, the extended lifecycle, and the high power density [2–4]. However, thermal issues related to Li-ion batteries need resolution, as they require a narrow operating temperature range for effective performance [5].

The strategies developed for thermal management (TMS) are based on the primary strategies that include air, fluid, and phase change material (PCM). The most cost-effective is the air-based TMS; however, air has limited cooling capabilities due to its low thermal properties [6]. Fluids generally have higher cooling capacities compared to air; therefore, they are capable of improved cooling and can effectively maintain the required thermal environment for the operation of the Li-ion cells [7]. The simplest form of liquid-based

TMS is designed using electrically insulating fluids. These fluids extract heat from the cell's surface through direct contact [8–10]. Usually, there is an increased viscosity associated with these fluids, which results in the requirement of a large pumping capacity. Moreover, in large battery packs, there is an increased risk of the leakage of the fluid that can cause unfavorable effects on the electrical battery pack [11]. To counter these issues, electrically noninsulating fluids are utilized. They have lower viscosities compared to insulating fluids; therefore, they require less pumping power. In order to incorporate them into the battery pack, cold plates are used that contain these fluids. The cold plates are connected to the cells' surface to extract heat. However, the application of the cold plates is restricted to rectangular cell geometries because the degree of manufacturing difficulty for cylindrical cells is higher [12–14]. An alternative to a cold plate is to use serpentine liquid channels developed to run through the cylindrical cell rows. These liquid channels contain electrically noninsulating fluids. In these cases, the part of the cell in contact with the channels cools at a higher rate than the part not in contact, causing temperature nonuniformity. Zhao et al. [15] developed multiple configurations of serpentine channels, and temperature uniformity occurred incrementally in the streamwise direction of the contact surface area with the liquid channels. Furthermore, to eliminate the restrictions of using cold plates for cylindrical cells, Rao et al. [16] used variable-sized blocks of aluminum placed in contact with the cylindrical cells, and the length of the block was varied to achieve uniform cooling through the battery pack. Liquid channels ran within these aluminum blocks to extract heat from the blocks. Liquid cooling is effective at extracting heat from the cells; however, the high pumping power causes a hindrance in its application within the battery packs [17].

PCMs are also effective in extracting heat through latent heat capacity. Initially, its use was suggested by Al-Hallaj and Selman [18], and later on, Javani et al. [19] showed that a high thermal uniformity is achievable in battery packs. Jiang et al. [20] applied a composite phase change material (CPCM), and maximum temperatures were limited to 44 °C. This CPCM was developed by combining a paraffin base with expanded graphite (EG) to increase the thermal conductivity. In comparison, through a 1 m/s airflow, the temperature reached 72 °C. He et al. [21] conducted an investigation with the EG/paraffin CPCM, and the results concluded with a 2.82 °C of temperature uniformity with 7% of EG. This was considered as the optimum EG proportion within the CPCM. Furthermore, a novel CPCM composition was investigated by Hussain et al. [22]. For the CPCM, the nickel foam was coated with saturated paraffin and graphene. Based on the results, when compared to pure paraffin, the thermal conductivity of the CPCM increased 23-fold. A novel CPCM consisting of dual regions of two different phase change temperatures was designed by Ye et al. [23]. The lower temperature region was activated at regular ambient temperatures. This enabled the TMS to maintain the battery module temperature between 25.9 and 34.9 °C. Similarly, at high ambient temperatures, the higher temperature region was activated. Through this, the battery module temperature was kept below 49.2 °C. In addition to dual phase change temperature regions, EG was added to the CPCM, and 1.98 W/m K of thermal conductivity was achieved. Additionally, a flexible CPCM was studied by Wu et al. [24]. The flexibility of the CPCM allowed it to encase the cylindrical geometry of the cell completely, which increased the surface area in contact with the cell, resulting in increased heat transfer. The novel CPCM had a thermal conductivity of 1.64 W/m K and exhibited better flexibility between −5 and 20 °C. Compared to a rigid CPCM, the results concluded a 4 °C reduction in the maximum temperature. Furthermore, Faraji et al. [25] developed a CPCM by adding hybrid nanoparticles, and the results concluded a safe operation of the electronic components with a significantly high temperature uniformity. Arshad et al. [26] added copper nanoparticles to the PCM and the results showed that the heat transfer rates were increased by ~2.9%, which resulted in a reduction in the melting time of the PCM by ~2.6%. Faraji et al. [27] also studied the insertion of nanoparticles in the PCM by adding SiO_2–MWCNT hybrid nanoparticles, and it was found that the melting time of the PCM increased by 92% compared to pure PCM, and the maximum temperature was reduced by 11 °C on the electronic components. The PCM application in the TMS exhibits promising results. However, its low overall heat transfer coefficient results in a hindrance to its mass adoption. When the PCM is utilized and completely melted, a secondary TMS is needed to charge the PCM.

Since there are a few drawbacks to using the cooling mediums individually in a TMS, hybrid TMSs are introduced that combine two or more cooling mediums. A simple hybrid system combines air and liquid cooling by incorporating mist into the airflow. This was investigated by Saw et al. [4], and it was concluded that 3% of vapors within 5 g/s of airflow is required to maintain the temperatures below 40 °C. Moreover, the natural evaporation of water was utilized by Wei and Agelin-Chaab [6,28]. Water channel strips within the battery pack were exposed to airflow. Through this, the water vapors were formed using natural evaporation and were incorporated into the airflow stream. Additionally, the air particles regained their heat capacity once they interacted with the water strips. This resulted in an improvement of 56% in the uniformity and 20% in temperature reduction. Wang et al. [29] studied a combination of cold plates and airflow. The cold plates were stationed below the cells, and the airflow circulated through cold plate surfaces to extract heat from the cold plates. Multiple strategies and concepts were tested. The temperature uniformity increased 2.42 times, and a further 3.45 °C maximum temperature reduction was achieved. In the authors' previous study [30], a novel design was developed by attaching liquid jackets to cylindrical cells. The liquid jackets contained fluid, and since it was isolated to each cell, the design was capable of employing direct cooling by using electrically noninsulating fluids.

In this paper, a novel hybrid thermal management strategy is developed by incorporating air, liquid, and PCM cooling, and analyzed experimentally and numerically. In this strategy, the PCM is the primary coolant, which is in direct contact with the cells, whereas the liquid and air are secondary coolants, which extract the heat simultaneously from the PCM, resulting in the increased heat extraction capability of the PCM and the improved thermal performance of the battery module. In order to eliminate the fluid pumping power requirements, a novel cold plate is designed which houses a single body of stationary fluid and can be placed between the cell rows and columns to improve the thermal performance of the battery module. Additionally, a novel modular battery pack design is developed which is capable of scalability in power and energy. The principal objective of this paper is to maintain the maximum temperature of the battery module within the optimal ranges of 25–45 °C and attain high thermal uniformity within 5 °C. Moreover, secondary objectives are to eliminate the pumping power for the fluids and reduce the airflow velocity, which reduces the power requirements of the TMS.

2. Description of the Proposed Hybrid Strategy

2.1. Battery Module Configuration

Based on the objectives, novel hybrid thermal management strategy is designed and developed. The strategy consists of a combination of air cooling, liquid cooling, and PCM cooling systems. In order to develop the strategy, the following designs are considered:

- The battery module should have a high temperature uniformity.
- The fluid pumping power should be eliminated.
- It should have modularity for scalability.

In the proposed hybrid strategy, PCM (paraffin) is used for direct cooling and airflow within the air duct and water in the cold plates is used for indirect cooling, as shown in Figure 1.

Rectangular cold plates of a 1.5 mm width and 0.5 mm wall thickness were added in between each cell's rows and columns, which were completely filled with fluid, and which protrudes in the air duct. This cold plate was sealed and closed from all sides so that there was no leakage of fluid. The cold plates were all connected to each other, and the fluid within this was considered a single body of fluid. The working principle behind this is divided into three stages. In the initial stage, the heat will be extracted using the PCM through direct contact. In the intermediary stage, the fluid within the cold plates will extract heat simultaneously from the PCM. In the final stage, the air flows over the protruded cold plate and will be used to cool the fluid within it. Due to a single fluid body, the protrusion of the cold plate will allow the fluid in between the columns to cool as well. Since the fluid in this strategy is sealed, the requirement for a fluid pump and reservoir is eliminated, as shown in Figure 2.

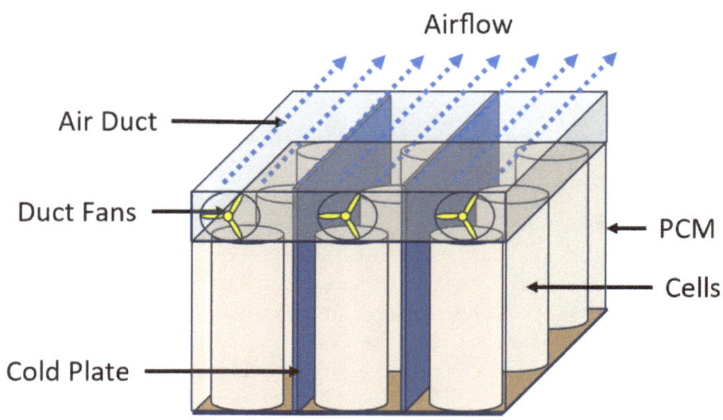

Figure 1. Three-dimensional schematic of the proposed strategy.

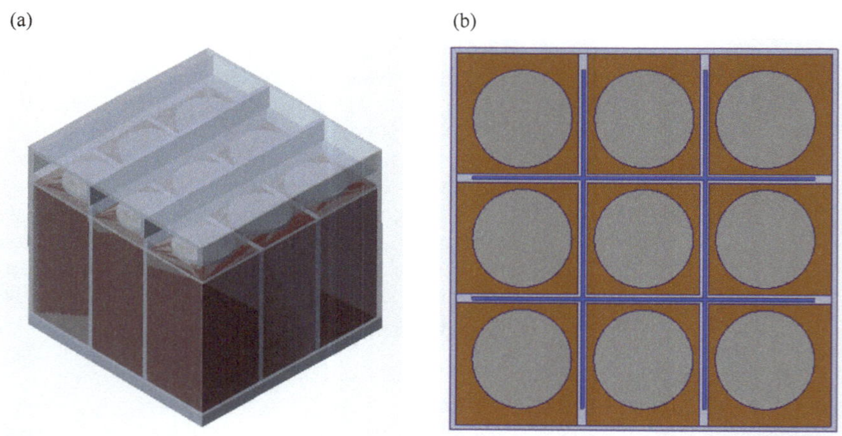

Figure 2. (**a**) Isometric view and (**b**) top cross-section view of the proposed strategy.

Moreover, for the experimental and numerical validation, the module housing was made of wood, and paraffin was used as the PCM. Once the model was validated, for all the numerical studies, an all-aluminum housing was used, and a CPCM with a higher thermal conductivity was used.

Pure paraffin usually has a low thermal conductivity. In order to improve the cooling of the Li-ion batteries, researchers have worked on increasing the various properties of paraffin, including the thermal conductivity, using different types of property-enhancing materials (PEMs), such as expanded vermiculite, graphene oxide, carbon foam, copper foam, porous graphite foam, and expanded graphite. A review study by Mishra et al. [31] reviewed a paraffin-based PCM equipped with various PEMs for the past 17 years that were developed by researchers. For this study, a paraffin base combined with copper foam was selected as the CPCM, as it has a higher thermal conductivity of 3.11 W/m K and has a feasible phase change temperature range for Li-ion batteries. The material details are provided in Table 1.

Table 1. Physical properties of the materials used [31–33].

Materials	Density (kg/m^3)	Specific Heat (J/kg K)	Thermal Conductivity (W/m K)	Melting Heat Capacity (J/kg)	Phase Change Temperature (°C)
Aluminum	2719	871	202.4	-	-
Wood	700	2310	0.173	-	-
Air	1.225	1006	0.0242	-	-
Water	998.2	4182	0.6	334,000	100
Paraffin	880	2150	0.21	245,000	42–44
Paraffin with Copper Foam	880	2150	3.11	170,400	42–43

2.2. Increasing the Thermal Conductivity of Water through Nanoparticles

The fluid within the cold plate is required to remove heat from the CPCM. Hence, thermal conductivity is one of the defining factors for heat removal. A high thermal conductivity will result in increased heat removal from the CPCM. To study this, the fluid within the cold plates was changed from pure water to water combined with nanoparticles. In order to select the nanoparticles, recent studies in open literature were reviewed [34–39], and it was found that 0.04% of silver and 0.16% multiwalled carbon nanotubes combined with a water base (H_2O–AG–MWCNT) provided the highest thermal conductivity of 0.92 W/m K at 42.24 °C [36]. The temperature of 42.24 °C was selected, as it is the phase change temperature of the CPCM (the paraffin base with copper foam).

2.3. Addition of Fins in the Battery Module to Improve Heat Transfer

This study attempts to further improve cooling by increasing the heat transfer from the fluid in the cold plate to the airflow. This was done by adding fins to the cold plate region within the air duct. Two separate configurations were developed. In the first configuration, one fin of a thickness of 0.5 mm was added parallel to the airflow in the middle of the air duct, as shown in Figure 3a. In the second configuration, two fins were added parallel to the airflow, as shown in Figure 3b. In order to keep the same Reynolds number of 1950, the velocity increased at the inlet, and the details for the two configurations are provided in Table 2. The Reynolds nummber is calculated using the following equation:

$$Re = \frac{\rho \vec{v} D}{\mu} \quad (1)$$

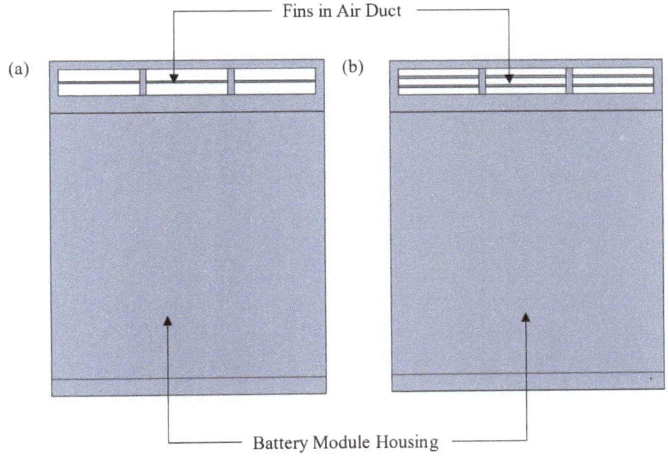

Figure 3. Schematic of the front view of (**a**) configuration 1 and (**b**) configuration 2.

Table 2. Number of fins and corresponding inlet velocity for the developed configurations.

Battery Module Configuration	Number of Fins	Inlet Velocity (m/s)
Configuration 1	1	5.93
Configuration 2	2	9.28

2.4. Scalability of the Battery Module

The battery module with no fins and water as the fluid was used to scale it to the entire battery pack. For reference, a Tesla 75 kWh battery pack with the dimensions of ~1660 mm in length, ~964 mm in width, and 174 mm in height and a total of 5880 cylindrical cells was used [40]. To accommodate the battery modules within the exact dimensions, 25 modules were connected in a series (in terms of the airflow), which amounted to 1600 mm of length, as shown in Figure 4a. Each series setup of modules (BMA) was connected 15 times in parallel (in terms of the airflow), which amounted to a width of 960 mm, as shown in Figure 4b.

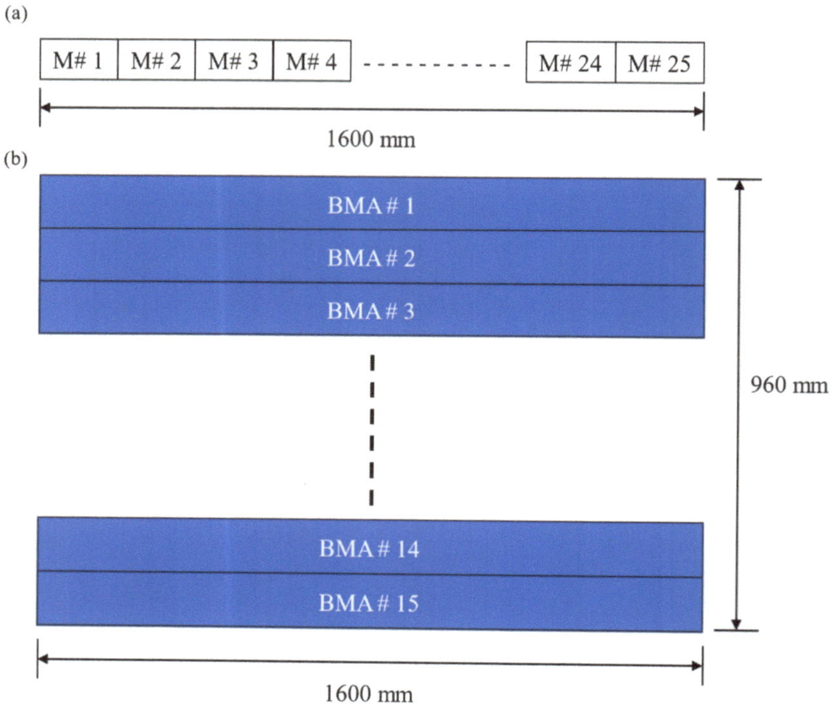

Figure 4. (**a**) Schematic of the battery module assembly (BMA) in the series; (**b**) schematic of the BMA connections in parallel.

Finally, two arrays of series and parallel modules were placed vertically on each other to amount to a height of 154 mm. Therefore, the total number of battery modules in the battery pack is 750, which contains a total of 6750 cylindrical cells. The modules in the width and height directions are in a parallel airflow; therefore, one series configuration of 25 modules, as shown in Figure 4a, was used for the evaluation of the scalability of the module. Since the spacing between the cells and cold plate is compact, it was not feasible to conduct the simulation for all 25 modules at once due to the high computational requirements. Therefore, the results were obtained for 2, 3, and 4 modules in a series, and the maximum temperature of the airflow and the battery modules was extrapolated. Moreover, a simulation was conducted for the airflow only across the 25 battery modules to obtain the pressure drop through the air duct. The schematic for the airflow domain is shown in Figure 5. The domain is based on a single airflow channel within the air duct.

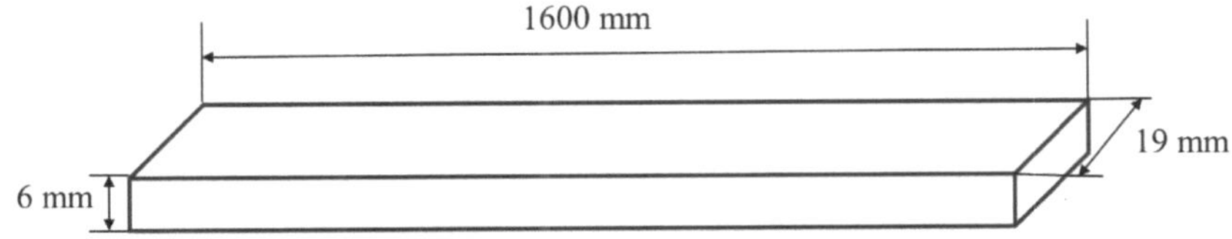

Figure 5. Airflow domain for a single series air duct channel of the battery pack.

3. Experimental Setup and Procedure

In this study, two separate experiments were conducted. In Experiment I, the heat flux profiles were obtained at the surface of a Samsung INR18650-25R cell. The heat flux profile was obtained as a boundary condition in the numerical modeling. In Experiment II, the battery module was manufactured, and the temperature measurements were obtained. The detailed experimental setup and procedure are described in the subsections below.

3.1. Experiment I—Heat Flux Measurement

Experiment I was performed to acquire the heat flux measurements. This will then be used in the numerical modeling as the transient boundary. The experimental setup is shown in Figure 6a, and the schematics are shown in Figure 6b. The Samsung INR18650-25R cell with aluminum tabs attached to the positive and negative terminals was used for this experiment. The length of the cell is 65 mm, and the diameter is 18 mm. The nominal discharge capacity is 2.5 Ah and 3.6 V [41]. The electrical insulation at the surface of the cells was detached to obtain accurate measurements, as the noninsulated cells were used in the battery module. For the charging of the cells, a Turnigy charger was used. It can charge 4 cell groups simultaneously up to 300 W and 20 A. The charger was powered through a power supply (Jesverty SPS-3010N) which was sourced from Toronto, ON, Canada. To discharge the cells, a load discharger (TDI RBL488) was used. It has high power draining load capacities of up to 800 W and 120 A. It incorporates variable-speed forced-air cooling to ensure a quiet environment. To measure the heat flux profiles, the Omega HFS-5 was used. In order to measure the readings from the HFS-5 sensor, the data acquisition device (DAQ) NI-9211 was used.

As shown in Figure 6a, alligator clips were connected to the electrode tabs of the Li-ion cell. The cell was charged in two stages. The first stage of the charging process comprised a constant current (CC) charge at 1 A. During this stage, the voltage of the cell increased gradually. Once the voltage reached 4.2 V, it was charged at constant voltage (CV) of 4.2 V, and the charging current was reduced from 1 A to 0 A. Once the cell was completely charged, it was kept at an ambient environment of 25 °C for 1 h. The discharge cycle of 2.5 A CC discharge was then initiated (which corresponds to 1 C) until the voltage reached the cut-off voltage. During the discharge process, the data were logged using the LabView.

The experiment was repeated three times for a 1 C discharge rate, and the error analysis was conducted using the technique recommended by Moffat [42]. The details are tabulated in Table 3. The average percentage uncertainty in the heat flux measurements is ±3.59%.

Table 3. Calculated values of the relative bias and precision errors and total uncertainty of the heat flux measurement.

Parameters	Reference Value (W/m^2)	Absolute Bias Error (W/m^2)	Relative Bias Error (%)	Relative Precision Error (%)	Total Uncertainty (%)
Min. Heat Flux	18.99	1.27	6.69	1.90	6.95
Max. Heat Flux	584.16	1.27	0.22	0.05	0.22

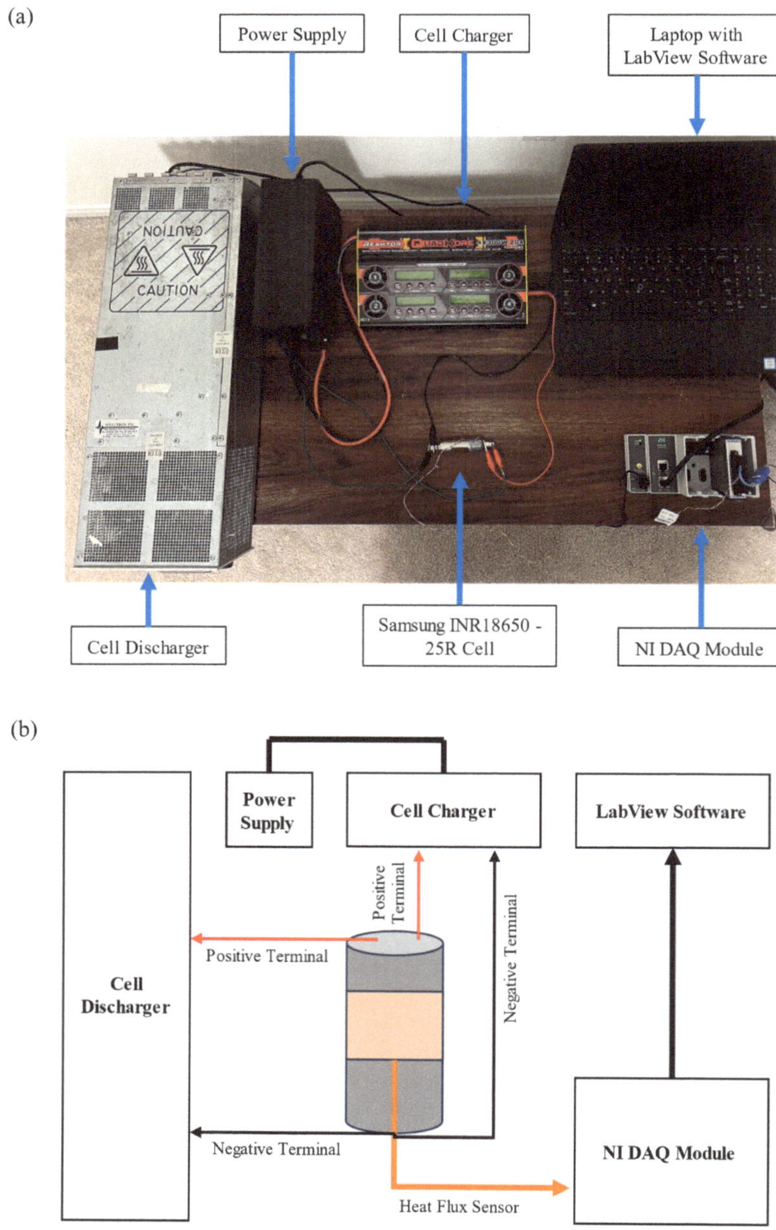

Figure 6. Experiment I—(**a**) setup and (**b**) experimental schematics.

3.2. Experiment II—Temperature Measurement

Experiment II was performed to attain the battery module's temperature profiles. The setup is shown in Figure 7a, and the schematics are shown in Figure 7b. The specifics of the equipment are already provided in the previous section. For this experiment, the module was manufactured. The outer casing was made from wood. The cells were placed within the battery module with a series connection. Aluminum liquid channels were filled with water and placed between the PCM material. An axial fan was used to force air through the air duct in the battery module, and the velocity of the airflow was measured using the Proster Anemometer (sourced from Toronto, ON, Canada. Finally, one T-type thermocouple was attached to each cell and the temperature was measured.

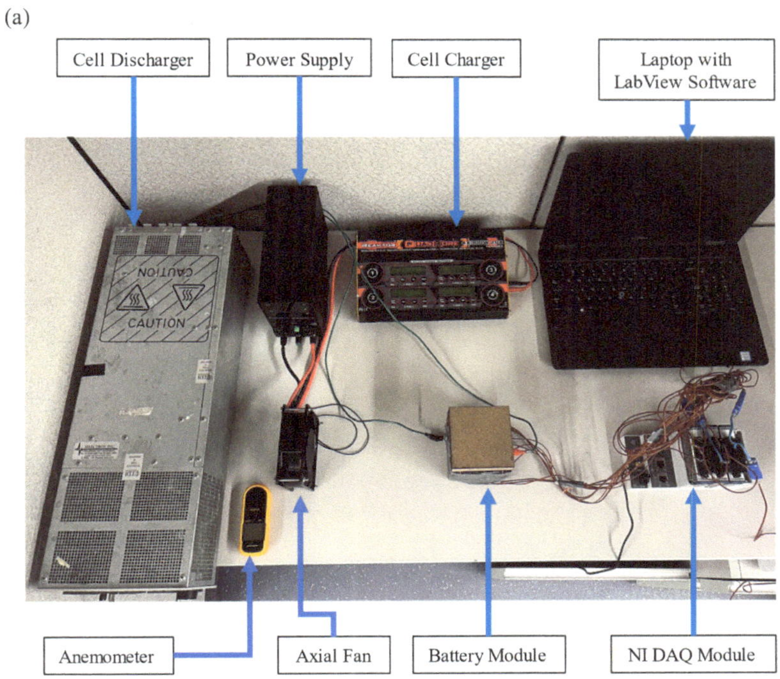

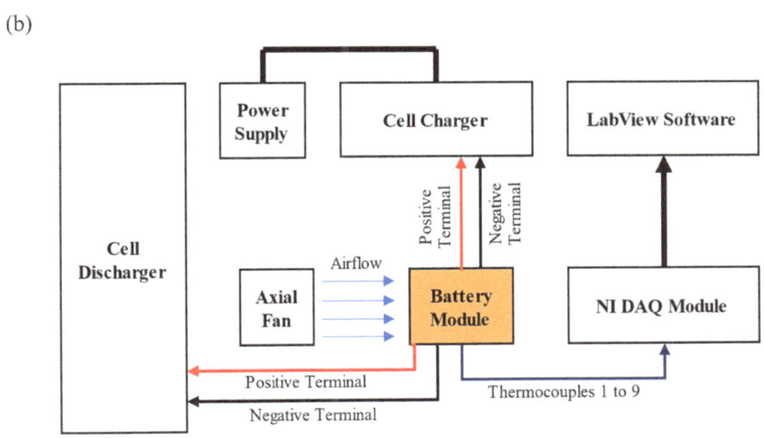

Figure 7. Experiment II—(**a**) setup and (**b**) experimental schematics.

To manufacture the battery module, the panels of the wooden housing were laser cut, including the holes in the bottom and the top plates to hold the cells and liquid channels in place. The wooden panels were then glued together. All nine cells were then placed inside the housing. Four aluminum liquid channels were placed inside the housing and filled with water. Thermocouples were then attached to each cell using adhesive tape. Then, the PCM blocks (paraffin wax) were melted using a wax melt warmer, and the melted PCM was then poured into the battery pack. The wires were then connected to the tabs of the cells to connect them all in a series. Finally, the air duct was placed on the module.

The procedure for Experiment II consisted of charging the cells using the Turnigy charger. The cells were charged with the CC-CV process. The total voltage at the end of charge was 37.8 V (the voltage of each cell was 4.2 V). The cells were discharged at the CC once the measurements from the thermocouples reached 25 °C. The readings were logged using LabView. Once the voltage of the battery module reached the cut-off voltage of 22.5 V, the discharge process was terminated.

The experiment was repeated three times for a 1 C discharge rate, and the error analysis was conducted using the technique recommended by Moffat [42]. The details are tabulated in Table 4. The average percentage uncertainty in the temperature measurements is ±3.59%.

Table 4. Calculated values of the relative bias and precision errors and the total uncertainty of the temperature measurement.

Parameters	Reference Value (°C)	Absolute Bias Error (°C)	Relative Bias Error (%)	Relative Precision Error (%)	Total Uncertainty (%)
Min. Temp.	25	1	4	2.22	4.57
Max. Temp.	47	1	2.13	1.49	2.6

4. Numerical Modeling

4.1. Governing Equations

The numerical model was conducted using ANSYS Fluent. The solidification/melting, energy, and flow models in Fluent were coupled to achieve the numerical model for this concept. The following assumptions were considered for the development of the numerical model:

- Airflow through the duct is incompressible and the temperature of the air at the battery module inlet is at an ambient temperature.
- The fluid in the liquid cooling is incompressible and the temperature of the fluid at the start of the discharge process is at an ambient temperature.
- There is no leakage of fluid from the liquid cooling components.
- The battery module is kept at an atmospheric pressure.

The governing equations for the energy and fluid flow models are as follows [32].
Energy Equation

$$\frac{\partial}{\partial t}(\rho E) + \nabla\cdot\left(\vec{v}(\rho E + P)\right) = \nabla\cdot\left(k_T \nabla T - \sum_j h_j \vec{J}_j + \left(\overline{\overline{\tau}}_{eff}\cdot\vec{v}\right)\right) + S_h \quad (2)$$

Momentum Conservation Equation

$$\frac{\partial}{\partial t}\left(\rho\vec{v}\right) + \nabla\cdot\left(\rho\vec{v}\vec{v}\right) = -\nabla P + \nabla\cdot\overline{\overline{\tau}} + \rho\vec{g} + \vec{F} \quad (3)$$

where:

$$\overline{\overline{\tau}} = \mu\left[\left(\nabla\vec{v} + \nabla\vec{v}^{T}\right) - \frac{2}{3}\nabla\cdot\vec{v}U\right] \quad (4)$$

Continuity Equation

$$\frac{\partial\rho}{\partial t} + \nabla\cdot\left(\rho\vec{v}\right) = S_m \quad (5)$$

The solidification/melting model was only applied to the PCM region. The enthalpy–porosity technique was used to model the phase change process. In this technique, the interfaces where the solid melts are not explicitly tracked. Instead, a liquid fraction (β) quantity is introduced, which provides the fraction of each cell volume that is in a liquid form. This liquid fraction is based on the enthalpy balance [32]. The enthalpy of the PCM is based on the latent heat and sensible enthalpy, as shown below [32].

$$H = h + \Delta H \quad (6)$$

where:

$$h = h_{ref} + \int_{T_{ref}}^{T} c_p dT \quad (7)$$

The liquid fraction is 0 if the temperature (T) is less than the solidification temperature ($T_{solidus}$) of the PCM. Additionally, the liquid fraction is 1 if the temperature (T) is more than the melting temperature ($T_{liquidus}$) of the PCM. The equation for the liquid fraction during the change in the phase is provided below [32].

$$\beta = \frac{T - T_{solidus}}{T_{liquidus} - T_{solidus}} \quad (8)$$

The latent heat equation in terms of the liquid fraction and latent heat of the material (L) is shown below [32].

$$\Delta H = \beta L \quad (9)$$

Therefore, the energy equation is as follows [32].

$$\frac{\partial}{\partial T}(\rho H) + \nabla \cdot \left(\rho \vec{v} H\right) = \nabla \cdot (k \nabla T) + S \quad (10)$$

The enthalpy–porosity technique considers the partially solidified region (the mushy zone) as the porous medium. In regions that are completely solid, the porosity is equal to zero. The momentum sink due to the reduced porosity in the mushy zone is provided in Equation (11) and is utilized in the term $\vec{F}$ in the momentum conservation equation (Equation (3)) [32].

$$S = \frac{(1-\beta)^2}{(\beta^3 + \epsilon)} A_{mush} \left(\vec{v} - \vec{v}_p\right) \quad (11)$$

4.2. Mesh-Independence and Time-Independence Studies

The cross-section of the developed mesh is shown in Figure 8. Since the air flows within the duct region only, the inflation layers were added to the walls in this region to capture the boundary layer.

Figure 8. (**a**) Isometric view and (**b**) cross-section view of the developed mesh.

In order to verify that the results are not dependent on the size of the elements in the mesh, a mesh-independence study was performed. A coarse mesh, medium mesh, medium-fine mesh, and fine mesh were developed. In each subsequent mesh, the total elements were increased by approximately 50%. To compare the results, the airflow velocity profiles are provided in Figure 9a. In addition, the average velocity is provided in Figure 9b. According to the results, the medium mesh was selected.

Moreover, a time-independence study (TIS) was conducted to make sure that the results are not dependent on the time step. An optimum time-step size is required, as a large time step will produce inaccurate results and a small time step will increase the computational time significantly. The results are provided in Figure 9c for the TIS and a time step of 0.5 s was chosen based on the results.

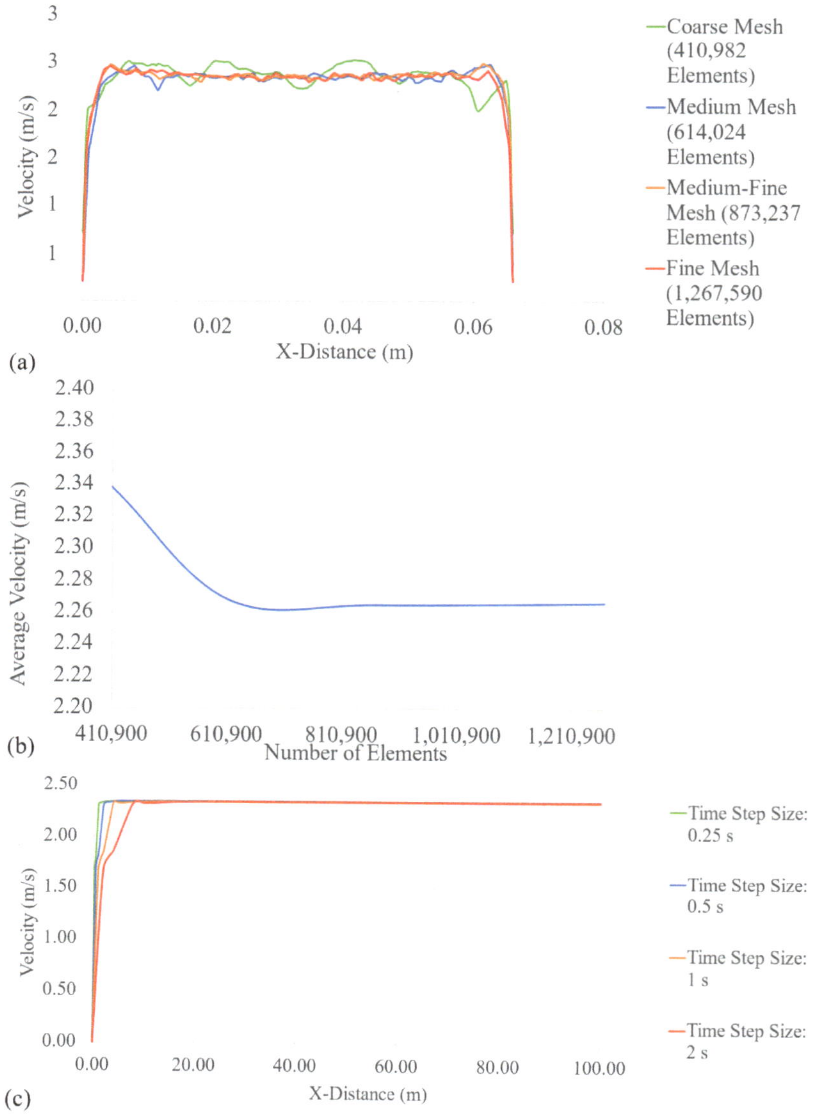

Figure 9. (**a**) Velocity profiles, (**b**) average velocity based on a number of elements, and (**c**) temporal velocity profiles for the different time-step sizes.

4.3. Boundary Conditions and Simulation Procedure

In order to assess the impact of the nanofluids, the ambient temperature was set to 42.24 °C to evaluate and compare the phase change in the CPCM with different fluid thermal conductivities. This is the temperature at which the CPCM will start to change its phase. The CPCM was solid at the beginning of the discharge cycle, and once the discharge process started, the CPCM started to melt due the heat absorption from the cells. The inlet velocity of the airflow was based on the Reynolds number of 1950. The heat flux boundary condition at the 7 C discharge rate was applied at the surface of the cells. Additionally, the configurations were also simulated at an ambient temperature of 25 °C to evaluate the reduction in the maximum temperature by the two configurations.

In order to evaluate the impact of the addition of fins, the ambient temperature was set to 25 °C. The heat flux boundary condition at the 7 C discharge rate was applied at the surface of the cells. The inlet velocity was changed for the three configurations, as per Table 2.

For the scalability analysis, the inlet air velocity was based on the Reynolds number of 1950. The heat flux boundary condition at the 7 C discharge rate was applied at the surface of the cells. The ambient temperature was set to 25 °C.

Moreover, a transient simulation was conducted using a pressure-based solver. The SIMPLE scheme was used for the pressure–velocity coupling. The convergence criterion was set to 0.000001 for the residuals of the continuity, velocities, and energy. Finally, a fixed time advancement of a 0.5 s time-step size was used.

5. Results and Discussion

5.1. Heat Flux Profile and Numerical Model Validation

The heat flux profiles were obtained at various discharge rates ranging from 1 C to 7 C, and are provided in Figure 10a. In comparison, the highest heat flux throughout the experiment was obtained for the 7 C discharge rate with 584 W/m^2; hence, 7 C was considered as the worst-case scenario, and the numerical simulations for all the strategies were conducted at this discharge rate.

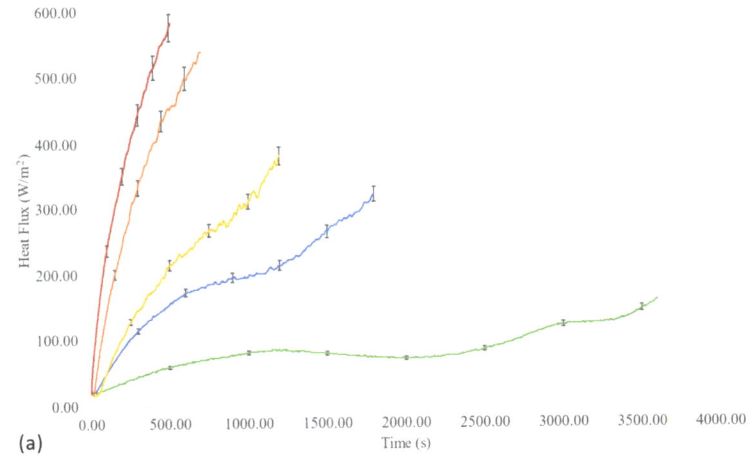

Figure 10. *Cont.*

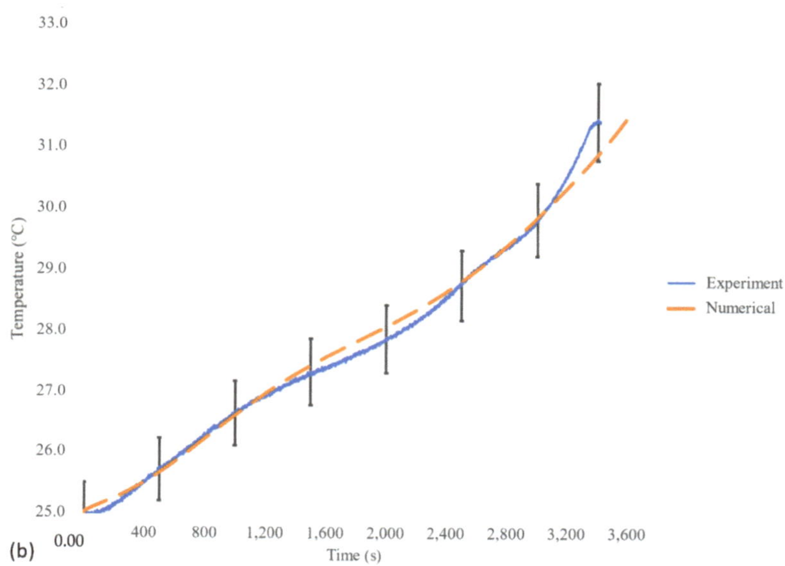

(b)

Figure 10. (**a**) Heat flux profiles and (**b**) numerical model validation results.

The numerical simulations provide effective results; however, it is important to validate the results of the numerical model with the experimental results. Since the numerical model is based on the battery module developed for Experiment II, the results of Experiment II were used for validation. The validation results are shown in Figure 10b. It can be seen that the numerical results are within the ±2% temperature error bars, which falls within the ±3.59% experimental uncertainty. Therefore, the numerical model is considered completely validated.

5.2. Thermal Analysis of the Proposed Hybrid Strategy

In order to comprehend the cooling phenomenon of the proposed hybrid strategy with the CPCM (the paraffin with copper foam), the contours for the temperature are provided in Figure 11. To evaluate the thermal performance, two separate simulations were conducted. The first simulation was conducted at a 25 °C ambient temperature, and the second simulation was conducted at the phase change temperature of the CPCM, which is 42.24 °C in this case. The temperature contours for the 25 °C ambient temperature are provided in Figure 11a–c, and the temperature contours for the 42.24 °C ambient temperature are provided in Figure 11d–f.

It can be seen that, due to the presence of the cold plates, a high amount of cooling was achieved. The lower-temperature surfaces are the ones that are in front of the cold plates, whereas the higher-temperature surfaces are the ones that are facing the module housing. Since the housing of the module is cooled through the ambient environment and the airflow only, as compared to the liquid cooling effect for the other cell surfaces, these temperatures were higher. This shows the effectiveness and importance of the presence of liquid cooling within the proposed strategy.

Moreover, according to Figure 11b, the cold plates also act as a heat sink for the CPCM. Adding a cold plate eliminates the high-temperature region within the composite PCM between the neighboring cells as the fluid removes heat from the CPCM. Since the high-temperature regions are eliminated, the temperature of the cells does not reach the phase change temperature of the CPCM, and the melting does not occur.

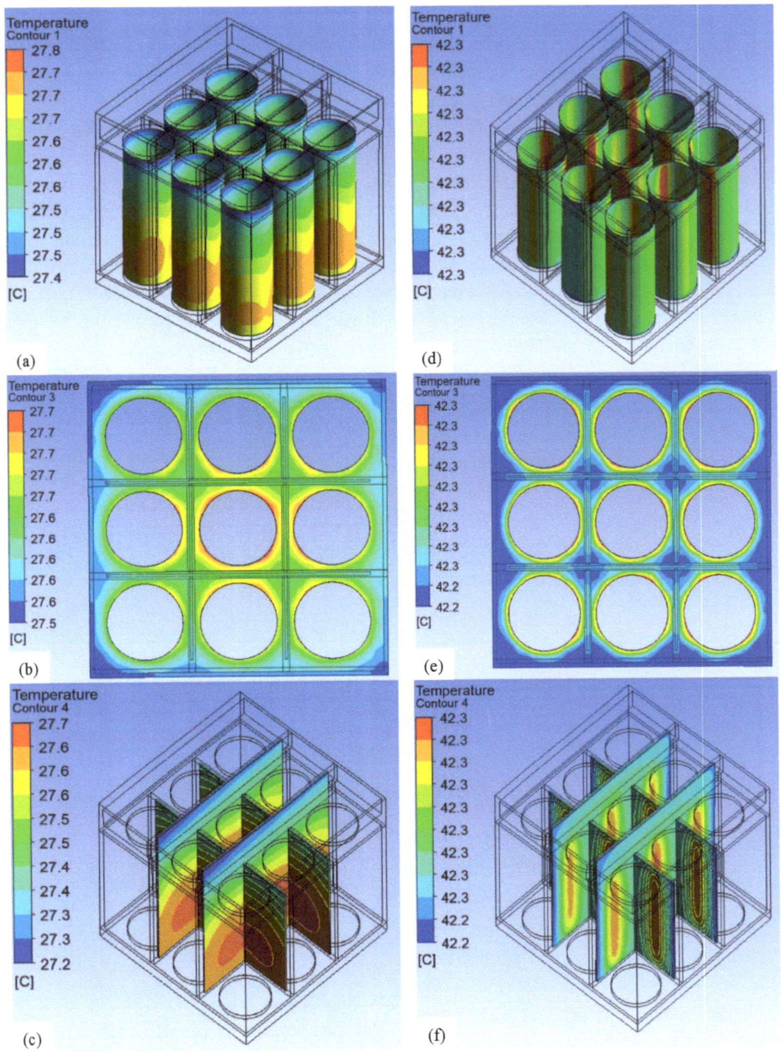

Figure 11. Temperature contours of the (**a**) cells, (**b**) composite PCM, and (**c**) fluid at the 25 °C ambient temperature; temperature contours of the (**d**) cells, (**e**) composite PCM, and (**f**) fluid at 42.24 °C.

Additionally, according to Figure 11c, the lower portion of the fluid has the highest temperature, and it gradually reduces to the minimum towards the top portion of the fluid. This is because of the air-cooling effect on the cold-plate region that is protruding within the air duct. Additionally, the lowest temperature of the fluid is exhibited near the inlet of the air duct, and as the air flows through the cold plate, its heat extraction capacity reduces, and the temperature of the fluid increases towards the outlet of the air duct. The impact of the single fluid body can be seen in Figure 11c, as the same airflow allows the fluid to cool between the cell columns.

Finally, at the ambient temperature of 42.24 °C, the phase change in the composite PCM begins. It can be seen from Figure 11d–f that all temperatures for the entire battery pack are within the range of 42.2 to 42.3 °C, as the phase change in the composite PCM limits the temperature of the cells, and the PCM and the airflow at the same temperature limits the cooling of the fluid to the ambient temperature of 42.24 °C.

5.3. Increasing the Thermal Conductivity of Water through Nanoparticles

As discussed in Section 2.2, the thermal conductivity of the fluid is increased to 0.92 W/m K using H_2O–AG–MWCNT from 0.6 W/m K of water. To compare the effect of increased fluid thermal conductivity using nanoparticles, two different sets of numerical simulations were conducted. In the first simulation, the amount of CPCM usage through the phase change was investigated. Since the phase temperature of the CPCM was 42.24 °C, the simulation was conducted at an ambient temperature of 42.24 °C. Moreover, in the second simulation, the increase in the maximum temperature was investigated, and for this, the ambient temperature was set to 25 °C. The result of the phase change evaluation is provided in Figure 12a. It can be seen that approximately 7% of the CPCM changed its phase using both fluids, and the graphs completely overlap each other. Additionally, the result for the maximum temperature increase is shown in Figure 12b. It can be seen that the maximum temperature in both instances increases to 27.8 °C, and there is no difference in the results, as the graphs completely overlap each other. This shows that the cooling from the fluid is not sensitive enough to capture the change of 0.32 W/m K thermal conductivity in the fluid.

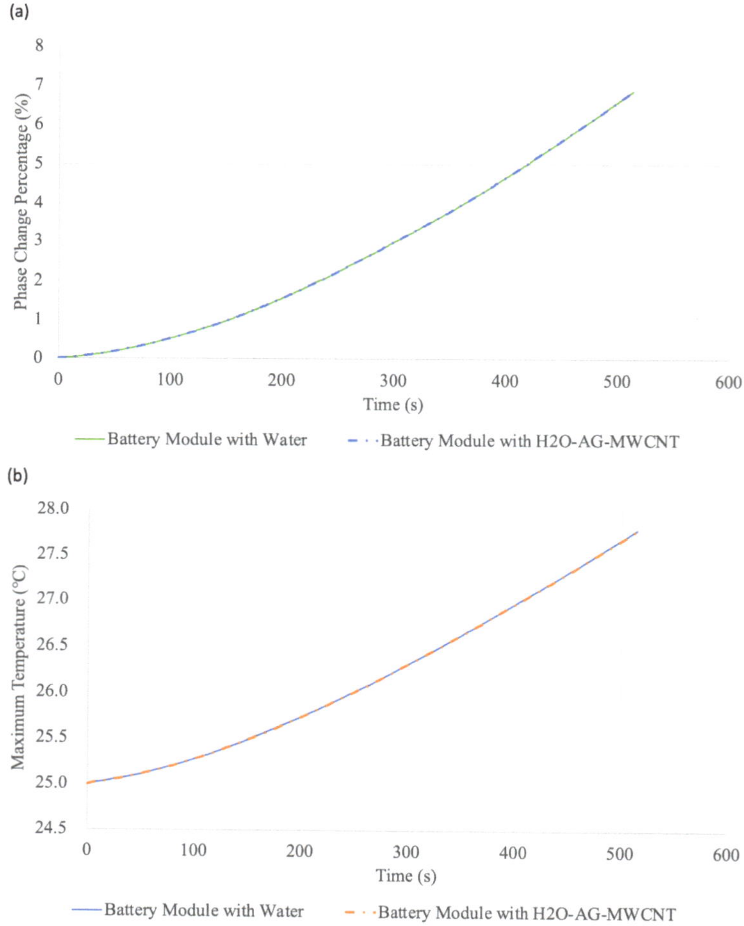

Figure 12. (**a**) CPCM phase change comparison and (**b**) maximum temperature increase comparison of the battery module with water and H_2O–AG–MWCNT.

5.4. Addition of Fins in the Battery Module to Improve the Heat Transfer

As discussed in Section 2.3, in the first configuration, one fin was added to the cold plate within the air duct, and in the second configuration, two fins were added. The fins were incorporated to assist in improving the heat transfer to the airflow, as, through the fins, the contact area with the airflow increases. However, this increase in the heat transfer will come at the cost of a higher pressure drop. The results of this study are shown in Figure 13. It can be seen from the results that the cooling achieved is not significant enough when compared to the pressure drop increase. With one fin, the pressure drop increases from ~11 Pa to ~38 Pa, whereas the maximum temperature reduces by only ~0.6 °C. Similarly, with two fins, the pressure drop increases by ~70 Pa, and the maximum temperature reduces by ~0.5 °C when compared to the results of the one-fin configuration. Therefore, the addition of fins does assist in improving the cooling. However, the pressure drop increase is much higher, which will result in higher fan power requirements.

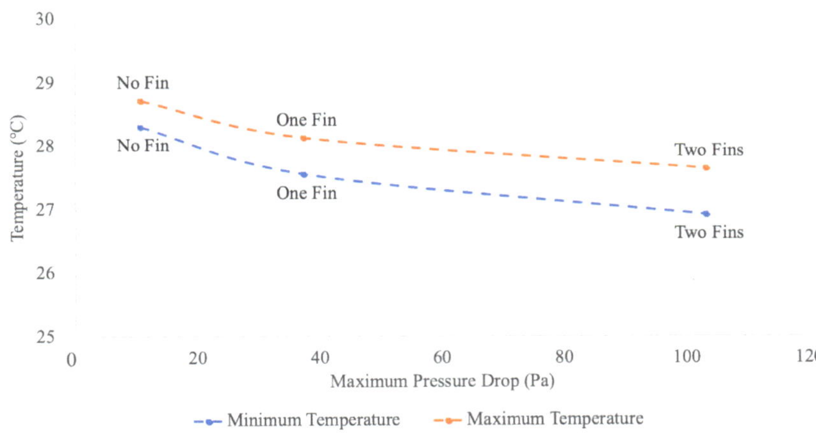

Figure 13. Variation in the maximum and minimum temperatures with the maximum pressure drop for the proposed configurations.

5.5. Scalability of the Battery Module

In order to evaluate the scalability of the battery module, four different numerical simulations were performed. The first simulation consisted of one battery module only, the second consisted of two battery modules in a series, the third consisted of three battery modules in a series, and the fourth consisted of four battery modules in a series. The maximum temperature of the airflow and battery modules was obtained and is provided in Figure 14. As expected, it can be seen that the temperature of the airflow and modules increases as the number of the modules increases in a series. This is because the air extracts a higher amount of heat from the initial battery modules, and, as it moves along in a series, its heat extraction capacity reduces, which results in an increase in the temperature. However, due to the hybrid design of the developed battery module, the impact of this phenomenon has been significantly reduced, as the temperature difference for the battery modules is approximately 0.31 °C across the four battery modules. This is mainly because the CPCM is used as the primary coolant, and it extracts the heat and maintains the uniformity. This can also be seen in the temperature contours provided in Figure 15.

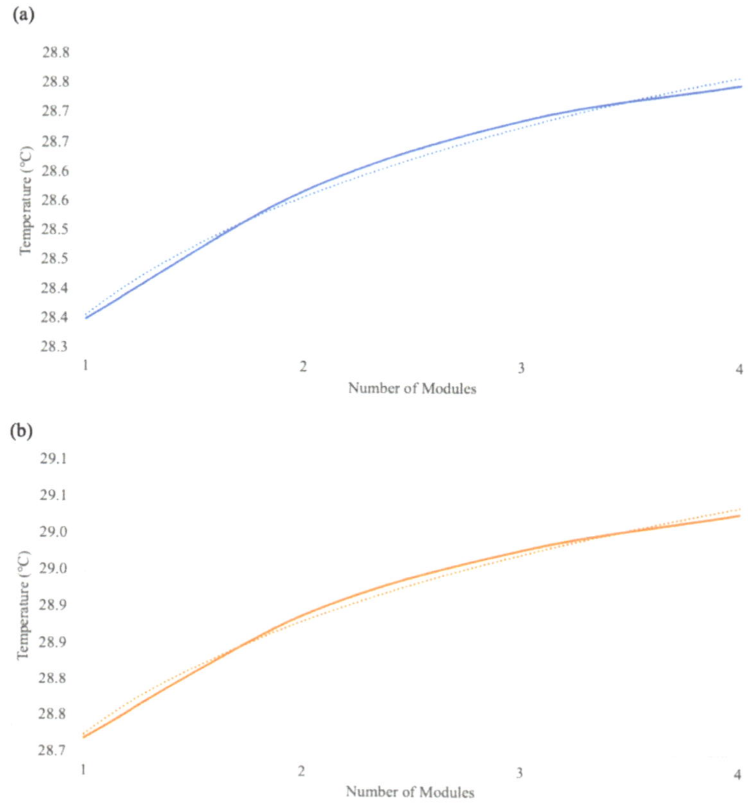

Figure 14. Maximum temperature with the increasing battery modules for the (**a**) airflow and (**b**) cells.

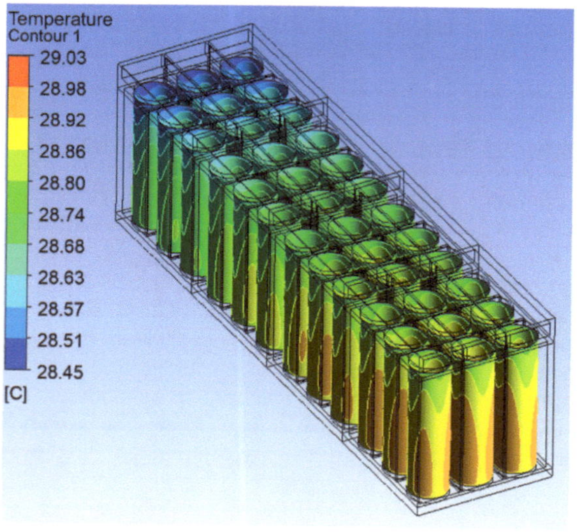

Figure 15. Temperature contours of the cells in four battery modules connected in a series.

In order to extrapolate the results over the 25 battery modules in a series, the equations obtained from the trendlines for both the airflow and cell temperature were used and are provided below. The R^2 value for the airflow trendline is 0.9953 and the battery module trendline is 0.9963.

$$\text{Air Flow Temperature} = 0.293122289793738 \ln B + 28.3571103960402 \quad (12)$$

$$\text{Battery Module Temperature} = 0.226234840293359 \ln B + 28.725253374811 \quad (13)$$

where B is the number of modules attached in a series.

The result of this extrapolation is provided in Figure 16. It can be seen that the temperature difference between the maximum temperature of the final module and the initial module is 0.72 °C, whereas the difference in the airflow is 0.94 °C. Moreover, the overall temperature uniformity across the entire battery pack is 1.14 °C. This shows that, even with 25 modules stacked in a series, the temperature uniformity is still high. Therefore, the developed thermal management strategy can be scaled up or down according to the vehicle requirements.

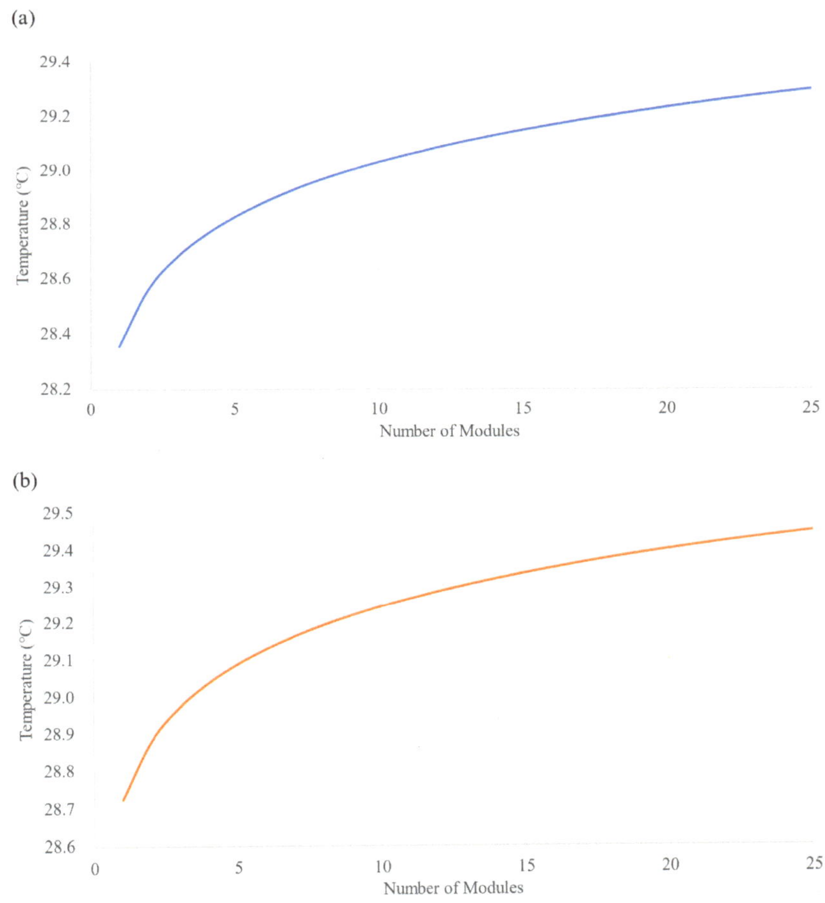

Figure 16. Maximum temperature across 25 battery modules for the (**a**) airflow and (**b**) cells.

Similarly, the pressure drop was also obtained using the numerical simulations, and the results for up to four battery modules are provided in Figure 17a. A trendline was obtained, and the results were extrapolated for the entire 25 battery modules in a series, as shown in Figure 17b. The trendline equation is provided below.

$$\text{Pressure Drop} = -0.0459999999999923(B^2) + 2.89520000000002(B) + 7.92099999999991 \quad (14)$$

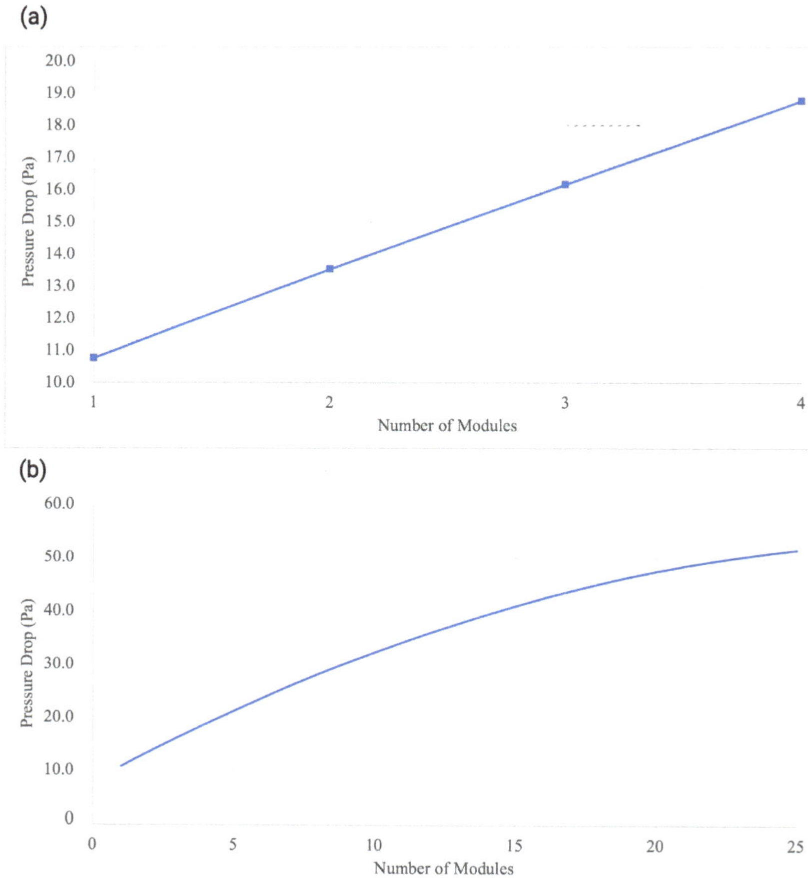

Figure 17. Pressure drop across the increasing battery modules (**a**) through numerical simulations and (**b**) through extrapolation.

Furthermore, for the verification, a numerical simulation was conducted over the entire 25 battery modules in a series, and the pressure contours for the single air duct channel are provided in Figure 18. For a single battery module, the maximum pressure drop was ~11 Pa (Figure 13), whereas, for the 25 battery modules, the maximum pressure drop increased to ~51 Pa. Additionally, for a single channel of 19 mm in width and 6 mm in height, the airflow in terms of the cubic feet per minute (CFM) amounted to ~0.75 CFM. Hence, for the complete battery pack, the total CFM requirement will be ~68 CFM.

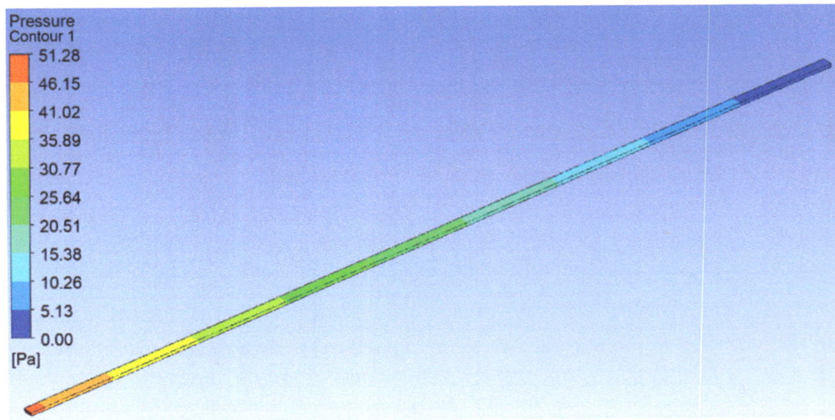

Figure 18. Pressure contours for the air duct channel for 25 battery modules in a series.

5.6. Comparison with the Open Literature

The results obtained from the scalability study were compared with the maximum temperature and temperature uniformity of various thermal management strategies developed by researchers in the open literature. The comparison is provided in Figure 19. Zeng et al. [43] developed a hybrid strategy by combining liquid cooling and microheat pipe arrays. Liu et al. [44] developed a hybrid strategy by combining a CPCM and copper tube liquid cooling. Xu et al. [45] developed a liquid cooling strategy using a wrench-shaped cold plate with a bifurcation structure. Xiong et al. [46] developed a novel liquid-cooling bionic flow channel in the shape of a spider web, and the results were compared with honeycomb and helical flow channels. Ling et al. [47] designed various configurations of minichannel cold plates and optimized the designs.

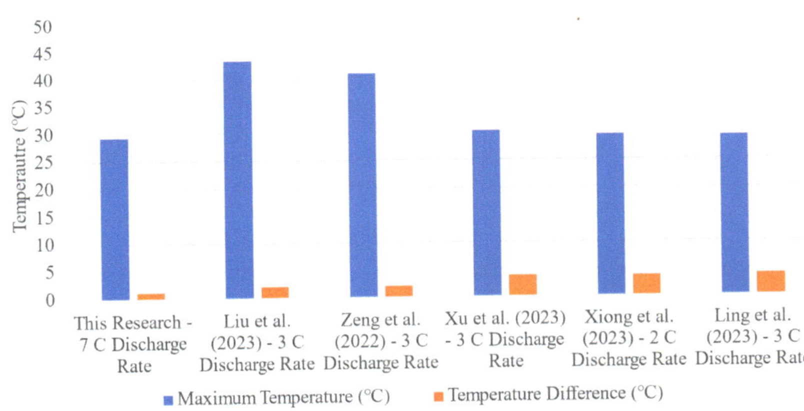

Figure 19. Comparison of results with the thermal management results in the open literature [43–47].

It can be seen from Figure 19 that the maximum temperature of the scaled battery pack at a 7 C discharge rate is similar to the results obtained in the studies in the open literature with a smaller battery module and discharge rates up to 3 C. Moreover, the temperature uniformity of 1.14 °C of the scaled battery pack is higher compared to the studies in the open literature. Additionally, the open literature studies require flowing fluid through the entire battery modules, which results in power consumption. However, the improved thermal performance obtained from the strategy developed in the present study is achieved without the flow of fluid and pumping power requirements.

6. Conclusions

This paper proposed a hybrid thermal management strategy for electric battery modules and conducted experimental and numerical studies to evaluate the strategy. At the core of the proposed strategy are cylindrical lithium-ion cells in direct contact with the composite phase change material. An air duct was added to the top of the cells to allow airflow. Rectangular cold plates were added between each cell's rows and columns, filled with fluid and protruding in the air duct. The cold plates were all connected, and the fluid within this was considered a single body of fluid. Due to a single fluid body, the protrusion of the cold plate within the air duct allowed the fluid in between the columns to cool as well. Two experimental studies were conducted. In the first experimental study, the heat flux profiles were recorded, and in the second study, the battery module was developed, and temperature profiles of all the cells were obtained at a 1 C discharge rate. A numerical model was also developed based on the proposed strategy and validated using the second experimental study. The numerical data stayed within $\pm 2\%$ of the experimental data. The numerical results showed that the maximum temperature was limited to 300.95 K, and a high temperature uniformity of 0.4 K was obtained. Additionally, when comparing the addition of nanoparticles to increase the thermal conductivity of water, the results concluded that the cooling from the fluid was not sensitive enough to capture the increase of 0.32 W/m K thermal conductivity through the use of nanoparticles. Furthermore, to enhance the airflow's cooling, the proposed strategy was modified by adding fins to the protruded cold plate section in the air duct. The results showed that the pressure drop through the addition of fins increased by ~245%, which will result in higher fan power requirements, whereas the maximum temperature of the battery module reduced by only 0.6 K. Additionally, when scaled up to an entire battery pack, the overall temperature uniformity across the pack was 1.14 K, with a maximum temperature of 302.6 K, which was within the optimal operating temperature and uniformity ranges. In terms of the airflow requirements for the 25 battery modules, the maximum pressure drop was ~51 Pa, which amounted to ~0.75 cubic feet per minute (CFM) of airflow. Hence, for the complete battery pack, the total CFM requirement was ~68 CFM. Finally, when compared to data in the open literature, the temperature uniformity of 1.14 K of the scaled battery pack at the 7 C discharge rate was higher compared to the smaller battery modules with discharge rates of 3 C. Therefore, the developed thermal management strategy provides significant cooling and temperature uniformity, eliminates the requirement of a pump and reservoir, and can be scaled up or down according to the energy and power requirements.

7. Limitations and Future Research Outlook

As concluded above, the hybrid thermal management strategy is capable of maintaining the battery module temperatures within the feasible operating range while significantly reducing the energy requirements for the operation of the thermal management system. However, gaps still remain which require further investigation and research in order to apply this strategy to practical applications.

The limitations of the research performed in this thesis are as follows:

- The research was conducted on 18650 Li-ion cells and it is not applicable to larger cells without further investigation.
- The research was conducted on cylindrical Li-ion cells and it is not applicable to different cell geometries without further investigation.
- The research is not applicable to discharge rates higher than 7 C without further investigation.
- The structural integrity of various developed components has not been investigated.

Furthermore, several recommendations can be made for future research, as listed below:

- Experiments should be conducted to obtain heat flux profiles of lithium-ion cells at high charging rates and at various drive cycles, and used to obtain and improve the thermal performance of the hybrid battery module.
- Experiments should be conducted by developing a prototype of the scaled-up battery pack to improve the numerical simulations for the scaled battery pack and further validate it.
- A study should be conducted for the developed hybrid battery module with larger lithium-ion cells, such as the 26650 and 42120 cylindrical cells.
- An experimental and numerical study should be conducted to evaluate the performance of the developed hybrid thermal management strategies and battery module in instances of thermal runaway.

Author Contributions: Conceptualization, S.S. and M.A.-C.; methodology, S.S. and M.A.-C.; software, S.S.; validation, S.S.; formal analysis, S.S.; investigation, S.S.; resources, M.A.-C.; data curation, S.S.; writing—original draft preparation, S.S.; writing—review and editing, M.A.-C.; visualization, S.S. and M.A.-C.; supervision, M.A.-C.; project administration, M.A.-C.; funding acquisition, M.A.-C. All authors have read and agreed to the published version of the manuscript.

Funding: This research received no external funding.

Data Availability Statement: The raw data supporting the conclusions of this article will be made available by the authors on request.

Conflicts of Interest: The authors declare no conflict of interest.

Nomenclature

ρ	density (kg m^{-3})
t	time (s)
$\vec{v}$	velocity (m s^{-1})
ϵ	is a small number (0.001) to prevent division by zero
D	characteristic length (m)
S_m	user-defined function for the continuity equation
P	pressure (Pa)
$\bar{\bar{\tau}}$	stress tensor
A_{mush}	is the mushy zone constant
$\vec{g}$	gravitational acceleration (9.81 m s^{-2})
$\vec{F}$	external forces, porous-media source terms, and user-defined sources
μ	dynamic viscosity (Pa s)
T	temperature (K)
U	unit tensor
E	energy (J)
k_T	effective thermal conductivity (W m^{-1} K^{-1})
H	enthalpy (J)
h	sensible enthalpy (J)
ΔH	latent heat (J)
β	liquid fraction
c_p	specific heat capacity (J kg^{-1} K^{-1})
m	mass (kg)
Q	heat generation (W)
Re	Reynolds number (dimensionless)
v	volume (m^3)

References

1. US EPA. Fast Facts on Transportation Greenhouse Gas Emission. Available online: https://www.epa.gov/greenvehicles/fast-facts-transportation-greenhouse-gas-emissions (accessed on 4 August 2023).
2. Omar, N.; Monem, M.A.; Firouz, Y.; Salminen, J.; Smekens, J.; Hegazy, O.; Gaulous, H.; Mulder, G.; Bossche, P.V.D.; Coosemans, T.; et al. Lithium iron phosphate based battery—Assessment of the aging parameters and development of cycle life model. *Appl. Energy* **2014**, *113*, 1575–1585. [CrossRef]
3. Sefidan, A.M.; Sojoudi, A.; Saha, S.C. Nanofluid-based cooling of cylindrical lithium-ion battery packs employing forced air flow. *Int. J. Therm. Sci.* **2017**, *117*, 44–58. [CrossRef]
4. Saw, L.H.; Poon, H.M.; Thiam, H.S.; Cai, Z.; Chong, W.T.; Pambudi, N.A.; King, Y.J. Novel thermal management system using mist cooling for lithium-ion battery packs. *Appl. Energy* **2018**, *223*, 146–158. [CrossRef]
5. Rao, Z.; Wang, S.; Zhang, G. Simulation and experiment of thermal energy management with phase change material for ageing $LiFePO_4$ power battery. *Energy Convers. Manag.* **2011**, *52*, 3408–3414. [CrossRef]
6. Wei, Y.; Agelin-Chaab, M. Development and experimental analysis of a hybrid cooling concept for electric vehicle battery packs. *J. Energy Storage* **2019**, *25*, 100906. [CrossRef]
7. Shahid, S.; Chea, B.; Agelin-Chaab, M. Development of a hybrid cooling concept for cylindrical li-ion cells. *J. Energy Storage* **2022**, *50*, 104214. [CrossRef]
8. Park, S.; Jung, D. Battery cell arrangement and heat transfer fluid effects on the parasitic power consumption and the cell temperature distribution in a hybrid electric vehicle. *J. Power Sources* **2013**, *227*, 191–198. [CrossRef]
9. Patil, M.S.; Seo, J.-H.; Lee, M.-Y. A novel dielectric fluid immersion cooling technology for Li-ion battery thermal management. *Energy Convers. Manag.* **2020**, *229*, 113715. [CrossRef]
10. Dubey, P.; Pulugundla, G.; Srouji, A.K. Direct Comparison of Immersion and Cold-Plate Based Cooling for Automotive Li-Ion Battery Modules. *Energies* **2021**, *14*, 1259. [CrossRef]
11. Shahid, S.; Agelin-Chaab, M. A review of thermal runaway prevention and mitigation strategies for lithium-ion batteries. *Energy Convers. Manag. X* **2022**, *16*, 100310. [CrossRef]
12. Panchal, S.; Khasow, R.; Dincer, I.; Agelin-Chaab, M.; Fraser, R.; Fowler, M. Thermal design and simulation of mini-channel cold plate for water cooled large sized prismatic lithium-ion battery. *Appl. Therm. Eng.* **2017**, *122*, 80–90. [CrossRef]
13. Panchal, S.; Mathewson, S.; Fraser, R.; Culham, R.; Fowler, M. Thermal Management of Lithium-Ion Pouch Cell with Indirect Liquid Cooling using Dual Cold Plates Approach. *SAE Int. J. Altern. Powertrains* **2015**, *4*, 1184. [CrossRef]
14. Zhao, J.; Rao, Z.; Li, Y. Thermal performance of mini-channel liquid cooled cylinder based battery thermal management for cylindrical lithium-ion power battery. *Energy Convers. Manag.* **2015**, *103*, 157–165. [CrossRef]
15. Zhao, C.; Sousa, A.C.M.; Jiang, F. Minimization of thermal non-uniformity in lithium-ion battery pack cooled by channeled liquid flow. *Int. J. Heat Mass. Transf.* **2019**, *129*, 660–670. [CrossRef]
16. Rao, Z.; Qian, Z.; Kuang, Y.; Li, Y. Thermal performance of liquid cooling based thermal management system for cylindrical lithium-ion battery module with variable contact surface. *Appl. Therm. Eng.* **2017**, *123*, 1514–1522. [CrossRef]
17. He, F.; Ams, A.A.; Roosien, Y.; Tao, W.; Geist, B.; Singh, K. Reduced-order thermal modeling of liquid-cooled lithium-ion battery pack for EVs and HEVs. In Proceedings of the IEEE Transportation and Electrification Conference and Expo, Chicago, IL, USA, 14 June 2017. [CrossRef]
18. Al-Hallaj, S.; Selman, J.R. A Novel Thermal Management System for Electric Vehicle Batteries Using Phase-Change Material. *J. Electrochem. Soc.* **2000**, *147*, 3231. [CrossRef]
19. Javani, N.; Dincer, I.; Naterer, G.F. Numerical Modeling of Submodule Heat Transfer with Phase Change Material for Thermal Management of Electric Vehicle Battery Packs. *J. Therm. Sci. Eng. Appl.* **2015**, *7*, 031005. [CrossRef]
20. Jiang, G.; Huang, J.; Liu, M.; Cao, M. Experiment and simulation of thermal management for a tube-shell Li-ion battery pack with composite phase change material. *Appl. Therm. Eng.* **2017**, *120*, 1–9. [CrossRef]
21. He, F.; Li, X.; Zhang, G.; Zhong, G.; He, J. Experimental investigation of thermal management system for lithium ion batteries module with coupling effect by heat sheets and phase change materials. *Int. J. Energy Res.* **2018**, *42*, 3279–3288. [CrossRef]
22. Hussain, A.; Abidi, I.H.; Tso, C.Y.; Chan, K.C.; Luo, Z.; Chao, C.Y.H. Thermal management of lithium ion batteries using graphene coated nickel foam saturated with phase change materials. *Int. J. Therm. Sci.* **2018**, *124*, 23–35. [CrossRef]
23. Ye, G.; Zhang, G.; Jiang, L.; Yang, X. Temperature control of battery modules through composite phase change materials with dual operating temperature regions. *Chem. Eng. J.* **2022**, *449*, 137733. [CrossRef]
24. Wu, W.; Ye, G.; Zhang, G.; Yang, X. Composite phase change material with room-temperature-flexibility for battery thermal management. *Chem. Eng. J.* **2022**, *428*, 131116. [CrossRef]
25. Faraji, H.; El-Alami, M.; Arshad, A.; Hariti, Y. Numerical Survey on Performance of Hybrid NePCM for Cooling of Electronics: Effect of Heat Source Position and Heat Sink Inclination. *J. Therm. Sci. Eng. Appl.* **2021**, *13*, 051010. [CrossRef]
26. Arshad, A.; Jabbal, M.; Faraji, H.; Talebizadehsardari, P.; Bashir, M.A.; Yan, Y. Numerical study of nanocomposite phase change material-based heat sink for the passive cooling of electronic components. *Heat Mass. Transf.* **2021**, 1–15. [CrossRef]
27. Faraji, H.; Yıldız, Ç.; Arshad, A.; Arıcı, M.; Choukairy, K.; El-Alami, M. Passive thermal management strategy for cooling multiple portable electronic components: Hybrid nanoparticles enhanced phase change materials as an innovative solution. *J. Energy Storage* **2023**, *70*, 108087. [CrossRef]

28. Wei, Y.; Agelin-Chaab, M. Experimental investigation of a novel hybrid cooling method for lithium-ion batteries. *Appl. Therm. Eng.* **2018**, *136*, 375–387. [CrossRef]
29. Wang, S.; Li, Y.; Li, Y.Z.; Mao, Y.; Zhang, Y.; Guo, W.; Zhong, M. A forced gas cooling circle packaging with liquid cooling plate for the thermal management of Li-ion batteries under space environment. *Appl. Therm. Eng.* **2017**, *123*, 929–939. [CrossRef]
30. Shahid, S.; Agelin-Chaab, M. Development of hybrid thermal management techniques for battery packs. *Appl. Therm. Eng.* **2021**, *186*, 116542. [CrossRef]
31. Kumar, M.D.; Bhowmik, C.; Bhowmik, S.; Murari, P.K. Property-enhanced paraffin-based composite phase change material for thermal energy storage: A review. *Environ. Sci. Pollut. Res.* **2022**, *29*, 43556–43587. [CrossRef]
32. ANSYS. *ANSYS Fluent Theory Guide*; ANSYS, Inc.: Canonsburg, PA, USA, 2020.
33. Paraffin 42-44, in Block Form 8002-74-2. Available online: https://www.sigmaaldrich.com/CA/en/product/mm/107150 (accessed on 24 August 2023).
34. Amburi, P.K.; Senthilkumar, G.; Nithya, A. Novel use of CuO nanoparticles additive for improving thermal conductivity of MgO/water and MWCNT/water nanofluids. *J. Therm. Anal. Calorim.* **2023**, *148*, 10389–10398. [CrossRef]
35. Bioucas, F.E.B.; Köhn, C.; Jean-Fulcrand, A.; Garnweitner, G.; Koller, T.M.; Fröba, A.P. Effective Thermal Conductivity of Nanofluids Containing Silicon Dioxide or Zirconium Dioxide Nanoparticles Dispersed in a Mixture of Water and Glycerol. *Int. J. Thermophys.* **2022**, *43*, 167. [CrossRef]
36. Pourrajab, R.; Noghrehabadi, A.; Behbahani, M.; Hajidavalloo, E. An efficient enhancement in thermal conductivity of water-based hybrid nanofluid containing MWCNTs-COOH and Ag nanoparticles: Experimental study. *J. Therm. Analy Calorim.* **2021**, *143*, 3331–3343. [CrossRef]
37. Corasaniti, S.; Bovesecchi, G.; Gori, F. Experimental Thermal Conductivity of Alumina Nanoparticles in Water with and without Sonication. *Int. J. Thermophys.* **2021**, *42*, 23. [CrossRef]
38. Jabbari, F.; Rajabpour, A.; Saedodin, S. Thermal conductivity of CNT–water nanofluid at different temperatures, volume fractions, and diameters: Experimental investigation and molecular dynamics simulations. *Microfluid. Nanofluidics* **2021**, *25*, 102. [CrossRef]
39. Ghasemi, M.; Niknejadi, M.; Toghraie, D. Direct effect of nanoparticles on the thermal conductivity of CuO-water nanofluid in a phase transition phenomenon using molecular dynamics simulation. *J. Therm. Analy Calorim.* **2021**, *144*, 2483–2495. [CrossRef]
40. Tesla Model S/X Pack—75 kWh/214 Ah/350 V. Available online: https://www.secondlife-evbatteries.com/products/tesla-75 kw-pack (accessed on 27 September 2023).
41. Introduction of INR18650-25R. Available online: https://www.powerstream.com/p/INR18650-25R-datasheet.pdf (accessed on 17 November 2023).
42. Moffat, R.J. Describing the uncertainties in experimental results. *Exp. Therm. Fluid. Sci.* **1988**, *1*, 3–17. [CrossRef]
43. Zeng, W.; Niu, Y.; Li, S.; Hu, S.; Mao, B.; Zhang, Y. Cooling performance and optimization of a new hybrid thermal management system of cylindrical battery. *Appl. Therm. Eng.* **2022**, *217*, 119171. [CrossRef]
44. Liu, Z.; Cao, M.; Zhang, Y.; Li, J.; Jiang, G.; Shi, H. Thermal management of cylindrical battery pack based on a combination of silica gel composite phase change material and copper tube liquid cooling. *J. Energy Storage* **2023**, *71*, 108205. [CrossRef]
45. Xu, Q.; Xie, Y.; Huang, Y.; Li, X.; Huang, H.; Bei, S.; Wang, H.; Zheng, K.; Wang, X.; Li, L. Enhancement of thermal management for cylindrical battery module based on a novel wrench-shaped design for the cold plate. *Sust. Energy Technol. Assess.* **2023**, *59*, 103421. [CrossRef]
46. Xiong, X.; Wang, Z.; Fan, Y.; Wang, H. Numerical analysis of cylindrical lithium-ion battery thermal management system based on bionic flow channel structure. *Therm. Sci. Eng. Prog.* **2023**, *42*, 101879. [CrossRef]
47. Ling, L.; Li, L.; Xie, Y.; Wang, T.; Zheng, K.; Shan, S.; Zhang, L.; Bei, S.; Xu, Q. Optimal Design of Minichannel Cold Plate for the Thermal Management of Cylindrical Battery Modules. *Energy Technol.* **2023**, *11*, 2201484. [CrossRef]

MDPI AG
Grosspeteranlage 5
4052 Basel
Switzerland
Tel.: +41 61 683 77 34

Batteries Editorial Office
E-mail: batteries@mdpi.com
www.mdpi.com/journal/batteries

www.ingramcontent.com/pod-product-compliance
Lightning Source LLC
Chambersburg PA
CBHW041101270926
51850CB00025B/3174
9783725824212